T0138197

Reading the Skies

Reading the Skies

A Cultural History of English Weather, 1650–1820

VLADIMIR JANKOVIĆ

The University of Chicago Press

Chicago and London

Vladimir Janković is lecturer in the Centre for the History of Science, Technology and Medicine at the University of Manchester.

Manchester University Press, Oxford Road, Manchester M13 9NR, UK
The University of Chicago Press, 5801 S. Ellis Avenue, Chicago 60637

09 08 07 06 05 04 03 02 01 00   1 2 3 4 5

ISBN: 0–226–39215–5 (cloth)
ISBN: 0–226–39216–3 (paperback)

Library of Congress Cataloging-in-Publication Data

Janković, Vladimir, 1961–
    Reading the skies : a cultural history of English weather, 1650–1820 / Vladimir Janković.
        p.   cm.
    Includes bibliographical references and index.
    ISBN 0–226–39215–5 (cloth : alk. paper) — ISBN 0–226–39216–3 (pbk. : alk. paper)
    1. Meteorology—England—History.   2. England—Climate—History.   I. Title.

QC857.G77.J36 2000
551 6942—dc21   00–053250

This book is printed on acid-free paper.

To my parents

Contents

List of figures

List of boxes

Acknowledgments

It is a pleasure to acknowledge the many people and institutions who have enabled and assisted various stages in the research and completion of this book. I would first and foremost like to express my gratitude to Christopher Hamlin who has encouraged me from the beginning and has greatly helped me with his conceptual and editorial suggestions. Phillip Sloan has been most generous in commenting on early versions of the manuscript and has kept me alert to the larger historiographic issues. Special recognition is due to Jan Golinski whose attentive readings inspired important revisions and to Michael Hunter for his suggestions and support during my research in London. Roy Porter read the manuscript and was kind enough to advise on the overall strength of the argument. Other scholars and friends whom I would particularly like to thank are Ernan McMullin, Michael Crowe, Christopher Fox, Don Howard, and Leposava Vuskovic whose constant encouragement and invaluable advice on all aspects of the research gave me the strength to complete the study.

The present work has greatly benefited from the opportunity to visit several archives in Great Britain and the United States. For these visits I was fortunate to receive support in 1995 from the American Council of Learned Societies and the Social Science Research Council. The funds provided by the Ford and Mellon Foundations enabled me to conduct research in Great Britain where several scholars and archivists helped me access rare materials and advised on further investigations. For their help I am especially indebted to Mary Sampson at the Royal Society Library, Joanne Crane at the John Rylands University of Manchester Library, and Maurice Crewe and Mick Wood at the National Meteorological Office Library and Archive in Bracknell. I am also grateful to the expertise of the staff of the British Library, the Wellcome Institute Library, and the Manchester Central Library.

The illustrations in this book are reproduced with the kind permission of the British Library, John Rylands University of Manchester Library, Manchester Central Library, National Meteorological Office Library and Archive, Margaret I. Kind Library, University of Kentucky.

During the final stage of research and writing I greatly benefited from the support of the Wellcome Trust and the congenial atmosphere in the Wellcome Unit for the History of Medicine at the University of Manchester. I am grateful to faculty and students for their encouragement and support.

During the final stage of research and writing I greatly benefited from the congenial atmosphere in the Wellcome Unit for the History of Medicine at the University of Manchester. I am grateful to the faculty and students for their encouragement and support.

In the United States, my gratitude extends to the colleagues and archivists at the Lilly Library at the University of Indiana where, in 1995, I spent a brief period as a Ball Brothers Visiting Fellow. More recently, a research fellowship from William Andrews Clark Memorial Library enabled me to use the library's valuable holdings on eighteenth-century British science. I am grateful to their library staff for their warm hospitality and outstanding level of assistance.

At several stages during my research and writing I was fortunate to have parts of the text read and commented on by many colleagues. Their suggestions are most appreciated and here I would like to thank them for their time and assistance. I am particularly indebted to Bill Luckin who closely read the manuscript and provided invaluable advice on the matters of style and composition. I should like to express my deep appreciation for the *esprit* with which Anne Caldwell has edited several chapters of the book. My gratitude also goes to David Allen, Katherine Anderson, Glen Burgess, Bruce Eastwood, Jane Elliott, James Force, John Harwood, Lynn Joy, and John Pickstone. Audiences at a number of meetings over several years have heard portions of this work and this is a proper place to acknowledge the valuable help and advice I received from Geoffrey Cantor, John Christie, Jonathan Clark, John English, Patricia Fara, Theodore Feldman, and Graham Gooday. For their help in editorial work I am indebted to Vanessa Graham at Manchester University Press and Susan Abrams at the University of Chicago Press.

This work would not have been possible without the love and support of my mother Olivera, who has always believed in the sunny side of my study of clouds. Although geographically distant, she has always been present in my thoughts. I am deeply grateful to her for her faith and support. My greatest thanks go to my wife Kristina. Her energy, wisdom, and love were great sources of inspiration for this work.

Introduction

JACK: Charming day it has been, Miss Fairfax.
GWENDOLEN: Pray don't talk to me about the weather, Mr. Worthing. Whenever people talk to me about the weather, I always feel quite certain that they
mean something else. And that makes me so nervous.

Oscar Wilde, *The Importance of Being Earnest*

In the last year of the eighteenth century, the meteorological writer John Williams
came forward with a project to de-humidify the British atmosphere. In each
county, a pair of buildings similar to cotton mills were to be supplied with electrical machines consisting of twelve hundred rotating cylinders. The cylinders
would be connected to an insulated bar leading to the top of the building where
the bar would end in a series of lamps for the diffusion of electrical fluid in the
surrounding air. As each cylinder was capable of electrifying over 5,000 cubic feet
of air, Williams calculated that the projected number of buildings and cylinders
would suffice to electrify the atmosphere of Great Britain to a height of one mile
and thus dissolve its notorious fogs and rainclouds. This would result in drier
and more equable seasons such as the country, according to Williams, had not
enjoyed for more than half-a-century.

Williams claimed that the project would end five decades of a progressive
deterioration in the weather, of wetter summers and milder winters which had
caused a general decline of agricultural output and the scarcity of the early
1770s. The reason for this deterioration, he wryly observed, was neither France
nor the calendar reform in 1752, but an increased evaporating surface of the
land effected by enclosures, plantations, tree-felling and especially the replacement of agriculture by more lucrative pasturage. This last trend was in turn
spurred by heavy taxation on the labor and products of arable land, while the
taxation itself was the result of the inflated price of grain. Williams judged that
"the late American revolution was the cause of [this evil], and the dreadful
struggle in which England has since been involved, in consequence of the unpar-
alleled, and I may say, unexampled events which have followed the French
Revolution, have increased it." His electrical design would offset the climatological
consequences of political adventurism.[1]

The project fell through; the *Anti-Jacobin Review* compared it to Martinus
Scriblerus's schemes for finding the stellar parallax from the center of the Earth
and for solving the longitude problem by means of equatorial poles with light-
houses on their tops. In fact, there had not been any degeneration of the weather.

The English climate was indeed *varium at mutabile semper* and as such always the object of abuse; but Williams's notion that there was something special about the last fifty years was similar to the despondency which Jonathan Swift had described in his poem "Description of a City Shower": "Sauntring in Coffee-house is *Dulman* seen. He damns the Climate and complains of Spleen." The English found their weather "uncommon" and "extraordinary" whenever it wasn't what it could have been, even if the presumed norm was nostalgic rather than real. "When we had a warm summer it used to be called one of Queen Anne's summers," observed the reviewer, for whom Williams's electrical engineering, however well meaning, could not make up for its false premise and the untimely national criticism that it implied.[2]

This episode belongs to a tradition of social, political, and scientific commentary precipitated by the quirks of British weather. "Peculiar weather" has always been common in England, but the English had no ready answers as to what, if anything, was wrong with their weather, much less how to correct it. In fact, even in the late eighteenth century, little was known about weather laws at all, to the extent that many advised the use of traditional weather proverbs and rustic prognostic rules rather than physical accounts. Paradoxically, even some of those who kept daily records of weather – and who would appear to us to have believed in the possibility of atmospheric laws – considered their journals of little help in discerning the regularities of seasons, lamenting that the atmosphere was such an "irritable mixture" of airs that any attempt to comprehend it would prove futile. For the prominent meteorologist Richard Kirwan, the regularity of weather changes "has hitherto in great measure eluded all research."[3]

Such doubts are particularly visible against the aura of optimism which surrounded contemporary astronomy, electricity, natural history or chemistry within the context of which the works of Newton, Franklin, Linnaeus, Priestley, and Lavoisier announced the triumph of rational synthesis. Why, then, did the skeptics question the prospect of such an achievement in meteorology? And why was this questioning so pronounced at the end of a period which saw an immense rise in popularity of the barometer and an increasing number of people engaged in observations? Was there a discrepancy between the expectations of Kirwan and those of his predecessors? What was it about pre-institutional, pre-professional meteorology that makes it such a seemingly complex and "unscientific" endeavor?

To tackle these questions, this book addresses assumptions about the social, theoretical, and literary characteristics of British meteorology during the long eighteenth century. By considering early weather observations as narrative news about *unusual* weather, this study explores the extent to which such reportage imbibed vernacular, religious, esthetic, and emotive dimensions of human engagement with the environment. Miss Gwendolen's insight affirms that the English weather was not just about its rains, but also about national commerce, politics, religion or the esthetics of "skyscapes." Its history can be told from the perspective of bodily pain, financial loss, millenarian doom-mongering, or from the position of those who, like Thomas Baines in the seventeenth century, thought that the English air was the cause of the mutability of English thought and thus the source of national characteristics such as newfangledness, rashness and love of rebellion.[4] This book argues that eighteenth-century meteorology embodied most of these natural, social, and cultural characteristics. It shows that the study of skies was also a study of English society in which inanimate nature possessed an unpredictable power to shape the meteorological lives of people. Meteors, weather, and seasons were among the most powerful forces informing the physical, moral, financial, and political landscapes of the eighteenth century.[5]

This book centers on the public and scientific perceptions of *meteors* – the phenomena which had traditionally defined the subject matter of *meteorology* – and pays special attention to the understanding and uses of extraordinary *meteoric* events. It follows the practice of reporting – *meteoric reportage* – as a dominant form of naturalists' engagement with the weather until the late eighteenth century. The adjective "meteoric" is meant to capture the qualitative description of discrete meteorological events separated by "anonymous" interludes of atmospheric tranquility. Rather than monitoring a continuum of weather, eighteenth-century observers described the periods of sublunary commotion such as storms, earthquakes, fireballs, waterspouts, flying dragons, or northern lights. The term 'meteoric' thus refers not only to *meteors* – the fundamental concept of Aristotle's *Meteorologica* used until the late eighteenth century – but also to the fleeting, uncommon and unstable character of these phenomena. In this respect, pre-1900 investigations differed considerably from the research of modern meteorology.

Taking notice of such events was not just a meteorological practice. Weather in general and extraordinary meteoric activity in particular were the subject of private diaries, chronicles, newspapers, and sermons. Weather is mentioned in early modern poetry, drama, geography, pamphlets, almanacs, broadsheets, ballads, the popular press, and oral culture. Unusual weather was known to Pliny the Elder, who wrote about it in the *Natural History*, and to Aristotle, Seneca, Cicero, Albertus Magnus, and Paracelsus. The early modern clergy exploited it in cautionary tales about divine intervention and the coming of the millennium. Political commentary linked severe weather to military affairs, while astrologers connected it to starry malevolence. Unusual weather was a realm made up of emblems, warnings, disasters, and geographical symbols.

How can a science assimilate these life-shaping agencies into a dimension of mere "nature"? What sort of technique does it require to divest weather of its human face? These questions are especially pertinent for a science produced – unlike botany, mineralogy or plasma physics – *inside* its own subject matter. Meteorology shares this feature with other "field" sciences in which people traverse glaciers, descend into caves, or dive into oceans. In such activities, the ability to separate the immediate experience from its scientific representation – believed to be a *sine qua non* of the scientist's ability to observe the fundamental and ignore the irrelevant – requires alienation from the esthetics of the setting's sights, sounds, smells, or surprises. But when the principles of scientific practice are still in the process of being negotiated, as was the case during the period surveyed in this book, the process of separation is necessarily incomplete and its results ambiguous. The eighteenth-century observer's interaction with the environment was characterized by a life-shaping immediacy, not disembodied representations: then as today, being caught in a storm had little to do with its representations in Salvatora Rosa's canvases, James Thomson's verses, or Thomas Barker's meteorological tables. – which is not, however, to deny that these representations possessed a reciprocal power to inform everyday perceptions.[6]

Seventeenth-century attempts to construct a science of weather worked on the premise that grand meteorological events represented privileged philosophical "facts." Bacon considered these events exceptionally valuable because nature spoke more clearly when it sported itself in the "out-of-the-ordinary." Following this view, naturalists wrote about monstrous births, bizarre medical cases, and rare minerals. They also recorded strange meteorological appearances: lunar rainbows, waterspouts, odd hailstones, mock-suns, fireballs, and curious lightning effects. The works of Robert Plot, Professor of Chemistry at Oxford, chronicle oddities such as low-flying fireballs, the atmospheric music of invisible aerial creatures, and the barking of "Gabriel's Hounds" from the coal pits around Wednesbury.[7] But where paradoxographical genres described the marvelous for its powerful emblematics, writings like Plot's sought to capture the Baconian "extraordinary" and give evidence for a natural philosophy of the atmosphere. Within this approach, early modern naturalists followed the content of marvel ephemera, but refracted its meaning through the prism of inductivism.

This practice has attracted interest not least because it demands a revaluation of an image of Enlightenment science. Lorraine Daston, Michael McKeon and Simon Schaffer have explained the interest in uncommon nature as methodological, commercial and psychological. Andrea Rusnock has pointed out that regardless of their theoretical relevance, investigations like these inaugurated an impressive network of collaborators and thus contributed to eighteenth-century science. David Miller has argued that such observations must not be dismissed as mere reportage. Such practice, says Miller, was the crucible of modern scientific credibility, a forum in which the testimony of reliable witnesses could be heard.[8] Whether or not we accept these insights is less important than a realization that the "quality" of eighteenth-century science must not be measured either against the Augustan marvelous curiosities or Newtonian syntheses. Nor should meteoric reportage be considered as anything less than non-trivial natural knowledge.

This book argues that the natural knowledge of this period should not be categorized in accordance with either the new philosophy of nature or its popular offshoots. Meteoric reportage and its literary expression will be considered in the present study as one amongst other species of natural knowledge. Consequently, it will be argued that the "uncommon weather" of "scientific" observation differed from other types of weather as they were presented in religious, political, and popular literatures. Such reportage was neither more nor less comprehensive than other practices, nor did its answers bear upon issues of greater relevance than those addressed in religious, political, or millenarian tracts.

If preachers had reasons to describe storms as divine signs, and if politicians had reasons to translate them into national issues, so naturalists had reasons to see them as facts. The weather was too public, too overwhelming, and too complex to be grasped within a single discourse.

One of the problems with which the present study is especially concerned is the appeal and persistence of this genre. What kinds of epistemological, cultural, or social impulses conspired to sustain this engagement with the physical environment? To what extent was the practice shaped by methodological and theoretical considerations or by vernacular reactions to unusual weather? Beginning with biographical and bibliographical information about eighteenth-century meteorologists, this book raises questions about their formal education, social status, places of residence, and other factors that might have had a pivotal influence on their work. Were religious, political or professional motivations formative, peripheral, or extraneous to their pursuits and to what extent was their practice central/marginal to their nominal occupations?

This study, then, is an attempt to link the meteoric tradition to society at large within the context of eighteenth-century British provincial naturalistic and humanistic scholarship. These individuals – mainly Anglican clergymen and members of the gentry – were the largest single group involved in meteorological observation during this period. The biographical information relating to this community will not support the claim that the literary/epistemic features of meteoric reportage merely reflected the conditions and aspirations of its practitioners. Neither is the present analysis used as a basis for a sociological study of natural knowledge. Rather, seeing meteoric reportage as both reflective and performative, I propose that such practice was less an embodiment of the consensus about the cultural role of eighteenth-century naturalists, than it was a force which informed and defined such a role. To argue for such a two-tiered role of natural knowledge, I have located the meteoric tradition less within the hackneyed "empirical meteorology" – which it was trivially – than within a larger body of ideas concerned with the nature of *place* and the place of nature. This study thus argues that meteorological reportage – and natural history in general – constituted an aspect of English geographical thinking during a period concerned with the creation and sustenance of national, regional, and parochial identities with respect to the moral topography of the land.

The emphasis on "chtonic" placement of people and knowledge plays an important role in the present study. Historians have recently argued that our understanding of science depends on an understanding of the local settings in which it is produced and that the conditions of knowledge depend on the naturalist's location in social and physical space. For example, physical access to sites of natural knowledge in the early modern period was mandated by the premium placed on direct rather than second-hand experience. Steven Shapin's work illuminates the social underpinnings of such a view. It also intimates that the legitimacy of natural philosophy can be interpreted as the institutionalization of settings already invested with material, social, and symbolic meaning. As a result of this the rise of the 'new' natural philosophy was linked to a process in which special sites – by virtue of their being the only sites for the appearance of scientific objects – were transformed into the exclusive epicenters of scientific activity. Hence the privileged spaces of the laboratory, the observatory, the clinic, and the museum.[9] Valuable as it is in the analysis of such paradigmatically scientific sites, the localization of scientific knowledge should also include the spaces *outside* the usual sites of knowledge: the public square, the field, the beach, or the rectory garden. If science moved indoors with the emergence of the laboratory and instrumentation, what was the scientific legitimacy of outdoors? If science was to be found in a gentleman's residence or a theatre, did that very fact entail a reciprocal effect on the science produced in a forest, cottage, or mill? If experimental knowledge existed in *urban interiors*, what kind of science was allowed to exist in *rural exteriors*? In other words, while historical studies of science have explored its positioning with respect to differences in occupation, social status, religious affiliation, and political allegiance, there has been little work done on whether the philosophy of nature had much to do with the setting – physical, esthetic, and social – in which nature itself had been encountered.[10]

This oversight cannot be justified in the light of an increasing recognition of the eighteenth-century rural/urban divide and the fact that until the end of the early modern period the dominant cultural and material forms of life varied with topographical setting. To ignore this dimension is to follow Enlightenment *philosophes* in turning a blind eye to a clash between provincial and cosmopolitan society. When the eighteenth-century republic of letters prided itself on erasing the bias of local knowledge through correspondence, lectures, and libraries, it did so because its members were all too ready to assume the existence of a provincial intellectual void. The London antiquarian Roger Gale, having received some letters from provincial correspondents, wrote: "who could have expected such a learned correspondence and so many curious observations to have been communicated to, and made by a set of virtuosi almost out of the world."[11] Do similar perceptions inform historical studies which continue to emphasize the processes of diffusion and popularization, and the social uses of cosmopolitan natural knowledge in provincial settings?

This book does not question the vigor of the process which Peter Borsay called the "English urban renaissance."[12] But it does stress that an analysis of these processes must come to terms with the impact of scholarship undertaken in non-urban settings. It will be shown that eighteenth-century non-metropolitan scholarship was in a crucial sense an *autochtonous* enterprise: it not only originated in a regional setting, but it pertained to it. Such scholarship might therefore have been geographically *local*, but it was socially and politically *central* to those who participated in its developments. Critical to understanding this dual character of the provinces is the question of how regional confines informed intellectual perceptions and determined the scope, profile, and priorities of eighteenth-century natural and humanist investigation. How did rural society perceive urban culture? How did the rural gentry look at science pursued in the capital? What were the differences and convergences between the rural rectory and the Royal Society? What did it mean to be in a field rather than a laboratory, in a cave rather than a lecturing hall? Finally, did the gulf between metropolis and province shape the uses, definitions, and understanding of "nature" in these two settings? This book engages with these issues by exploring the political and cultural arguments used to legitimize knowledge produced in the countryside (as opposed to the city), in the field (as opposed to the laboratory) and by provincial parishioners (as opposed to indoor experimentalists and urban *philosophes*). It discovers science-sites where they are not usually looked for, but were nevertheless equally powerful in legitimizing natural knowledge.

The abruptness of the social and cultural transition between the capital and the provinces during the eighteenth century suggests a radical change in the degrees of environmental impact on life in these settings. I argue that if the rural landscape had a remarkably human face – both defining and defined by boundaries of landed property, habitation, agricultural practices, and the limits of mobility – its inscription in the work of provincial scholars acquired a tone conflicting with that of their urban compatriots. This difference is visible in the customary tendency on the part of local communities to establish their identities through a record of the unusual and the exceptional. The plausibility of such material was bolstered if local observers could assign such information to a specific locality. Conversely, such records bestowed a unique identity on surrounding regions. "'The *Romans* were here,' says the antiquary; 'the Romans were *here*,' says the local historian." The emphasis on *here* supplied a social rationale for the very popular seventeenth- and eighteenth-century genres of English local history and regional geography.[13]

What was the meaning and purpose of these genres? The historian of English geography, Richard Helgerson, has argued that with their historical, genealogical, and topographical undertakings, seventeenth-century regional geographers – usually referred to as chorographers – and their patrons sought to transfer national identity from loyalty to the monarch to loyalty to the land, thereby securing for themselves political immunity and customary privilege. In such studies chorographers enacted a region-oriented cultural quest for identity, for the study of which local annals, antiquities, and geographical physiognomy were prime vehicles. Landed property made itself visible in chorographic works and maps, in exchange for which the land granted its transcendental authority to its chorogaphers and their patrons. Thus chorographic scholarship not only embodied the social ambitions of its practitioners and their audiences, but served as a medium and norm of their material, religious, and political individuation. Chorography was particulary prominent among late Stuart scholars, whose parading of the 'disinterested' evidence about natural events and local monuments never sought to conceal the gentrified patronage and local pride which underwrote such studies. Vicar White Kennet from Ambrosden stated in 1695 that next to his official duties, he did not know "how in any course of studies I could better serve my patron, my people and my successors than by preserving the memoirs of this parish."[14]

By the end of the seventeenth century, chorographers were appropriating Bacon's principles of natural study. At this juncture, "empirical" investigation of local nature turned into a powerful means for evoking regional eminence and drawing conspicuous attention to it on cultural maps of the nation. Regions acquired identities through their local nature, while local nature acquired the identity of its regions: Cornish storms, Yorkshire echoes, and the salubrity of Norfolk's air were not only natural, but also historical peculiarities of these areas. Especially potent markers of local identity were *unique* natural events: the "prodigious," "rare," "uncommon," "extraordinary," and "strangest phenomena ever seen" transferred regional ontology into the sphere of curiosity and gave it an impressive intellectual cachet.

It was this association with the "unique" rather than the "weather" which placed the eighteenth-century interest in uncommon meteors within the project of regional topography and local history. Unusual, prodigious and destructive weather was not interesting merely as such, but because it occupied a specific place in a region, a parish or a village. It was observed because it was important to those associated with these topographic and administrative units, for whom such weather existed beyond and above the gaseous atmosphere. Eighteenth-century "observational" meteorology was thus subordinate to thinking about the meaning of landed society and it both served and defined that culture during the British *ancien régime*. Yet that is not to say that chorography and natural history acted as a prosthesis of landed interest; they were one of its very resources.

Regional study can help us move away from epistemic grounds of meteoric reportage and appreciate its *cultural* relevance. When eighteenth-century provincial naturalists recorded remarkable meteors, they didn't do it because they saw such phenomena merely as Baconian data, but because they practiced chorography and sought visibility within the national republic of letters. It is crucial to understand this development in view of the fact that chorographic writers – many of them provincial correspondents of the Royal Society – had in common a provincial and mainly rural, residence, level of income, university education, and parochial authority. Provincial clergy played an important role in these activities. In the early eighteenth century the country parson had become representative of a new class of natural historians. His provincial residence not only stimulated his studies of local nature but at the same time added to the social visibility of the lower provincial clergy who offered spiritual guidance, engaged in local administration, and participated in the vernacular culture of the community. This book demonstrates that, in their role as proxy administrators and local scholars, clerical naturalists sustained a chorographic ethics of research. Thus their authority derived not merely from codes of gentlemanly civility, but from their residing in a place to which metropolitan naturalists had no access and over which they could not exert their authority. This study argues that provincial meteorological reportage was an autochtonous scholarship of the countryside: it translated the physical, social, and epistemological placement of its practitioners into a genuine engagement with nature. This placement explains why early meteorological facts recorded "important" parochial weather rather than the progress of esoteric parameters which could never fully reflect a vivid experience of extraordinary meteoric appearances and severe weather.[15]

With its stress on qualitative and narrative framing of "important" weather, this view offers an alternative approach to understanding early weather research. It suggests a new way of looking at the relevance of instrumental recording in eighteenth-century meteorology. It provides a challenge to the often held view that instrumental recording represented the cornerstone of modern meteorology. For instance, historians of science have generally associated the "beginning" of meteorology with regular weather records and the decline of the Aristotelian doctrine of *meteora*. For Kirwan, the birth of meteorology coincided with the work of Toricelli, Cornelius Drebbel (the inventor of the thermometer), Charles Theodore, and the Elector Palatine of the Rhine (founder of the first academy devoted to an international collection of weather observations). Later historians have linked these organized efforts with the end of qualitative description of meteors and maintained that until the 1780s meteorological theories remained "extremely vague and unsatisfactory" lacking the "necessary fundamental physical theory for it to be otherwise." After this period, meteorology changed from a "qualitative pursuit" to a "quantitative subject based on abstract, mathematical theory and precise, systematic experiment." It is argued that in this process naturalists turned away from Aristotle's *Meteorologica*, and centered on new techniques for measuring the atmosphere and debating its chemical, electrical and physical properties.[16]

Instead of historical changes and shifts, the present reading of meteorology emphasizes a cultural and epistemological continuity spanning the period from the end of seventeenth to the end of the eighteenth century. Most meteorologists during this period used Aristotelian *meteora* to make sense of the aleatory character of weather phenomena; narratives about unusual weather situations in this sense instantiated *meteora* and so perpetuated a classical understanding of meteorology. This book does not therefore offer an account of how several early anticipations of modern meteorology negotiated their wider acceptance against the neo-scholastic survivals of the Aristotelian ideas, nor how these "seminal" breakthroughs eventually came to be institutionalized during the nineteenth century. Rather, it explains what the *dominant* view of the matter was for the *greater* part of the period under survey, maintained by a *larger* number of naturalists, their informants, and their audiences.

This stress on continuity of theory and practice is at odds with the tradition which sometimes treats eighteenth-century science as an unassuming transition from the Scientific Revolution to exuberant nineteenth-century professionalization. Historical accounts of these periods would be untenable without the use of notions such as "change," "shift," "decline," "rise," "new attitudes," or "ascendancy." This change-oriented historiography rests on evidence emphasizing the definition of natural knowledge as grounded in new methods, new forums of institutionalization, new vehicles of accreditation, and new social implications. If our interest is in these trends, then a history of Hanoverian meteorology would indeed emphasize discontinuities and change. It would confirm the view that Aristotelian theory declined after the electricians explained lightning, cloud-formation and earthquakes. Simultaneously, meteoric reportage could be described as challenged by measurements and networks making the natural historic idiom look ludicrous when juxtaposed against laboratory simulation and analysis of the atmosphere.

It is beyond doubt that an increasing number of late eighteenth-century naturalists began to argue for different methods and aims in meteorology. Yet this change was not the result of rational argument winning provincials over to the coordinative enterprise of quantification. As this book shows, such arguments were put forward several times during the eighteenth century without produ-cing any far-reaching change in the provincial practice of qualitative reportage. Symptomatically, however, many potential observers already kept naturalist calendars of seasons with information of intrinsic local interest. Why did local naturalists not participate more enthusiastically? The reasons may have included the scarcity of reliable instruments, seasonal mobility, or a lack of discipline. But we may go further and question the good judgement of London's meteorological projectors when they hoped to elicit support from the people who had no stake in joining a venture which would deprive them of authority and intellectual criteria grounded in regional identity. When qualitative reporting eventually began to disappear in the scientific press of the early nineteenth century, it was not because provincials adapted to a new discipline and embraced the dictates of centralization, but because, more trivially, the editors of the scientific press ceased to accept their contributions. This was an end in visibility, rather than a triumph of rationality.

What reasons did the new meteorologists and editors of the early nineteenth-century journals propose against meteoric reportage? The most frequently quoted argument against the chronicles of unusual weather had to do with methodo-logical inadequacy. What purpose could there be in describing a single jack-o-lantern or a single lunar rainbow? How could such rare instances add to meteorological knowledge? Was there a methodological tool available to connect singular data into a coherent body of knowledge? When, by the end of the eighteenth century, such questioning became more insistent, especially among new groups of experimentally trained naturalists, provincial reports on indi-vidual weather events began to appear out of touch with the practices of natural science. Indeed, by the 1810s, the most influential scientific periodicals ceased to publish observations on 'most uncommon' meteorological events and started to devote themselves to far less exciting tables of daily and monthly means of temperature and atmospheric pressure. But the change did not occur overnight: there were at least three factors whose cumulative effect accounted for the de-clining public visibility of meteoric reportage.

The first was the impact of natural *theology* on the meteoric tradition as an agency of "normalization" of its "prodigious" and "numinous" dimensions. The emphasis on knowable cycles of benevolent weather represented a tendency to diminish the significance of individual meteorological events on both theological and epistemological grounds. From the theological point of view, the meteoric tradition was criticized as a survival of Puritan enthusiasm and political millenarianism, both of which were unacceptable to Georgian intellectual orthodoxy. From the epistemological point of view, descriptive accounts of meteors were criticized as non-comparable and hence useless as instances of disciplined induction. The second factor was the rise of practical agrometeorology. Mid-eighteenth century agricultural writers adopted a natural theological perspective and argued for the possibility of predicting the weather based on the traditional weather rules. They distanced themselves from *causal* explanations of meteors and focused on the *statistical* correlation between signs and signifieds. Distrusting instruments, they favored the empirical authority of the ancient writers and modern shepherds. They argued for a science based on long-established praxis, not on instruments and newfangled hypotheses. Most importantly, they stressed the agricultural relevance of seasons and explained abnormalities as either normal, or irrelevant to the vegetative cycle. The result of this shift of interest was a situation in which meteoric reportage seemed increasingly redundant either as a prognostic tool or as practical knowledge.

The third factor comprised new theoretical developments. Most important here was the strong bias toward chemical investigation of the atmosphere. Meteorologists who subscribed to the idea that meteorology represented no more than a species of chemical philosophy argued that no atmospheric process ought to be left unrecorded, however insignificant it might initially appear. Thus, where the agrometeorological meteorology of seasons normalized the problematical weather of meteoric tradition, chemistry problematized the normal. In asking their contemporaries to view problematical as normal and normal as problematical, late eighteenth-century meteorologists put an end to the meteoric tradition and, by extension, to classical *meteorologica*.

The result of these processes was that the chorographic impulse lost its cultural rationale in sustaining local natural history. With the enquiry into *local* weather becoming a methodological prerequisite for a knowledge of *globally* evolving atmospheric systems, the meaning of locality changed from its status as an exclusive end of investigation to a specimen in a larger entity, a point on a grid. Scrutiny of local weather, whether in the form of rules of prediction or series of observations, mattered only to the extent to which atmospheric "unity" manifested itself in a locale. Rather than seeking meteorological attention to place, observers began to monitor the atmospheric vagrancy displayed in the motions that now had a global significance. Their local relevance was incidental. And even though the emphasis on observation in distant and more numerous *places* was as strong as ever, demand was determined by methodological, rather than cultural norms. In this conversion of emphases, the weather was about to lose its human relevance, and its investigation its appeal in terms of the topophilic culture of the Georgian countryside. The culture of "country airs" lost out to the physics of planetary circulation.

These new emphases did not reflect a disembodied rationality of chemical and laboratory expertise. This expertise was promoted by a community of practitioners which shared the same expectations about what meteorology should be and how it should change. Importantly, this community also shared a distrust in the traditional meteorological experience because such experience could not meet new priorities, nor make sense of new knowledge that had become available during the last decades of the eighteenth century. These new priorities were publicized by those whose training, practice, and prestige evolved around ideals fashioned in a specific geographic and didactic environment: the city and the laboratory. By changing the scale of meteorological phenomena from life-shapers to parameters, metropolitan experts worked to replace the curiosity-driven authority of rural parsons with indoor simulations of atmospheric "tides" and storm paths.

This book follows this series of transitions by examining major seventeenth- and eighteenth-century meteorological books, papers, and weather-related ephemera. The first two chapters look at how classical meteorology – from Aristotle to the mid-eighteenth century – appeared in popular, religious and philosophical treatments of severe weather and how these treatments shaped provincial meteoric reportage. Chapter 3 complements these analyses by paying closer attention to the early eighteenth-century tension between religious and secular uses of weather. It shows how the Great Storm of 1703 and the northern lights of 1716 gave rise to arguments on the origins and meaning of meteorological phenomena and the limits of natural explanation.

The two succeeding chapters engage with meteoric reportage from a social perspective. For example, chapter 4 argues that the regional study of history and nature – chorography – provided cultural resources for the practice of local natural history, and specifically, for the investigation of regional weather. The chapter looks more closely at the methodological aspects of this tradition and shows that early reporters put more premium on the common sense of informants than on the social accreditation of witnesses. In chapter 5 the career of the Rev. William Borlase from Cornwall serves as an example of the regional study of nature which I consider formative to eighteenth-century meteoric reporting. Here it is also shown that theoretical developments made in the "high" (metropolitan) tradition of natural philosophy only marginally influenced practices of country clergy and gentry. Consequently, meteoric reportage is seen as reflecting concerns about social identity and religious duty rather than a growing "urbanization" of rural intellect.

In the last two chapters, attention shifts to the marginalization of the meteoric tradition. Chapter 6 examines how the factors and priorities formulated outside the chorographic culture of individuation undermined classical meteorology and the important role which this had on the rise of the agricultural prognostics of weather. Chapter 7 expands on the displacement of the meteoric tradition (with its focus on the analogy between the atmosphere and the laboratory) which by the end of the century defined debates on nature and limits of meteorological knowledge. Metropolitan chemists persistently argued against the relevance of qualitative meteorological data and in the early decades of the nineteenth century created a community whose reliance on analytical methods required an experimental education. By associating meteorology with experimental training and chemical analysis, they presented all other meteorological writers as outdated dilettantes and their unsystematic records as unworthy of consideration. Meteorology ceased to come from the countryside in the form in which its dwellers considered it appropriate: from now on it was to be imposed on to the countryside and delivered in the form required by its metropolitan reformers.

Did this displacement cause the local enthusiasts to protest? We do not find much by way of complaint, but there does continue to be a strong interest in parochial weather. Nineteenth-century newspapers, magazines, popularizations and the publications of natural history clubs persisted with articles on extraordinary colds and heats, spectacular rainbows and sublime northern lights, romantic storms and mischievous whirlwinds. Extraordinary weather was perhaps relegated to the "footnote" of meteorology but it survived in the titles of the newspaper stories and in topographic pamphlets, a pale reflection of Stuart chorographic folios. During the last decade of the nineteenth century, John Cairns Mitchell published a booklet which, in addition to praising the salubrity of non-industrial Chester, described an unusually mild winter with the explanation that "the weather of 1893 has been so remarkable as to call for special notice, being most memorable, and well worth a place of our annals." Notwithstanding the experts, remarkable weather survived in annals, memoirs, weather talk, and the daily press. This book is about an era in which all these cultural conduits coalesced into a pursuit of autochtonic meteoric reportage.[17]

I

Imperfect mixtures

A great error possesses mortals: men believe that this universe is an accident, revolving by chance, and thus tossed about in lightning bolts, clouds, storms, and all the other things by which the earth and its vicinity is kept in turmoil.

Seneca, *Naturales Quaestiones.*[1]

In a science so very difficult as that of the weather, it is not to be supposed that anything like a certain and established theory can be laid down: our utmost knowledge in this respect goes no farther as yet than the establishment of a few facts.

George Adams, *Lectures on Natural and Experimental Philosophy*[2]

Meteorology is about meteors: this makes it unique and precarious. For meteors – classical meteors – are not plants, minerals, mammals, or stars. They are not things to be measured, observed, or displayed, and a science of such entities naturally invites doubt, uncertainty, and criticism. Modern generations are not the first to jest at the experts' professed ability to predict the weather, nor are we the first to doubt the efficacy of anti-hail projectiles launched into stormy clouds. It had already been noticed, at the turn of the nineteenth century – and more than two millennia after the first meteorological work had been written – that "observations we are in possession of are too few and too inaccurate for the purpose of forming a [meteorological] theory." Meteorologists were thus at a serious, perhaps unredeemable, loss in handling the vicissitudes of atmosphere. Some called for aggressive data accumulation, some insisted on more theory and hypotheses, some argued for the value of folk weather-wisdom, and some pursued astrometeorology. Following the paralyzing fog of 1813, a London writer observed that nothing was so "striking a proof of the little progress hitherto made in meteorology, than the difficulty of proposing a legitimate explanation of a phenomenon so common and familiar as a thick fog during winter." What emerged in lieu of these laments was a consensual picture of meteorology as profoundly different from other sciences. Those sciences, some argued, could be viewed as the species of Linnaeus's *Horloge Botanique*, that is to say at a specific stage of intellectual development. Among these, however, there was one "which owing at once to the bodies that . . . compose its members, and to the nature of the agencies, combined with the inadequateness of the senses or ordinary faculties of man to grapple with its parts, must be considered as an exception to the rule, as though, in mockery, placed without the pale of human attainment; and is it necessary for me to add, that the science alluded to is *Meteorology*."[3]

Something about the weather eluded the scientific mind when no correlations, laws, or theories could challenge a belief in its inscrutability. To appreciate such attitudes it is necessary to look at how naturalists in the past tried to come to terms with the weather. To do so, we will begin with early meteorological ideas and explore their evolution up to the period with which the present book is concerned. It can be shown that eighteenth-century meteorology was based on a conjunction of ideas drawn from several cognitive provenances: Greco-Roman natural philosophy, Elizabethan and Tudor paraphrase of that knowledge, and an interrelated set of beliefs derived from astrology, magic, and weather folklore.[4] Despite theoretical developments during the mid-eighteenth century – notably, research into atmospheric electricity, hygrometry, and chemistry of gases – a majority of eighteenth-century meteorologists perceived their subject within a framework established by the ancients. For example, in his *A Rational Account of the Weather* (1738), John Pointer juxtaposed contemporary opinions on the causes of wind with those drawn from the canonical works of Aristotle, Pliny, and Seneca. Samuel Horsley, in an article for the *Philosophical Transactions* in 1775, discussed the prognostics of Theophrastus and Aratus while William Jones, in his *Physiological Disquisitions* (1781), quoted Cleomedes and Horace on lunar influences on weather and tides. The references to Greek and Roman authors was not made merely in passing. It will be demonstrated below that the definition of meteorology – the order of presentation of its subject matter and a predominant part of its causal structure – all reflected ancient conceptions of meteors, their behaviour, and their position in the universe. As a result, theoretical meteorology at the turn of the nineteenth century may be said to have been the result of the decline of classical theory which, by the 1850s, had lost its original meaning, scope, and methods.

The term *meteoros* has an ancient genealogy; here we are concerned with only two of its important senses and derivative forms. *Meteor – iso* means to "raise to a height," which in the passive, *meteoros*, is "to be raised up," "to be suspended," "aloft," "elevated," even "sublime," "noble," and "magnificent." In this sense *meteor* may refer to "rising" (of the smoke or dust, water or vapor), "rising up" (as from one's bed), or of wind "rising" from the stomach (*meteor – menos* means "suffering from flatulence," where "flatulence" may stand for both an intestinal disorder or windy and empty speech). In this sense *meteoro – logeo* is best translated as "talk of high things" and *meteoro – logia* as discussion of *ta meteora*, or elevated natural phenomena. A *Meteorologikos* is a person skilled in meteorology. The second sense refers to the specific *character* of human thought and speech. *Meteoro – kopeo* means to "prate (chatter) about high things." *Meteoro – leshes* refers to "star-gazer," or "visionary." In this context *meteora* signify "abstruse" or "lofty" speculations. In addition, a mind might be said to be *met – eoros*, or "excited," "in suspense," or "in doubt." With the same context of meaning we also encounter *meteor – geomes* which denoted "mental trouble," or "disturbance," as well as "vain, wild imagining." Within a similarly negative framework it might also mean "to buoy up," or "elevate" with false hopes.[5]

Antiquity also knew of the term *meteoro – fenaks* (a "quack" in matters of *meteora*) in opposition to *meteoro – sofistes* (a "philosopher"). "Talk of high things" might therefore become a "high (or tall) talk of high things." This suggests that two of the above meanings might blend in such a manner as to suggest a profound uncertainty about the status of these phenomena, to the extent that an understanding of this status could be seen as a contrasting image of more orderly natural phenomena. The history of meteorology may thus be conceived as an effort to resolve this uncertainty, or, better yet, as a series of recurring failures to do so.[6]

Meteorologica

Aristotle was not the first to treat meteors in an extended way, but was the first to make their explanation part of an all-inclusive doctrine of the natural world. In the doxographic parts of the *Meteorologica* he presented the views of Hippocrates on comets, Democritus on the Milky Way, Anaxagoras on hail, Anaximenes on earthquakes, and Empedocles on lightning. According to Aristotle's definition, *meteorologica* – the study of "things on high" – examines phenomena which occur less regularly than the movements of the primary element (i.e. the "aether," the fifth element, which is a substance of the heavenly bodies and their spheres).[7] Accordingly, his investigation of *meteora* continues to follow the enquiries into the causes of natural motion (*Physics* v–vii), the movements of the stars (*De Caelo*, ii), the transformation of the four elements (*De Caelo* iii, iv), and the phenomena of growth and decay in general (*De Generatione et corruptione*). This makes *Meteorologica* one of the so-called *libri naturales*, the "natural books," which endeavored to put the first principles of Aristotle's natural philosophy into practice. "Natural books" discuss sciences which did not enquire into the essence of things, but rather into "the essential attributes of the genus of phenomena dealt with in these books."[8] *Meteorologica* thus explores essential attributes of sublunary bodies, generated in the combination of earth, water, air and fire.

These elements represent the material cause of meteoric production; the efficient cause was the celestial sphere whose motion, transferred into the sublunary sphere, produces meteors as mixtures of elements. It is certainly significant for the scope and possibility of meteorological knowledge that the earliest formulations defined it as a study of "imperfect" and transient mixtures of elements. Meteors, in other words, either did not contain all the four elements or contained them in a compound of different ratios. In addition to this vague characterization, Aristotle deliberately omitted the discussion of the formal and final causes of meteors which he justified with an explanation that when the transformation of matter was still in a rudimentary stage – and meteors were at just that stage – the causes of this transformation remained unknown.[9]

More generally, Aristotle addressed these issues by denying that inductive reasoning (*epagoge*) could yield certainty when applied to the realm of sublunary corruption and change. The best that could be achieved was speculation based on analogy. In *Meteorologica* I.7, this view is linked to difficulties involved in writing an account of comets. In such cases, "the matter has been sufficiently demonstrated according to *logos* if we bring it back to what is possible."[10] Other meteors "we find inexplicable, others we can to some extent understand." (339a 1). Overall, the *Meteorologica* rests on the assumption that the subject matter requires one to accept an analogical possibility rather than demonstrable interpretations of natural bodies.

The central question, then, is the nature of the matter which fills the region of meteoric events, the region between the earth and the moon. Aristotle explains that this region contains two principal agents which he terms *exhalations*. These are produced by the sun's action on the surface of the earth. The first – the hot and dry exhalation – is produced when the sun's rays fall on dry land, from which it rises toward the fiery region of the sublunary sphere. It consists of earthy particles which are kindled in the fiery layer by celestial motions or some other cause.[11] The second – the cool and moist exhalation ("vapor") – occurs when the sun's rays fall on water. It is composed of water particles. Ascending, it mixes with the dry exhalation and produces the moist and hot "air" (*aer*), or the "inner" stratum, from which cloud, rain, snow, and so on are produced. This "air" differs from the element air in that it is thought of as a compound of exhalations, not a combination of qualities. Consequently, no exhalation can exist without the other, but one always predominates in any given mixture. The fiery stratum is thus almost exclusively composed of dry inflammable exhalation, while the moist vapors preponderate in the lower strata.

What sort of meteors are produced by these exhalations? Aristotle begins with those created in the fiery layer which consists of the dry exhalation under the influence of a celestial sphere.[12] The phenomena produced here are shooting stars, comets, the Milky Way and "burning flames" (aurora borealis (northern lights), in modern terminology). The next layer – the "lower atmosphere" – contains rain, cloud, mist, dew, snow, hail, and the causes of rivers, springs, climatic changes, seas, and coastal erosion. In this same region, meteors defined by the action of air include winds, earthquakes, thunder, lightning, hurricanes, firewinds and thunderbolts. The sequence ends with halos, rainbows, and mock-suns.

Aristotle next defines a meteor in terms of exhalation dynamics. Meteors in the high sublunary layer are produced when the celestial sphere sets the layer in motion, causing the inflammable exhalation to burst into flames. The kind of meteor which is formed depends on the place, distribution, quantity, and homogeneity of the material available (341b 22–35).[13] In the lower regions, the cool and moist vapor rises up from the water surface, cools down and condenses into the clouds, which may either give rain or remain in the form of mist (depending on the availability of vapors, 346b 22–36). The process of "condensation" (*pikneosis*) does not, however, carry a narrower meaning, but refers to an elementary change: mountains thus have the ability to *generate* (condense) rain from air that is potentially water (352b 1 ff). The same holds for the term "evaporation": dew and hoar frost are said to appear when the evaporated moisture ("elevated by the sun's heat") has not risen enough during the day and falls again when it is cooled during the night.

These are non-controversial issues. Production of hail, on the other hand, presents a difficulty because of the counterintuitive appearance of hail during the summer. Aristotle rejects Anaxagoras's view that this occurs when a cloud freezes at high altitudes. He denies this on observational grounds – hailstones are rarely observed on high mountains (sic) – and explains the meteor by using the notion of *antiperistasis*, a "mutual reaction" between elementary qualities. It may also mean "to oppose by surrounding" or to "be replaced by another substance," as the water yields place to any body immersed in it.

The use of *antiperistasis* illustrates Aristotle's contention that analogical thinking could secure knowledge in the absence of demonstrative proof. The analogy used in the theory of hail production is that subterranean heat – underground places are colder in hot weather and warmer in frosty weather – serves to suggest that the cause of hail ought to be sought in a contrary reaction somewhere in the sublunary region. In warmer seasons cold is concentrated within the surrounding heat. When the cold cloud descends into a layer of hot air – a process contrary to Anaxagoras's ascending theory – the "mutual reaction" is strong enough to effect freezing and form hail (348b 2–20).

Similar agonistic analogies informed meteorology until the late eighteenth century and were especially appealing when used in the theories of wind, thunder, lightning and earthquakes. *Meteorologica* describes winds as hot and dry rivers of air, each with their individual qualities, direction of movement, and medical properties. Such a view allows that at certain times two or more winds may simultaneously blow over the same region (364a 27 ff). The wind is, therefore, not the air in motion, as Hippocrates contended, as it would imply that all winds are one, "without differing in reality." This would be equivalent to thinking of all rivers as one (349a 25–32).

In addition, having positioned the wind in the lower region of sublunary space, Aristotle uses it in his account of earthquakes – which take place when an exhalation agitates the earth (366a 5) – and thunder, which is the ejection of the dry exhalation caught in the clouds during the process of condensation. In the latter case, the ejected exhalation strikes the surrounding clouds and makes the familiar sound of thunder. Thunder can also be occasioned by the collision of a wind and a dense cloud in which case "[t]he ejected wind burns with a fine and gentle fire, and it is then what we call lightning" (369a 10–4). Lightning is produced after the impact and so later than thunder" (369b 5–9). Finally, hurricanes, typhoons, firewinds and thunderbolts are all brought about by the abundant flow of dry exhalation; they differ from thunder mainly in the quantity of the exhalation.

It is clear that *Meteorologica* presents a theory quite different from the modern understanding of meteorology. It also includes a larger number of phenomena. On the one hand, it covers all phenomena that, for different reasons, linger "elevated" in several sublunary layers. It includes entities associated with the terrestrial globe with the exception of living creatures and stones, although both of these are touched upon in the fourth book.[14] On the other hand, *Meteorologica* does not, in any obvious sense at least, suggest that the sum total of meteors (rain with wind and occasional thunder, for instance) can represent something that could be described as the "weather." Meteors are understood *primarily* as individual and individually explicable items in the sublunary sphere – perhaps in the sense of *meteoro – fanos* ("appearing in the air") and only *incidentally* as weather phenomena. This distinction makes sense of Aristotle's interest in enumerating, ordering, and explaining *the totality of phenomena* produced in a specific cosmological realm, rather than a concern with *weather patterns* in a geographic area. Put simply, *Meteorologica* is concerned with meteors, not the weather.

From a structural perspective, Aristotle's explanations are narratives with an intrinsic diachronic dimension: the explanation of a meteor is equivalent to the process by which it comes into existence. In other words, each of the narratives construes a meteor as a contrived *event* which involves protagonists (elements, exhalations, vapors and other meteors) undergoing transformations and culminating in the production of the meteor in question. When Lucretius later wrote about things "happening in the earth and sky," and "events" and "transactions" taking place in the heavens,[15] he was voicing the notion that meteors occur in the same way as battles, epidemics, and theatrical plays. This early recognition of the temporal yet quantum character of meteoric appearances would continue to inform subsequent discussions and acquire prominence in the seventeenth and eighteenth centuries, when the strictures of meteorological reporting demanded narrative form. Aristotle's doxography, theory, and style represent the basis for much of this later meteorological tradition. The *Meteorologica* had been summarized in his *De Mundo* which had a significant circulation after its first Renaissance edition in 1497. Indeed, the Hellenistic and later meteorology derived from it was an elaboration of Aristotle's theory of exhalations. It was, however, never simply a restatement of it.[16]

One contentious issue was whether meteors participated in the world's order or not. Aristotle thought of meteors as being subject to less-than-demonstrable theory. But *Meteorologica* does not present meteors as either inexplicable or ontologically indeterminate. When the Hellenistic commentators took up the subject, however, *epistemic* concerns sometimes became *ontological*. Theophrastus and Lucretius stated that meteors followed no rational order. Seneca argued to the contrary, and Pliny distinguished between regular and accidental meteors. Not only were meteors considered difficult to account for, but that difficulty was now believed to originate in the world, not the mind.

While a recent reconstruction of Posidonius's natural philosophy suggests minor additions to Aristotle's ideas,[17] Theophrastus's *Meteorology* differs from Aristotle's in respect to its content, form and method. It is organized according to elemental order (fire, water, air, earth), rather than "presentational" development from the two exhalations. Significantly, however, Theophrastus introduced explanations with multiple causes.[18] For thunder, for instance, Theophrastus lists seven causes: collision of clouds; rotation of the wind in a hollow cloud; clouds extinguishing falling fire; wind striking an icy cloud, and so on.[19] Lightning and earthquakes have four causes; the formation of clouds, two. But even if Theophrastus's interpretation features neither the notion of "opposing tendencies" – the *antiperistasis* – nor the idea of mixed exhalations, his conception of meteorology remained Aristotelian to the extent that it continued to use exhalations and vapors as primary explanatory vehicles. Thus, winds are formed from a fine vapor, clouds from a moist one, and earthquakes from the winds enclosed in the bowels of the earth. Theophrastus also enriched Aristotle's use of analogy by introducing a compendium of familiar instances intended as approximate explanations. The same strategy was followed by Epicurus.[20]

Theophrastus, however, radicalized Aristotle's assumption of the indeterminacy of meteorological knowledge in arguing for an ontological separation between the cosmic order and the irregularity of meteors. He writes that regularity may only *appear* as the result of several simultaneous causes, but that, ultimately, meteors cannot originate in God and are not endowed with order: "It is thus not right to say [about] hurricanes that they come from God; [we may] only [say the following] about something that happens to us to our harm or that diminishes divine power: *it happens without any order.*"[21] Thunderbolts are for this reason known to have fallen in uninhabited areas, in seas, on trees and animals, and yet "God is not angry with those!" Moreover, thunderbolts are also known to have struck "the best people and those who fear God, but not those who act unjustly and propagate evil." Thunderbolts and natural disasters are thus not of divine origin.[22]

In the encyclopedic *Naturales quaestiones* Seneca denies this assertion. As recurrent and orderly, storms and winds are knowable and are the subject of a science relating to events "occurring between the sky and the earth."[23] This places meteorology (*sublimia*) between uranology (*caelestia*) and geography (*terrena*). Seneca's reasons for making meteors a part of knowable order reflect his theology and ethics: in his view, an investigation of celestial phenomena helps humans transcend their limitations through an understanding of lofty things such as stars, comets and meteors. The order of exposition and some of the theoretical treatments differ from Aristotle's, but a more substantial difference is the lengthy discussion of the theory of divination, which, when taken up by Pliny, serves as a pretext for his extensive compilation of miraculous meteors.[24]

Pliny distinguishes between regular and accidental meteors.[25] Local winds, storms, gusts, and bursts of subterranean air come from the exhalations of the earth and "their onrush is quite irregular" (*NH*, II, 49). The wind in general, however, is a mere motion of air whose steady blowing obeys the seasonal change and influence of stars (*NH*, II, 45). This distinction rests on a difference in causes. Purely sublunary and spatially local phenomena are accidental, as opposed to those caused by the regular motions of the celestial sphere, which are necessary and undergo regular cycles. Noting the Tuscan belief in terrestrial thunderbolts, Pliny writes that they were not "the ordinary ones and [those that] do not come from the stars but from the nearer and more disordered element [i.e. the air]" (*NH*, II, 43). This differentiation between sublunary disorder and celestial regularity reflects Aristotle's distinction between the material cause (mixing of exhalations), and the efficient cause of meteors (the movement of the celestial sphere). It can be also accommodated within Seneca's understanding of the airy region as a medium of communication between the starry vault and terrestrial processes.

Pliny's work gives us an insight into the kind of interest in sublunary *accidentalia* that would recur in early modern meteorological discourses. *Natural History* abounds with the meteorologically uncommon. The work reports on unusual meteors mentioned in earlier sources (*res*, *historiae*, and *observationes*) and the grotesque incidents and paradoxes (*curiosum* or *mirabile*) which can in part coincide with *historiae*.[26] These meteorological *mirabilia* include rains of blood, flesh, milk, and iron, accounts of predictions of earthquakes and falling of sun-stones, as well as the principles of auguries from thunderbolts. Typical of Pliny's reportage is a story about a lady at Rome struck by lightning when pregnant, and "though the child was killed, she herself survived without being otherwise injured" (*NH*, II, 52). There are also reports on apparitions in the sky. During Marius's consulship, inhabitants of Ameria saw two heavenly armies advancing from the East and West to meet in battle. Phenomena like these could be of interest both for their own sake and as natural signifiers. Thus in the study of the earth, earthquakes, rivers and springs are considered subject to a natural law, but also appear as *signs* (or portents) of impending political, military, or social turmoil. This double causation came to be widely adhered to in later periods.[27]

Before we proceed to classical meteorology in medieval and early modern scholarship, another group of ancient works deserves a brief mention. These are the works on the prognostic signs of weather, i.e. the weather-signs. There is a considerable difference between theoretical meteorology and the content of these treatises. If Aristotle defined meteors as the elements in a system of sublunary change – that is, in terms of exhalations, *antiperistasis*, condensation, and ignition – writers on weather-signs dealt with the concrete "species" of weather outside theoretical systems. They were less interested in the "essential attributes of the genus" of meteors, than in the knowledge of weather patterns in a concrete location and at a specific time of year.

The first Greek text of this nature is the *Phaenomena* by Aratus of Soli, written early in the third century BC. It is an astronomical poem to which a section is added on prognostic signs, entitled *Diosemiai*. This and the anonymous *De Signis Tempestates* (attributed to Theophrastus), represent two chief sources for the most influential work in the genre, the first book of Virgil's *Georgics* (37 BC). Poetry rather than a didactic work,[28] the *Georgics* was the most reliable source of weather signs and agrometeorological knowledge in the ancient period.[29]

The poem discusses seasonal changes of weather as far as they affect the farmer and sailor, that is, as far as they concern ordinary people rather than philosophers. The *Georgics* lists the signs of stormy and fair *weather*, dry and wet, windy and calm, warm and cold, clear and foggy. These kinds of weather are foretold by plants, animals and other phenomena: by the behavior of ants, swallows, frogs, and ravens, or by characteristic appearances of the moon, sky, clouds, and stars. It should be noted that, first, these weather-signs were presented as purely "phenomenological" correlations: they would have been considered more reliable merely because they were causally explained. In fact, the majority could have had their origins in oral tradition. Second, the *Georgics* and its later paraphrases do not pretend to an exhaustive treatment of all sublunary phenomena. Because they reflect and elaborate on a non-philosophical knowledge of nature, these works deal with what may be termed the "existential" weather, which is not only geographically and temporally specific, but is also relevant to everyday life. In these works, therefore, the weather possesses an anthropocentric, rather than theoretical, meaning. The concern of the weather-signs tradition is the *weather*, in contrast to theoretical meteorology, which discusses causes of *meteors*.[29]

Early modern meteors

Instrumental in the propagation of classical meteorology from the Elizabethan era to the Restoration, was the circulation of translated ancient and Arabic texts, scholastic and early modern commentaries, and a group of specialist writings by English and Continental authors that drew upon Greco-Roman authorities.[30] Besides these sources, in view of the explosion of vernacular print, three other traditions are worth noting. The first included topical writings on unusual and portending meteors, representative of popular *and* learned interest in relation to marvelous and providential aspects of the natural world. Such publications played a role in the theological and political arguments attending major historical events, such as, in the case of England, the Reformation, the Thirty Years War, the English Civil Wars, the Restoration, and the Glorious Revolution of 1688. Secondly, extraordinary meteors were also discussed in astrometeorological ephemerides, almanacs of weather prognostications, and calendaric and agricultural ephemera based on Theophrastian, Virgilian and other collections of weather-rules.[31] Finally, in the wake of Francis Bacon's natural history project, some seventeenth-century students of nature opened the door to what was to become the "natural history of meteors," an empirical study informed by and often overlapping classical cosmographical and travel writings. It is worth briefly examining these several traditions.

We may begin by observing the impact which the dissemination of the Aristotelian corpus had on the content and exposition of natural knowledge in the early modern academic setting. The scholastic version of Aristotelian natural philosophy had, since the late medieval period, assimilated new theoretical and doctrinal elements and eclectic uses of classical and contemporary sources. This led the commentators of the early modern period to develop what Patricia Reif has called a "superficial type of syncretism." This was a tendency to multiply authorities for the purpose of establishing the highest authority, "the common opinion of philosophers."[32] Thus modified, Aristotelian ideas "eventually became so deeply ingrained in the European consciousness as to be accepted unquestioningly and their original source lost sight of." Even staunchly anti-Aristotelian thinkers of the high Renaissance could not disentangle their criticisms of the "philosophy of schools" from Aristotelian thought. Charles Schmitt termed this inability the "problem of the escape from the Aristotelian predicament."[33]

The resilience of the scholastic tradition and the problem of "escape" from it suggest that doctrinal premises taught within the academic sphere remained considerably more homogenous in the seventeenth century than is commonly assumed. The indebtedness to Aristotle, on the one hand, of the "modern" research of Andreas Vesalius, Girolamo Cardano, William Harvey, and René Descartes, and, on the other, the inclusion of this same research by orthodox "second scholastics" such as John Henry Alsted, Daniel Sennert, Bartholomew Keckermann, Robert Sanderson, and Thomas Barlow, affirm a continuing vitality of the Aristotelian world view among many scholars writing in the exuberant decades of the so-called Scientific Revolution.[34]

In England in particular the situation was described as the "revival of scholasticism." This tendency reflected the pre-Laudian "middle-of-the-road Protestant" fear of both Puritanism and Arminianism which, during the reign of James I, made the scholastic synthesis overwhelmingly appealing to the professoriate of Cambridge and Oxford universities.[35] A curriculum composed of humanist and scholastic subjects remained unchanged for several subsequent decades, despite attempts at Baconian reforms made by Samuel Hartlib and Johann Comenius. The anonymous *Directions for a Student* (c.1648–50), for instance, used at Emmanuel College, Cambridge, recommended logic and ethics for the first year, physics and metaphysics in the second, more ethics, logic and physics in the third, and in the fourth Seneca and Lucretius and Aristotle's *De Caelo* and *Meteorologica*.[36] Students' notebooks of the 1650s were indistinguishable from those of the 1640s and even the so-called "Wadham group" of Oxford did not seem to question the pedagogical conservatism of Thomas Barlow, whose *Library for Younger Scholars* did not go beyond the early seventeenth-century standards.[37]

Within this framework, meteorological texts published on the Continent and England at the turn of the seventeenth century were mainly commentaries on the first three books of Aristotle's *Meteorologica*.[38] During the seventeenth century at least eighteen works presented its major ideas. In these documents, Aristotle's discussions on causes, exhalations, and *antiperistasis* were kept within their original formulation, while minor alterations existed in relation to names and classification. In 1592, Theodoricus Piripachius described the *Meteorologica* as the doctrine of *corporibus imperfecte mixtum*; in 1590, Vitale Zuccolo wrote the *Dialogo delle Cose Meteorologiche* to explain "all the marvelous things generated in the air, and other wonders regarding springs, rivers and sea," founded on the doctrines of Aristotle and other illustrious writers.[39] The exhalation theory was universally accepted; those who attempted to formulate alternatives, such as William Gilbert in England, failed to escape from the scholastic framework. Gilbert sought to detach himself from a theory of elements but nevertheless resorted to a version of Aristotle's imperfectly mixed bodies and, ironically, introduced another element – the pure earth – as a source of sublunary effluvia.[40] In all other respects, Gilbert followed the first three books of *Meteorologica*.[41]

In English universities after the 1650s, the most influential among Aristotelian interpretations was the work of Libertus Fromondus, Professor of Philosophy at the University of Louvain. His *Meteorologicum libri sex* was published in Louvain in 1627 and the first Latin edition in England appeared in 1656.[42] Fromondus worked on the etymology of *meteorum* ("sublime, aut suspensum") and defined it as a "corpus imperfecté mixtum, ex vapore vel exhalatione ortum in sublimi." Meteors were placed in three regions of air (the upper, middle and lower) and their *causa materialis* was "vapore, calido & humido, & exhalatione, calido e sicca."[43] Fromondus's presentation mimicked Aristotle's in beginning with fiery meteors, proceeding with comets (sublunary and stellar) and winds, and concluding with watery productions (rivers, springs, clouds, rain, snow, and so on) and *meteoris apparentibus* (rainbow, halo, parhelia, and so on) (FIGURE I).[44]

Apart from such commentaries, perhaps the most widely consulted works on meteors during the seventeenth century were the tracts written by William Fulke, a Puritan divine and former Master of Pembroke Hall, Cambridge. In

FIGURE I   Early modern practical calendars contained information on agricultural works, seasons, health, astrological advice and meteorological phenomena. This engraving represents a stylized representation of the spectacular fiery meteors, the flying dragon and the leaping goat, caused by what was believed to be the ignition of fatty exhalation in the atmosphere. *Shepherd's Kalendar* (1585).

Cambridge, Fulke studied law, mathematics, and languages, wrote a polemic against astrological prognostics in 1560, and in 1563 and 1564 published two meteorological tracts, *A Goodly Gallerye* and *Meteors*.[45] In these works, Fulke divided meteors according to a theory of four-fold elemental origin. His commentary on the nature of meteors is based on a syncretic reading from at least twenty-eight classical and medieval sources.[46] Despite this background, Fulke claimed to have started from nothing, or rather from "no-thing": as far as he knew, no writer had explained causes of bodies (*impressiones*) generated in earth or on the "height" (*fossilia* and *meteora*). Although the usage of the term "impression" was common in Fulke's time, meaning "atmospheric influence, condition or phenomenon," Fulke argued that no earlier writer could demonstrate *what* those impressions might be. No earlier definition was helpful, and indeed, Aristotle's own was derived from "doutfullnes."[47] The terminological problems which Fulke tried to resolve were only complicated by his usage of the notion of "affections," signifying "a temporary or non-essential state, condition, or relation of anything" (as in "the coldness or other affection of Air".)[48]

Fulke thus defined meteors as bodies "compounde without lyfe naturalle" and classified them according to three criteria. First, they could be either *imperfectly* mixed, (if reducible to the elements), or *perfectly* mixed (if they resisted change). Second, according to the qualities of matter they could be either moist or dry and, finally, in the elemental division, they could have either fiery, airy, watery or earthy origins. Exhalations and vapors were the material cause of meteors, while the efficient cause was two-fold: God – the primary efficient cause, and the heavens – the "proximate" efficient cause.[49]

Fulke's was certainly not an academic work and if its religious messages attracted a lay audience, it is plausible to assume that a wider non-university public of the seventeenth century could have been familiar with the basics of Aristotelian meteorology. The next chapter will examine similar publications whose distribution might have had the same effect. But, even if it is safe to say that both the academic and non-academic spheres shared a largely Aristotelian conception of the sublunary world, it would be a mistake to assume that no new ideas came on to the horizon. For example, by the seventeenth century, the chemical philosophy of Paracelsus had appeared. In all fairness to its originality, however, it can be argued that Paracelsus's theory, rather than replacing the existing ontological framework, worked more toward enriching it and making it even more resistant to radical change.

Paracelsus's aerial chemistry was discussed in his *Meteora* (1556) where he sought to replace Aristotle's treatment of the exhalations material with a chemical nomenclature. Paracelsus adapted the Stoic notion of "sophic fire" to "a sulfur of fire" and used it to explain thunder and lightning.[50] The idea of "vital sulfur" was then integrated into the concept of aerial nitre, which under the name of the "Master Ingredient" had appeared originally in Cardan's *De Subtilitate* (1550). Early in the seventeenth century, the "sulfur-nitre" theory was further developed by Daniel Sennert, a German neo-scholastic and the sulf-nitrous theory of thunder, lightning and earthquakes prominently featured in the natural studies of John Mayow, Robert Hooke, John Wallis, Petms Gassendi, William Clarke, and George Cheyne.[51]

What were the principles of the Paracelsian sulf-nitrous doctrine? In *Meteorologica*, Aristotle claimed that lightning and other fiery meteors came about when the dry exhalation caught fire. This fire was caused either by the movement of the celestial sphere or by another unspecified sublunary agency. The problem with this account was that the cause and timing of ignition were arbitrary. What exactly was the composition of the volatile material responsible for producing certain meteors at certain times and certain places? Was the exhaled material explicable in chemical terminology and was it possible to associate it with specific characteristics of the soil at places where meteors occurred?

In 1674, John Mayow wrote that the probable cause of the ignition was the moment at which particles of air struck those of aerial nitre, throwing them into a fiery motion and producing a flame perceived as lightning. The flame produced was "very impetuous for the same reason as in the case of gunpowder, [for] it has been shown elsewhere that the force of gunpowder is caused by nitro-aerial particles bursting out in densest crowd from the ignited nitre."[52] The motion of air was the result of nitrous and sulfurous exhalations set on fire and the similar underground explosions effected earthquakes and volcanoes.[53] The moment of ignition had now been chemically fixed by a replacement of the Aristotelian undefined material – described as "unctuous," "clammy," or "inflammable" – with the "vital air, "Nitre," "Sulphur," or simply "Gunpowder." The gunpowder analogy gained so much prominence that Browne ventured to claim that "a main reason why the Ancients were so imperfect in the doctrine of meteors, was their ignorance of Gunpowder and Fire-Works, which best discover the causes of many thereof."[54]

The inflammability of sulfurous mixtures was applied to the phenomena involving sudden release of heat and light, such as lightning, fireballs, shooting stars, northern lights, and Ignes Fatui (the lambent, flickering fires observed in graveyards). In summer thunderstorms, the "stormy" matter was thought to be made of sulfur, and its clouds as reservoirs of vapors gathered from the earth. In the *Examen of Antiperistasis* (1665), Robert Boyle argued that the subterranean heat observed during the winter was not caused by *antiperistasis*, but by subterranean steams of sulfur. These steams were particularly copious around mines, but "there may be probably very many [other places] that may supply the air with a store of mineral exhalations, proper to generate fiery meteors and winds."[55]

In the late seventeenth century, the writers working to consolidate these ideas in a comprehensive and more accessible form were John Woodward and Thomas Robinson. Judging by the numbers of references made to his work, Woodward's controversial treatise on Mosaic geogony (the theory of creation and the evolution of the earth) was extraordinarily well received during the first half of the following century.[56] In this work, in the section on meteors, Woodward laid stress on the frequently observed simultaneity of earthquakes, volcanic eruptions, and lightning, suggesting that a possible explanation might be looked for in the nature of the soil at the site of concurrent events. For instance, he associated frequent thunderstorms around Etna in Sicily with the sulfurous soil surrounding the volcano, eruptions of which might transport the earthy sulfur into the atmosphere and thus make it conducive to the production of lightning and other fiery meteors.

Making meteors the manifestations of earthly effluvia, Woodward laid down the guidelines for a majority of the eighteenth-century naturalists. His "interactional" meteorology of vapors argued for both the *interconnection* between meteors and for their *unity*. Meteors were considered mutually commensurable in virtue of their common physical origin – i.e. mineral and watery exhalations. Meteorologists thus required only quantitative factors to account for differences between meteors of the same class. For example, the quantity of water vapor determined whether it condensed into rain, dew, or mist.[57] In another instance, the intensity of fermentation and the quantity of aerial sulfur determined whether the product of ignition was lightning, aurora borealis, or ignis fatuus. More importantly, however, meteors were considered reciprocally predictive phenomena. Falling stars foretold thunder, for "they shew the Air to be inflam'd with much Heat, and consequently that Thunder and Lightning will ensue." Long droughts anticipated the approach of earthquakes, and earthquakes signaled the approach of pestilential seasons. Earthquakes also anticipated the northern lights as they both ensued from the same sulfurous vapors.[58] Interestingly, even after the theory of exhalation had lost its appeal during the second half of the eighteenth century, the weather was often correlated with the incidence of earthquakes and auroral lights.[59]

Woodward's *Essay* reiterated the peristatic theory with an enticing metaphor: when the nitrous and sulfurous vapors saturated the air to a critical level, he claimed that they constituted a kind of "Aerial Gunpowder," the cause of "dismal and terrible Thunder and Lightning which commonly, if not always, attend Earthquakes." The metaphor was taken up by several writers, its most elaborate and idiosyncratic version being in the work of the Northamptonshire natural historian, Thomas Robinson.[60] He wrote that stormy weather was a battle between the "armies" of fire and water, the conflict between which begins when fire sends forth his chariots to meet a detachment of those from the vaporous army, after which the observer hears "the Thundering Sound of the Battel."

> The Battel by this time growing very hot, the Main Bodies engage, and then nothing is to be heard but a Thundering Noise, with continual Flashes of Lightning, and dreadful showers of Rain, falling down from the broken Clouds. And sometimes random shots flie [sic] about, kill Both Men and Beasts, fire and throw down houses, split great trees, and tear the very Earth. The two irreconcilable Enemies still keep the Field until one of them be utterly destroyed. If the fiery Exhalations keep the Field, the East Wind blows still hot and sulphureous. If the Vapors get the Victory, the West Wind blows cold and moist, the Sky is clear, the Air is cold, the Battel is over and the Earth burries the Dead and gets the Spoil.[61]

The idiom was taken up by others. In 1715 Edward Barlow, a watchmaker and mathematician, wrote that the atmospheric battle becomes more dreadful by the addition of "Fire-Arms to the Fury of Clouds, especially amidst the heat of summer, while the Air is, on all sides, impregnated with the sulphureous and combustible matter." If the "Train of Etherial Gunpowder" gets enclosed between the clouds, it "displodes a Thunderbolt" and the battle lasts as long as the store of sulf-nitrous ammunition holds out.[62]

A dominion of exhalations

Slight variations of these ideas – if perhaps with less military imagery – can be found in most meteorological texts written during the first half of the eighteenth century and in some instances even after that time. This period of unequivocal consensus was premised on the Woodwardian theory of exhalation challenged only after the sudden increase of electrical research in the 1750s. This state of affairs contradicts the majority of historical accounts of the period's intellectual activity in that it suggests a continuity, consolidation, and refinement of quasi-Aristotelian views rather than change. In the omnibus Quaerie 31 of the second English edition of the *Opticks* (1717), Newton himself asked if the sulfurous steams abound in the earth where they ferment with minerals and occasionally "take fire with a sudden Coruscation and Explosion." He surmised that earthquakes came about as a result of explosions of released air which had been "pent up in subterraneous Caverns," and the released vapor started tempests, hurricanes, and a number of fiery meteors.[63]

Contemporaries concurred. The sulf-nitrous force of fermenting vapors and "belligerent" meteors is found in a remarkable number of eighteenth-century documents and among diverse circles of scholars, pamphleteers, theologians, encyclopedists and meteorological observers. It featured in numerous editions of major textbooks in natural philosophy and in an occasional paper in the *Philosophical Transactions*.[64] The lexicographer John Harris, explained it in his *Lexicon Technicum* (1704), and the chorographers Charles Leigh and John Morton subscribed to it in their regional geographies. Following John Flamsteed's explanation of an earthquake as an aerial explosion, William Derham thought the Great Storm of 1703 was caused by the long summer drought which raised the sulf-nitrous matter, "which when mixed together might make a sort of explosion (like fired gunpowder) in the atmosphere." Edmund Halley explained condensation as an effect of the frigorific power of nitre. He believed that exhalations were the true cause of weather diversity: were the earth covered with water alone, changes in the weather would cease as the mixture of heterogenous vapors would be removed, "which as they are variously compounded and brought by the winds seem to be the causes of those various seasons which we now find." Whiston used it to explain the northern lights of 1716 as a "mean state" of the fermentation of sulf-nitrous exhalations. Others explained fireballs and shooting stars. When, in the middle of this series of explanations, someone remonstrated that the exhalations could not possibly reach so high an altitude by simple convection – having to surmount in such an ascent the extreme cold and rarity of the upper regions – Halley responded: "the fact is indisputable and therefore requires a solution."[65] As late as 1798, John Tytler argued that, because of their local character, meteorological phenomena depended on the "changes which take place in the bowels of the Earth, whence meteorologists ought not only to be perfectly acquainted with geography but with mineralogy also."[66]

Popularizations followed. During the 1690s, the bookseller John Dunton and his associates in the Athenian Society disseminated the mineral theory of exhalations in the *Athenian Oracle*. They explained that thunder and light-ning resulted from coagulation of the moist exhalations which in turn ignited the nitric ingredient. The editors used the idea of the opposition between the qualities of cold and hot as a mechanism for the formation of snow, hail, and frost. In 1734, *Time's Telescope*'s brief "Discourse of all kinds of Meteors" cited Hooke's theory of unctuous vapors and even Fulke's explanation of fiery meteors. In 1762, Benjamin Martin wrote that *meteorography* accounted for fiery meteors and assumed them to be inflamed vapors of fatty matter raised from the earth.[67]

What kind of evidence supported this unanimity? It must be pointed out that the exhalation theory gained only limited support from quarters conversant with vanguard research in barometry, pneumatics (research into the nature of gases) and the mechanical properties of air. Robert Boyle, among others, was in part responsible for this situation when he distinguished between the "science of meteors" and seventeenth-century pneumatics:

> [W]hen I speak of the air, I do not in this place understand that air, which I elsewhere teach to be more strictly and properly so called, and to consist of springy particles; but the air in its more *vulgar* and *lax* signification, as it signifies the atmosphere, which abounds with vapours, and exhalations, and in a word with the corpuscles of all sorts, except the larger sort of springy ones.[68]

Similarly, the Oxford polymath Robert Plot spoke of the air, or rather of the "atmosphere" as the subtle body of exhalations, but "whether beside these exhalation, there be any peculiar simple body, called Air, I leave to the more subtile Philosophers, and consider it here only, as 'tis the subject of storms, of thunder and wind, of *Echo's*, and as it has relation to sickness and health." In a digest of Varenius's *Geography*, the Cartesian popularizer Richard Blome doubted whether "any Thing or Body else be contained in exhalations."[69] Such skepticism demoralized those who might otherwise have hoped to establish a link between meteors and the corpuscular philosophy.

As a consequence, empirical evidence for mineral meteorology remained a moot issue. The dependence of the theory on instrumental observation was practically nil: barometers, thermometers, and hygrometers were used to meas-ure the weight of air, its temperature, and the amount of water vapor in it, but no known instrument registered the presence of sulfur or nitre. The exhalation theory thus derived its strength mainly from analogies established by Wallis who correlated the aerial nitre and the properties of gunpowder. In addition, it was strengthened by Woodward, Lister, and Robinson who linked the subterranean vapors and supraterranean meteors, and Halley, Derham, Whiston, and others, who applied these ideas to the atmosphere.

This is not to say that there was no effort to produce experimental evidence. Naturalists experimented and undertoook calculations, such as the "frigorific" researches of Boyle, Lister, Nehemiah Grew, and Derham.[70] Edmund Halley estimated the global marine evaporation of water vapor, and the physician Leonard Stocke performed experiments on dew-fall in 1742.[71] An anonymous Dutch naturalist took a dark-brown hailstone and threw it in the fire where it "gave very great report,"[72] and Joseph Wasse measured the power of lightning by taking "a Cowhorn charg'd with three Quarter's of a Pound of very good Powder, wadded with thick Paper, and fired it against a Stone."[73] John Whiteside, Keeper of the Ashmolean Museum, showed by experiment that inflammable sulfurous vapors, by an "innate Levity will ascend in Vacuo Boileano" and collect at the top of the receiver, while most other fumes would fall down. By analogy with Whiteside's result – and only by analogy – Halley concluded that as atmospheric vapors fell as rain and dew, the vapors ascended to the top of the atmosphere and there became fiery meteors.[74]

More pertinently, Stephen Hales referred, in 1750, to an experiment from his *Statical Essays* (1733) as proof of the sulfurous origins of earthquakes and lightning. In the experiment, he produced sulfurous fumes in a bottle by putting its opening into a vessel containing *aqua fortis* and powdered pyrites. Letting the fresh air into the bottle, he observed a "violent agitation between the two airs," resulting in the production of a "redish turbid fume, of the colour of those vapours which were seen several evenings before the late earthquakes." The sultry weather preceding earthquakes was, by analogy, "the intestine motion between the air and the sulphureous vapours, which are exhaled from the earth," and lighting no more than a contact between pure and sulfurous atmospheric air: the former accumulated above thunderclouds, the latter below them.[75]

Experiments such as these, however, made up only a minor portion of the empirical support for the Woodwardian theory of mineral exhalations. Instead, the bulk of evidence came from everyday observation: olfactory perception, ocular observation, and the correlation of weather conditions with meteoric occurrences. The smell was by far the most frequently cited observation in meteorological reports prior to the 1750s. The sulfurous smell attended storms, earthquakes, lightning, aurorae boreales, falling stars, and fireballs. In the 1726 edition of his geogony, Woodward argued that it was caused by salts, mineral particles, and other "Exhalations of the Abyss," forced into the atmosphere by subterranean heat. Reinforced by the fermentation of vapors, the heat announced approaching storms and earthquakes.[76] Chambers's *Cyclopedia* explained that the presence of sulfurous vapor in lightning "is known by the sulphureous [sic] smell and sultry heat in the air that preceeds it."[77]

The smell of rising vapors had often been mentioned by earthquake witnesses. Marcello Malpighi communicated to the Royal Society an account of an earthquake in Sicily in which the author, an Italian nobleman, wrote about water springing from local fissures which had a distinct sulfurous smell.[78] Borelli, Malpighi, Woodward, and, later Buffon, observed similar clefts. Hales surmised that they were produced by droughts in vents in which the terrestrial vapors, rising to the earth's surface, took fire and produced the explosion – the "Earth-Lightning." Above the earth's surface, the explosion – the earthquake – was announced by dark-red mists filling the muggy atmosphere during prolonged hot weather. Indeed, after earthquakes, witnesses remembered the mists.[79] Individuals of more sensible constitution sensed the vapors even before they were visible: some suffered asthmatic attacks, dizziness, shortness of breath, or actually vomited before thunder and lightning.[80] *Aurorae* produced similar effects: they were preceded by crimson mists, explosions, and an odor of sulfur which sometimes "affected and disordered the Spectators to a great degree."[81]

These experiental engagements with the sublunary world – textual, experimental, and commonsensical – testify to unusual interpretative difficulties which that world generated. How, in other words, was it possible to account for the variety of fleeting phenomena which the dominant body of scholarship forced into the framework of exhalations, *antiperistasis*, and mineral turbulence? For even if the mineral exhalation theory of the early eighteenth century underwent a small number of experimental tests, it is clear that its theoretical underpinning did much to transcend the quasi-Aristotelian seventeenth-century chemistry. If Newton's or Halley's meteorological views can be used as guidelines in surveying developments before 1750, the trumpeted revolution in naturalistic thought had apparently failed to reveal itself in relation to an understanding of the sublunary world. After that time research in atmospheric electricity, the theory of gases, and atmospheric chemistry led some prominent naturalists to abandon terrestrial exhalations and undertake more specialized enquiries (see chapter 7).

Eighteenth-century naturalists, however, argued that a mechanical, Newtonian or pneumatic meteorology would be too ambitious to supersede the specific nature of the exhalation theory. Even more important was the fact that when Aristotle declined to discuss the issue of weather prognostics, he effectively divorced "meteors" from "weather." As a consequence, classical meteorology persisted as an enterprise distinctive in both method and purpose from traditional prognostics and a more modern conception of meteorology as the science of weather. This separation was maintained until the end of eighteenth century when the "weather" was invariably defined as "the state or disposition of the atmosphere, with regard to heat, cold, wind, rain, frost, etc," whereas "meteorology" still referred to a "doctrine of meteors." Neither the "weather" nor the "atmosphere" were necessarily associated with "meteorology."[82]

In classical meteorology, therefore, the weather could never have been considered an aggregation of meteors since a list of all meteors at any given place and time could not possibly be equivalent to a knowledge of the state of weather in that place and time. This disjunction reflected the fact that meteors and weather existed in incommensurable domains: meteors in a theory of sublunary elements, the weather outside theoretical consideration. The weather happened, meteors were explained.

With this distinction in mind, the following chapters will explore how and why seventeenth- and eighteenth-century naturalists developed a "natural history of meteors" as a systematic *empirical* study of phenomena previously considered only in theory. These chapters will describe the process by which meteors became concrete "matters of fact," a process which was peripheral to meteorological theory and central to other forms of knowledge and life.

2

Observing the extraordinary

> It is the pleasure of philosophy to attempt something by way of solution concerning every extraordinary fact which falls under its cognizance: and though it be not always so happy as to produce satisfaction, it may at least succeed in the way of amusement. Under the influence of these notions, let us see what offers respecting the philosophy of the curious appearance before us.
> William Cockin, "Account of an Extraordinary Appearance in a Mist."[1]

The first meteorological matters of fact were unusual meteors, not numerical observations. Unusual meteors were found in public spaces, not in barometric tubes. This empirical study of meteors – the "meteoric tradition," "meteorological reportage," or "meteorological matters of fact" as I will also call it – was a unique rapprochement between atmospheric emblematics, meteorological theory and a customary belief in God's intervention: it could be argued that seventeenth- and eighteenth-century meteorological authors transposed the idiom of marvel and providence literature into an inductive enquiry into natural and preternatural Creation.[2]

Several features distinguish the meteoric tradition from other empirical engagements with the natural world. First, in terms of domain, the meteoric tradition developed within the largely classical conception of meteors. As in classical categorization, seventeenth-century meteorological facts were events: observations were narratives about unusual meteorological activity. Outside the intervals in which activity took place, meteors did not occur and could not be observed. The meteoric tradition was in this respect a "chronicle" of individual Aristotelian *meteora*.[3] Consequently, many natural occurrences, considered today to belong to other disciplines, were labeled as meteorological during this period. This is illustrated, for instance, by the organization of the early nineteenth-century abridgment of the *Philosophical Transactions* of the Royal Society (1809), in which tides, earthquakes, meteors, volcanic eruptions, and the "unusual agitations of the sea," all belonged to the province of meteorology. The editors of this abridgment, the mathematician Charles Hutton and the physicians George Shaw and Richard Pearson, classified meteorology as a part of chemical philosophy, but considered it in a classical sense as a science of meteors. Today, however, seismology investigates earthquakes and volcanoes; astrophysics meteorites ("falling stars" and other "fiery meteors"); optics explains halos, mirages, and St. Elmo's Fire, and aurorae boreales are the subject matter of geophysical studies of terrestrial magnetism. Modern diversification suggests the extraordinary inclusiveness of the subject matter of classical meteorology.

Second, the meteoric tradition was *qualitative* and *descriptive*. For lack of a better term, it was emphatically "empirical": it sought both testimonial authenticity and eschewed theory. However, the meteorological theory of exhalations was compatible with meteoric tradition in one respect: the assumption that the subject matter of meteorology was *events*, rather than long-term *processes* – climatological changes and planetary movements – or tangible items such as stones, plants, or animals. In this, and only in this sense, can the meteoric tradition be interpreted in principle as an empirical investigation of classical meteors. It rendered concrete Aristotle's theory.

Third, meteoric reports were personalized *narratives* of extraordinary, striking, or rare phenomena. The contents of these reports were recently characterized as "abnormalities in weather," "unusual" rather than "normal" weather patterns, and "strange weather."[4] The meteoric tradition was not about discovery of a global conception of the weather – what John Herschel would later term an "ideal unity" of the weather. It was about *extraordinary* phenomena. It was the *excess* and *oddity* that interested meteorologists, not regularities. They thus described "prodigious" rainfalls, droughts, frosts or winds, or "awful" waterspouts, airquakes (mysterious explosions heard in the air), northern lights, and fireballs. Many of these phenomena were considered difficult, if not impossible, to explain, – for example, ignis fatuus (fool's fire), *Castor* and *Pollux*, or the ineffable ball of noise observed in 1742 by one Mr Vievar in Essex. Most were perceived as frightening, unexpected, or extreme in extent and intensity, like hurricanes, tornadoes, earthquakes, thunderstorms, and lightning.

Fourth, the authors of these reports preferred ordinary rather than "technical" language. The "domestic coloring" of these reports, one historian has remarked, "shows how much their authors were at home in the weather, how much it formed part of their daily life, and how little able they were to objectify the weather for the purpose of analysis."[5]

Finally, meteorological writers were predominantly provincial men of letters. Like geological and botanical studies, archeological, and antiquarian field-research, and natural history in general, meteorological observation of this kind was part of an "active, outdoor, riding culture," associated with university-educated professionals, clergy, improving farmers, and the aristocracy.[6] Meteorology required modest means and skills for its cultivation, and possession of instruments became increasingly fashionable among middle- and upper-class households. By the 1790s, for instance, the barometer was said to be "a widely owned piece of furniture," and often used as nothing more than a toy.[7]

Until the late eighteenth century, meteoric reportage loomed large in books on natural history and comprised about 57 per cent (255 out of 466) of all meteorological articles published in the *Philosophical Transactions*.[8] For example, Charles Leigh, Thomas Robinson, John Morton, and William Borlase dealt in their natural histories with items such as "sea-murmurs foreshadowing storms," remarkable hailstones, county echoes, whirlwinds, "hurricanes," and unusual agitations of the sea. The *Philosophical Transactions* regularly published contributions on inundations, mock-suns, earthquakes, lightning damages, fireball explosions, and numerous unclassifiable phenomena in the atmosphere. Recording "strange" weather was the most popular form of empirical program in eighteenth-century meteorology and the material thus collected, it was frequently pointed out, served as a legitimate contribution to a complete natural history of meteors.

In 1781, however, William Nicholson lamented the state of meteorological knowledge and in 1806, another commentator thought that the theory was not forthcoming "merely from diffidence in the observers to write down their observations."[9] But the vitality of meteoric tradition throughout the eighteenth century suggests that naturalists, on the contrary, wrote extensively about the weather. These observations, however, were presented in a form of which Alexander Tilloch and Nicholson did not approve. Early nineteenth-century naturalists became more robust – and eventually more successful – in their demands for a meteorology that took it as self-evident that "extraordinary facts *teach us nothing.*"[10]

From this time on, meteorologists were increasingly asked to view their science as a quantitative and laboratory enterprise. Laboratory pneumatics and storm maps moved meteorology indoors. Measurement was everything: data was collected and tabulated for the daily, monthly and annual course of temperature, wind direction, atmospheric electricity, pressure and humidity. Only organized disciplined, and standardized use of scientific instruments – not the narratives about strange weather – was now taken to legitimize a science of weather. Meteorology was conceived as a *synoptic* description of the states of weather at different places and a description of weather in time. The weather needed to be mapped on to a spatio-temporal grid, its nodes corresponding to the measurements made at remote observational posts.

The idea of continuous weather observation was formulated early by Blaise Pascal who wrote that the "control" barometer left at the foot of the Puy-de-Dome – indicating changes in the atmosphere's weight and so eliminating their influence in the final analysis of readings of the barometer taken to the summit – was the source of "une expérience continuelle," "because one observes it, if one likes, continually, and one finds the mercury at almost as many diverse points [on the scale] as there are different weather [conditions] where one observes it."[11] But although regular observation was not a new idea, enthusiasm for it before the late-eighteenth century was limited. The call for a network of meteorological observatories equipped with standardized instruments began in England as early as the 1680s.[12] Beginning with Robert Boyle, Robert Hooke, Martin Lister, and John Woodward, these schemes aimed at an exhaustive "natural history of meteors." In his *Method for Making a History of the Weather* (1663), and "Instruction for Seamen" (1667), Hooke gave instructions for daily observations and drew up a standardized record-sheet. The objective was either to establish a weather–disease relationship, or to confirm astral influences on changes in the sublunar regions of air. John Locke, Robert Plot, Lister, Richard Towneley, and others started registers but regretted a lack of enthusiasm among their correspondents.[13]

In 1723 the physician James Jurin published an "Invitation to meteorological observation" appealing to all interested in beginning a systematic, standardized, and long-term measurement of atmospheric variables. The observations were to be entered in a tabular form and communicated to the Royal Society for further analysis.[14] In 1728 Isaac Greenwood, a professor of mathematics at Cambridge, New England, proposed an extension of Jurin's scheme by inclusion of marine observations.[15] A number of journals arrived from European academies and from private sources such as travellers, diplomats, and professional people. But after Jurin's secretaryship of the Royal Society had ended in 1727, communications on the subject decreased. "These investigations," observed a contemporary, "require not only industry and inclination, but also leisure and means and opportunity, which you seldom find in private persons."[16] Half a century later, European governments proved more efficient in this respect – information was gathered through the channels of the Société Royale de Médecine (1778) in France and Societas Meteorologica Palatina (1780) in Germany.[17] But even then, only meteorological anomalies created an interest. As a French bureaucrat noted, the harsh winter of 1776 excited the learned societies' interest, but as the weather grew warmer, "their zeal cooled."[18]

The parallel existence of the two dissimilar empirical approaches – the qualitative reportage of extraordinary weather events and the quantitative measurement of a "weather-continuum" – requires an explanation of both the persistence of the former and of the recurring low profile of the latter. The eventual acceptance of the quantitative "continuum" program should not be seen as either natural or inevitable, but rather as an acceptance of its advantages *relative* to meteoric reportage. By the late eighteenth century one or more aspects of meteoric reportage – methodological, theological, and social – ceased to elicit support among meteorological writers. But as the acceptance of the instrumental method occurred more than a century after its promulgation, the problem was not the failure of the meteoric tradition, but its lengthy methodological *supremacy.* The history of early modern meteorology reflects, in several respects, the history of the meteoric tradition and one of the important issues has to do with the forces that sustained its appeal. Do the features of this meteoric tradition point to any unique intellectual and socio-cultural practices responsible for its remarkable presence in eighteenth-century science?

Wonders, marvels, and ominous meteors

Accounts of various sublunary excesses and unusual workings of organic nature comprised a large number of popular and learned printed works in the period that stretched from the London earthquake of 1580 through "Defoe's" devastating storm of 1703, to the London earthquake of 1750. Publications in all formats, from ephemera and single-sheet narratives to embellished folio compilations carried detailed descriptions of the marvelous, miraculous, and prodigious productions of nature of various kinds: the latest monstrous births, gruesome murders, prolonged fasts, the sudden sinkings of the ground, conflagrations, rains of blood and wheat, and apparitions of armies in the sky.[19]

The tradition was of an ancient origin and was universal in early modern Europe. It has been called "the early modern vogue for the preternatural," and linked with the late Renaissance revival of magic, occultism, witchcraft, and demonology.[20] It has also been associated with Elizabethan beliefs in portents, tokens, and signs and demonstrates that early modern people perceived unusual natural phenomena as divine concern for the moral fate of mankind, manifested as promises, forewarnings, or reprimands. "So was there never any great change in the world," wrote the Elizabethan Robert Recorde, "but God by the signs of heaven did premonish men thereof."[21] In 1661, *Mirabilis Annus* explained that

> Accidents of this kind do portend the futurition or manifestation of some things as yet not existent or not known, which usually carry in them some kind of agreement and assimulation to the Prodigies themselves, as raining of blood may signify much slaughter, the noise of guns and the apparition of Armies in the Air, wars and commotions, great innundations, popular tumults and insurrections, yet we still know that god is unreachable in his ways.[22]

Prodigious natural productions had a conspicuous semiotic dimension. For instance, following the "All Saints Flood" of November 1570 and the St Bartholomew's Day Massacre of Huguenots in 1572, the mathematical writer Thomas Hill wrote a treatise, *The Contemplation of Mysteries,* a tour de force on the variety of weather "impressions" that carried ominous political meaning. Hill provided a "manual" of correspondences between extreme meteoric events and their implications for humanity: eight types of thunder, for example, were said to portend eight different types of political turmoil. A century later, the instrument maker Thomas Willsford published a similar collection, *Nature's Secrets,* featuring an explanation of "the efficient and finall [sic] causes of comets, earthquakes, deluges, epidemicall diseases, and prodigies of precedent times; registered by the students of nature."[23]

Between Hill and Willsford the reports and compilations of extreme natural occurrences proliferated. Especially during the times of political and religious crises – the Thirty Years War, the English Civil Wars, and the Restoration of Charles II – their production soared: excesses of nature were held to both reflect and announce social disturbance (BOX I, FIGURE 2).[24]

In publications like these, the natural merged with the political. The history of the world here was as much about human affairs as it was about how nature impinged upon them. In 1640, the year after the regicide, the royalist James Howell wrote that in the sublunary world "*the sea is never still, the aire is agitated with winds, and new monsters and meteorological impressions are hourely*

BOX 1   Pamphlets and "wonderful news" about strange meteors
in Stuart England

*A Report from Abbington towne in Berkshire, being a relation of what harme
Thunder and Lightning did on Thursday last upon the body of Humphrey
Richardson, a rich miserable farmer. With Exhortation for England to repent.*
London: William Bowde, 1641

*A Strange Wonder or The Cities Amazement. Being a Relation occasioned by a
wonderful and unusual Accident that happened in the River of Thames, Friday
Feb. 4, 1641.* London: printed for John Thomas, 1641

*A Strange and Wonderful Relation of the Miraculous Judgment of God in the late
Thunder and Lightning.* London: Bernard Alsop, n.d. [c. 1620–50]

Charles Hammond, *A Warning-peece [sic] for England. By that sad and fearfull
Example that happened to Men, Women and Children, all sorts of cattle and
fowles; by Stormes, Tempests, Hail-Stones, Lightning and Thunder, June 25
1652.* London, 1652

*Mirabilis Annus (Eniagtos Terastios) or the year of Prodigies and Wonders, being a
faithful and impartial Collection of several Signs that have been seen in the
Heavens, in the Earth, and in the Waters . . . all which happened within the
space of year last past, and are now made publick for a reasonable Warning to
the People of these Kingdoms speedily to repent and turn to the Lord . . .* London,
1661

*This Winters Wonders: or A True Relation of a Calamitous Accident at Bennenden
in the County of Kent; how the Church and several Houses were destroyed by
thunder and Lightning, on the 29th of December last, being Sabbath day.*
London, 1673

*Strange and Terrible News from Sea. Or, A True Relation of a Most Wonder-
ful Violent Tempest of Lightning and Thunder on Friday, the 18th of this
Instant Jan. 1678.* Westsmithfield: A. P. and T. H. for John Clarke, 1678,
6 pp

*Memento to the World: Or An Historical Collection of Divers Wonderful Comets
and Prodigious Signs in Heaven, that have been seen, some time before Birth of
Christ, and many since that time in divers Countries, with their wonderful and
dreadful Effects.* London: T. Haly, 1680

T. T., *Modest Observations on the Present Extraordinary Frost: Containing I A
Brief Description thereof, and its Natural Celestial Causes inquired into. II An
Account of the most eminent Frosts that have happened [in the past]. III Philo-
sophical presages of what may be feared now to ensue [scarcity, sickness, pestilence]
IV The cries of the poor, and an easy way proposed how there may be 20 000
pounds a week raised for their relief . . .* London: George Larkin, 1684

engendered."[25] The following year, a broadsheet explained that that year's
unnatural return of the tide in the Thames, if the past chronologies were not
mistaken, would be followed by "dismall heavi issues [sic], either of deaths of
unmatchable and Peerlesse personage, of Battaile, sicknesse, or Famine."[26] Late
in the century, *The English Chapmans and Travellers Almanack* inserted "a brief

chronology" (in "years ago") of the world in which historic public events were interspersed with meteorological and medical events of consequence (BOX 2).

During the seventeenth century, documents like these had a millenarian message. As the Kingdom of God was at hand, doomsday was to be announced by a series of spectacular events – the end of Catholicism, Judaism and Mohammedanism – preceded by grand meteors, apparitions, and uncommon productions of organic nature.[27] The storm of 17 January 1678 came with "such dreadful thunder and lightning, [th]at some started up half distracted, thinking it to be the day of judgment;" the thunder was "so extremely violent, and after such a

FIGURE 2   Parhelia (multiple suns) observed during the Thirty Years War. Engraving from R[ichard] B[urton]'s *The Surprizing Miracles of Nature and Art* (1683).

BOX 2   "A brief chronology" (in "years ago") of the world in *The English Chapmans and Travellers Almanack for the Year of Christ 1697*

623	The conquest of England by Duke William
317	The invention of guns
247	The art of printing
246	The infectious sweating sickness
122	The whole heaven seemed to burn with fire
117	The general earthquake in England
92	The gunpowder treason
66	The plantation began in New England
53	The fight at Newbury
53	The fight at Marston Moor
52	The fight at Naseby
37	King Charles returned to London
35	The strong tempestous wind (1662)
33	The blazing star appeared (1664)
23	Last great snow 11 days
13	A great frost containing 10 weeks[28]

FIGURE 3   The engraving accompanying the *The Uxbridge Wonder*, a single sheet account of a violent storm in Middlesex in July 1738. Narratives like these usually combined several meteorological phenomena to convey the extent of atmospheric turmoil. Shown here are thunderclouds with lightning, a comet, and missile-shaped fiery meteors.

dreadful manner, as some in fright were ready to judge the approach of the day of doom."[29] (FIGURE 3).

Pamphlets forecast such events. *A True Relation of the Wonderful Apparition of a Cross in the Moon* (1688) asked if it foreboded "the destruction of all the lunated crosses in Germany, Hungary etc which the haughty Turks have set up there in token of their victories?"[30] *The Age of Wonders* from 1710 described "a remarkable, and fiery apparition [of] a man in the clouds with a drawn sword." In 1763, it was reported that at Riga, Livonia, "a Multitude of People have seen in the open Sky, a Coffin, fiery rods, three Dead-heads, a Serpent and a Pyramid." In Kirschberg, Prussia, after twelve hours of thunder, the sky opened to display three singing angels, a fiery sun, and a great cannon: "The clouds turned as red as fire, the winds did blow, Woe, Woe, did the three Angels cry: 'Oh! People leave your pride.' "[31]

Where millenarian ephemera exploited marvelous meteors for spiritual ends, chapbook histories (pamphlets of popular tales hawked about by chapmen) of preternatural exotica used them for profit. The wonderful easily turned into the titillating. This "tradition of Wonder," according to J. Paul Hunter, represented a backlash against the disenchantment of nature inaugurated by the new sciences, "a tradition of just celebrating – not trying to explain – the strange, the surprising, the awesome, and the wonderful."[32] The fascination with the unusual drew upon the need for that which eluded rationalization, and the wonderful in the natural world in particular appeared to undermine the mechanistic cynicism of the materialist philosophy. A political poem published a year after the death of Queen Anne, described unusual heavenly lights as the mystery which "Men of Art" could only *try* to explain as no scientific principles sufficed for a true account. The poet implied that scientific accounts of such phenomena were about power, not truth, as the excellence of the learned "lies in puzzling those they find less wise," whereby they gain the upper hand "in thing's that neither understand."[33] John Dunton, editor of the *Athenian Mercury,* wrote of how Nathaniel Crouch, a compiler who popularized the wonder tradition during the 1680s, "has melted down the best of our English histories into twelve penny books, which are filled with wonders, rarities and curiosities; for you must know, his title-pages are a little swelling."[34] Crouch wrote a series of popular histories of "Most Remarkable and Tremendous Earthquakes, . . . dreadful and Conflagrations and Fiery Irruptions," [sic] the "most Remarkable Transactions and revolutions [combined] with Variety of Excellent Speeches, Strange Accidents, Prodigious Appearances," and "wonderful and supernatural Delusions."[35] "Knowing strange things" of both natural and social provenance was the essence of Crouch's historiography.

Such was the interest of shorter and more *au courant* reports on the preternatural, like the sixteen-page *God's Marvellous Wonders in England in 1694.*[36] It brought, among other oddities, a circumstantial description – with listed witnesses – of a "strange and wonderful shower of wheat in Wiltshire," and a "terrible storm" that dropped coronet-shaped hailstones, six inches in length. "Strange are the wonders and products of the airy region," the author noted, "as well as airy Phantoms of Men and Horses, castles, ships, and the like, and many times representing fights and battels, with great lightnings and thunders, as well as curtaines of fiery and dusky clouds, to draw over the seeming tragick scene."[37] The *London Journal* of 1725 claimed that the catalyst for such sensational pieces as well as for contemporary journalism was in the universal "strange love of Novelty." Where there was no news to report, the *Journal* wrote:

How often do we see them driven to the last shift, by having Recourse to Prodigies and Omens? They are forced to alarm our Apprehension with Accounts of terrible unheard-of Monsters in one Country, and most surprising Apparitions in another; Earthquakes, Volcano's, Hurricans [sic], and Inundations, Rivers flowing with Blood, and Mill-stones swimming against Tyde, floating Islands, and Castles in the Air are frequent Articles in our Modern Diaries and other papers of Intelligence.[38]

It is crucial to emphasize that "strange news" ephemera reported the items which powerfully mobilized public energies, generated arguments, controversy, and a preoccupation with symbols and even heightened the sense of crisis in times of scarcity or brewing social unrest. Such events represented an aspect of the early modern public sphere. They were *public* meteors which transgressed the norms of the expected and possible and chaneled the collective feelings of discontent into foreboding, anxiety, fear, even panic. Public meteors epitomized pre-modern hybrids of nature and society.[39]

In contrast to the philosophical treatment of meteors – in which they were abstracted from place and time – "strange news" publications described only specific meteoric occurrences. In this sense, public meteors – whether political portents, providential signs or disturbing marvels – were experiental items *par exellence*. The empirical authentication of these items became a critical issue in mid-seventeenth-century theological discourse. This period witnessed more and more discussion about the authenticity of biblical miracles and their historicity. It was accepted that salvation by a transcendent truth required a prior act of being empirically verified. By the 1660s, this requirement was extended beyond biblical events: contemporary English writers of strange news reports also had to pay heed to the rules of empirical verification.[40]

These strictly epistemological developments should therefore be read in the context of some of the Restoration authors in order to endorse the proliferation of miracles as a means to "rebut freethinkers and atheists by showing that divine revelation is still being manifested, quite literally, before our eyes."[41] The certainty of revelation could therefore be demonstrated not only through an exegesis of the Testaments, but from the verifiable particulars about true miracles. One of the issues raised in the discussion was the nature of acceptable proof, inasmuch as authors sought to comply with the criteria in terms of "evidence, proof, and testimony."[42] Establishing the truth of miraculous particulars was equivalent to claiming the historicity of miracles. Those who, in this way, upheld revelation against the materialism of the age, derived "their techniques of authentication from the very stronghold of skepticism which it [was] their purpose to refute." In 1691, Richard Baxter thought that his belief in revelation ought to derive from the "sure Evidence of [the] Veracity" of apparitions, because, as visible manifestations of invisible Spirits, these apparitions "might do much with such as are prone to judge by Sense."[43] Given this rationale, it is not hard to imagine the significance of meteorological prodigies in the demonstration of the supernatural. As public events, these prodigies seemed less controversial and less difficult to verify than manifestations of spirits; the witnesses of extra-ordinary frosts outnumbered those who witnessed battleships.

On the other hand, the quest for truth could be a rejoinder to purely natural explanations. By the mid-seventeenth century, the "strange news" genre became a target of derision: scholars denounced it as insidious, politicians as seditious, and divines as superstitious.[44] In his works on meteors, Descartes, for instance, denied reality to the lights appearing in the serene summer nights which "give idle people cause to imagine squadrons of ghosts who battle in the air, to whom they prophesy loss or victory for a particular group they admire, according to whether fear or hope predominates in their fancy."[45] To forestall such charges, pamphleteers listed trustworthy observers and supplied readers with details of the place and the time of the occurrences.[46] Alex Walsham sees this quest for circumstantial evidence as "a quasi-scientific verification" aimed to counteract the corruption of the well-authenticated reports by the "great mixture of false and foolish and unwitnessed fictions."[47] And even as she considers this quest for objectivity as a sales strategy – a trope which many "contemporaries regarded as tantamount to denial" – she also notes that the circumstantial evidence of such reports became an indispensable attribute of the wonder tradition and "strange news" genre.

By the end of the Stuart era, it seemed as if no meteorological marvel was left unauthenticated. Writers and ministers visited sites, noted effects of prodigious waterspouts, questioned local inhabitants, and listed testimonies. As early as 1583, William Averell produced signatures of four eye-witnesses in the appendix to his *Wonderful and Strange News* about a wheat rain in Suffolk county.[48] In 1628, a pamphlet described an apparition of a battle above Hatford in Berkshire, attended by the fall of several "thunderbolts," i.e., stones universally believed to be generated in the thunderclouds. "[O]ne of them was seen by many people to fall at a place called Bawlkin Greene, being a mile a halfe from Hatford: which thunderbolt was one Mistris Green caused to be digged out of ground, she being an eye witness amongst many other."[49] The cross on the Moon of 1688 was seen by no less a figure than the Lord Chancellor of Ireland; the author of a pamphlet describing the phenomenon looked for evidence in contemporary weather diaries (see note 30). Marvels and prodigies were becoming matters of fact in a manner resembling that of the new natural history and experimental philosophy.

Yet these strange meteorological events were well-documented not just because they were fascinating on their own terms, but because their authenticity determined the fate of the religious, social, or political readings they received in the press. In recording such events, writers' purposes were *trans*-factual: whatever "matters of fact" they sought to present, these facts served to attest a "higher" truth, the truth of the miraculous, portentous, millenial, or political significance. Resembling the status of plants and animals in the Renaissance compendia of natural history, these items were more than just the facts of a *natural* history – they also were emblems. The interest in the emblematics of public meteors was therefore not a pre-modern search for the causes of meteors, but an independent tradition that persisted even after the erosion of its theological and philosophical presuppositions.[50]

"Vulgar Baconians"

Francis Bacon put a strong emphasis on the study of meteors, but was skeptical as to which were legitimate candidates for his inductive method. The chief source of his beliefs on the nature of facts appears in *The Advancement of Learning*. Natural history investigates nature "at liberty," (the history of generations), "in her errors" (the history of pretergenerations), and "in constraint" (the history of arts). The second class is deficient, however, as the author found "no competent collection of the works of nature digressing from the ordinary course of generations, productions and motions; whether they be singularities of place and region, or strange events of time and chance."[51] Marvels and prodigies – the tokens of nature's digression – had previously been the subject of either paradoxography[52] or the tradition of wonder, but with Bacon they were the subject of a new natural history and were to be explained in the same way as ordinary phenomena of nature. In fact, Daston has argued, marvels and prodigies came to represent the first facts in Bacon's program of natural history where they were used to replace the "contentious knowledge" of classical empiricism, based on unconnected *exempla* and illustrations of old natural philosophies. Bacon's facts were "handpicked for their recalcitrance" in regard to the explanatory schemes of old systems. The first scientific facts were strange ones because "natural philosophy required the shock of repeated contact with the bizarre, the heteroclite, and the singular in order to sever the age-old link between 'a datum of experience' and 'the conclusions that may be based on it'; in other words, to separate facts from evidence."[53]

This analysis can provide an entry into a discussion of the origins of the meteoric tradition. If we sketched the meteoric tradition as a program of recording unusual meteors, it would appear that, on Daston's account, it fulfilled Bacon's expectations as to the aims of *preter*natural history: it recorded unusual, abnormal and extreme meteorological phenomena – exactly the kind that would perform "the repeated contact with the bizarre." But were these Bacon's expectations? While it is possible to agree with Daston on the methodological importance of the Baconian preternatural, it is also important to note that the creation of meteorological facts was not just the result of methodological uses of marvels, but also of changing criteria in accepting marvelous narratives. In addition, the strong impulse to understand the first meteorological facts – i.e., the meteoric tradition – came from the Restoration Baconians, not Bacon himself.

Bacon however is unclear about how to proceeed. Like Descartes, he also criticized "prodigious" narratives. In *Parasceve*, he reflected on the "stuff and subject matter of true induction," which he posited in opposition to classical canons. A new science should do away with antiquities, citations, and eloquence, and with the old natural histories, whose material on the "wanton freaks of nature" could be a pleasant recreation but valueless for the advance of knowledge. For Bacon, those "sports" of nature – the monsters of organic nature in particular – are of the "nature of individuals," i.e., "small varieties of species," and may be of interest only as such, for their own sake, but not as the "stuff for true induction."[54]

The hyperbolical stories in which these individual instances were found, all superstitious, unverifiable "old wives' stories," should be rejected. But in this condemnation, Bacon did not include "stories of prodigies, when the report appeared to be faithful and probable." These stories are assigned a prominent role within a separate study of the *unnatural* workings of nature, i.e., the natural histories of "pretergenerations," or the histories "of prodigies which were natural." The common experience, argued Bacon, did not probe nature as deeply as the "*[d]eviating* instances, such as the errors of nature, or strange and monstrous objects."[55] All such instances should therefore be collected in "a natural history of all prodigies and monstrous births of nature; of everything in short that is in nature new, rare, and unusual."[56] In the same context, Bacon returned to the superstitious histories of marvels which he had originally condemned, and proposed to remit them "to a quite separate treatise of its own; which treatise I do not wish to be undertaken now at first, but a little after, when the investigation of nature has been carried deeper."[57]

It is not immediately clear whether Bacon does or does not want to use superstitious accounts of natural phenomena. It is not clear whether the "superstitious history of marvels" is a history of *marvels* or a meta-history of superstitions *regarding* marvels. Should marvels be also considered prodigies? And why wouldn't one count the freaks of nature as prodigies?

It may be suggested that there are two separate issues under analysis. First, there is the ontological status of preternatural instances, which Bacon tends to see as superior to ordinary phenomena. In the study of nature they bring forth the most recondite aspects of the world's ontology. The second issue is the epistemological status of the reports about such instances. These reports Bacon makes a subject of close scrutiny and, like the defenders of miracles, argues that his goal is not a demonstration of theological truth, but a revision of the world's inventory. The semiotics of marvels must be replaced by induction from privileged instances.

In this context, Bacon explains that the distinction in philosophical value between the "history of monsters" – "the collection of heteroclites, irregularities of nature" – and books of "frivolous impostures, for pleasure and strangeness," is in how thoroughly the latter have been purged of fables and errors. If only the observers' trustworthiness could be assured, Bacon implies, all would be well, and the difference between the histories of monsters and those of frivolous impostures would only be a matter then of the respective authorial intentions.

This explains why Bacon reserved a niche for natural extravagances at the end of the *Parasceve*, where he provides a long, if only tentative, list of subjects which he desires to see investigated in his natural history project (BOX 3)[58].

Given the fact that Bacon follows ancient philosophers in assuming that the region of meteoric production is one of genuine "tumult, conflict, and disorder,"

BOX 3   Francis Bacon's "Catalogue of Particular Histories by Titles"
in *Parasceve*

History of all other things that fall or descend from above, and that are generated
in the upper region
History of lightnings, thunderbolts, thunders, and corruscations
History of fiery meteors
History of winds and sudden blasts and undulations of the air
History of showers, ordinary, stormy, and prodigious; also of waterspouts
History of sounds in the upper region (if there be any), besides thunder
History of the seasons or temperature of the year, as well according to the
variations of regions as according to accidents of times and periods of years;
of floods, heats, droughts, and the like

---

it is understandable that this list favors extraordinary, severe, and even marvelous
meteors. A Baconian history of meteors can fairly be approximated with a history
of meteoric excess, for example: thunderbolts, waterspouts, droughts, prodigious
showers, and so on. For Bacon, the uppermost region of sublunary realm – the
arena for most unusual meteors – resembles the border between two kingdoms,
which "is harrassed by continual incursions and violence, while the interior of
both kingdoms enjoys peace, security, and profound tranquillity."[59] This is well
illustrated in the preface to *Historia ventorum* where the origin of wind has an air
of mystery: winds are "not primary creatures, nor among the works of the six
days; as neither are the other meteors actually; but produced according to the
order of creation."[60] Expelled from *Genesis*, meteors are the refuse of nature:
hard to describe, predict, and understand. It thus appears ironic that the first
natural history Bacon prepared as an illustration of his new organon was the
History of Winds (*Historia Ventorum*, 1622).

Why did Bacon choose the wind as the subject for the first installment of
the history of nature out of all other possible candidates? The reason might have
been that winds have traditionally been granted a privileged place in natural
study from antiquity to the modern era. Because of their universality; regularity
(as well as its irregularity); local characteristics with respect to humidity, heat, and
strength; and most of all because of their importance for seafaring, winds had
always been a subject of intense interest. *Historia ventorum* follows this tradition,
being a compendium of etymologies, frequencies, and properties of winds. The
treatise asks questions about the times, durations, and regions of specific winds,
but, generally, it does not offer any significant first-hand observations. The only
"inductive" facts Bacon considered epistemologically acceptable were the non-
controversial and well-authenticated second-hand reports about winds.[61]

Historia ventorum does not exactly stand as an example of Bacon's ideal of
the inductive philosophy based on new facts. The reason for this may be sought
in the *absence* of any earlier study of the same character. Additionally, Bacon's
questions about winds that underscored their being specific to a particular place

and time would have been difficult in the context of fiery meteors or lightning. Such phenomena were rarer and largely independent of local climates which made them less susceptible to geographical analysis. This was the unacknowledged problem as to the kind of descriptive procedure which could, at the same time, be applied to non-windy meteors *and* which would satisfy the requirements of Bacon's inductive program.

How did Bacon's view of meteorological facts develop into the concept that served as the basis for the meteoric tradition? It can be argued that seeing meteors as Baconian facts – rather than as Aristotelian examples or Renaissance portents – owed much to the discussion of the nature of historical knowledge generally and, more specifically, of what constituted acceptable evidence in natural history. This discussion abolished Bacon's identification of the prodigious with the fictitious, and supplied a rationale for inclusion of the "suspicious" evidence into the program of natural history. The works of Joshua Childrey and Thomas Sprat exemplify the beginning of this process, while its further expansion into the "mainstream" meteoric tradition can be followed in the works of Robert Plot, John Aubrey, and other late seventeenth-century geographical authors.

To understand these developments, it should be noted that Bacon's revision of natural philosophy and the magnitude of his goal presented practical and theoretical snags from the outset. For instance, it was difficult to determine in advance when and where the enquiry should terminate. For Bacon, the inductive natural histories, *per definitionem*, were at least several generations away from a *philosophy* of nature which they were proposed to service. Acolytes would have to make observations with "a most religious care," shunning speculations other than those of the tentative first vintage. Furthermore, such practice could end in piles of haphazard data. As Bacon believed that *any* information had a potential to be useful in natural enquiry, the latter could – and did – become increasingly hard to differentiate from the compilations of spurious reports, such as Crouch's sensational chronicles. Joshua Childrey's *Britannia Baconica, or The Natural Rarities of England, Scotland & Wales* (London: the Author 1660) illustrates these problems and sheds light on what K. Theodore Hoppen has called the "vulgar Baconianism," that is to say, the practice of Restoration naturalists who perceived the Baconian program "in an extremely eclectic and catholic light." The seventeenth-century Royal Society of London could thus support an expedition in search of dragons in the Alps, discuss spirits appearing during anatomical dissections, or analyze a tooth removed from a testicle.[62]

The *Britannia Baconica* represents an early example of such eclecticism (FIGURE 4).[63] Childrey (1623–70), born in Rochester, received his B. A. in 1646 at Magdalen College, Oxford.[64] After the Restoration he made his living as a clergyman, holding ecclesiastical posts throughout England. Although *Britannia* had been published before Childrey made contact with Henry Oldenburg and the Royal Society, it embodied some of the characteristics of the venture that the Royal Society was to begin under the guidance of Oldenburg, Robert Boyle, and Robert Plot. In 1658 Childrey began to collect and arrange his work in the manner suggested by Bacon by using over one hundred folio notebooks – one for each species of natural history listed in *Parasceve* – and two journals, containing the dates of historical droughts, comets, earthquakes, and other natural convulsions. In another journal *Geographia Naturalis*, Childrey began a geography of natural rarities.[65]

Britannia Baconica, the result of these compilations, epitomized narrative natural history in that it moved towards "losing the balance between induction and mere accumulation."[66] Contemporaries harbored misgivings about its merits. In a letter to Oldenburg, Dr. Nathaniel Fairfax observed that in Childrey's book he had found "no weight: I have found so many stories coming either to quite nothing, or changing into quite another thing . . . it is barely founded on ye relation of an Historian."[67] Childrey's work, however, proved influential, not

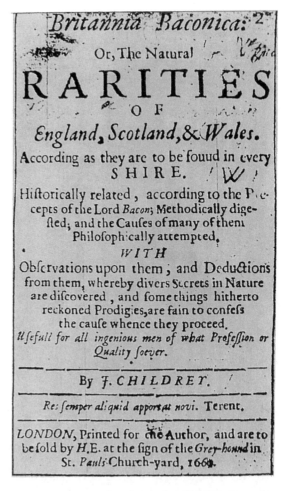

FIGURE 4   The title page to Joshua Childrey's *Britannia Baconica* (1660).

least because its audience could have enjoyed the capacity of its stories to change "into quite another thing," a marvel, that is.

The title-page juxtaposition of "rarities" and "Baconian" suggests that Childrey could have considered Bacon less a model of a new natural philosophy, and more of a patron of encyclopedic curiosity. The book is therefore a collection of brief – and mainly borrowed – accounts of the air, soil, stones, and county topographies. But the main stress is on the out-of-the-ordinary.[68] For instance, there are reports about the two-hour blood-shower in the year of 1176 in Hampshire, of Berkshire's blood-boiling well, of that "very violent tempest of shower and rain, at Nottingham," after which, on August 4, 1585, the ground suddenly sank. At Pool in the year 1653 it rained warm blood. "I once had a conceit that the blood might be engendered of some Vapour drawn up by the Sun from that part of the Sea where the cruel sea-fight was fought between the English and Dutch."[69] Other information included the description of the earth near Woburn – known among the local people to turn wood into stone, an account of a three-acre piece of ground mysteriously moved for forty rods, and the famous street ponds in Dunstable that never dried.

Reflecting on the status of this information, Childrey commented that they were without a doubt natural, but how they turned out to be so strange, he thought was "a secret we dare not meddle with til we have satisfied Sir F. Bacon's mind, by writing a compleat History of Nature and Art."[70] This caution could justify the seventeenth-century natural histories of the "miscellaneous" and give an encouragement to large-scale collecting projects: Childrey, for instance, amassed material for a larger work which was to deal with rarities of the world but the book was never finished.[71] He made clear in the preface that the vulgar should learn from the book

> not to mis-believe or condemn for untruths all that seemes strange, and above their wit to give a reason for, who are the least able of all men to do it. Not that I will undertake for the truth of all the Relations in Mandeville, and other credulous Writers; but so much may be said in their behalf, that all is not as the most is; that they have many truths interserted with their fables and falshoods, and some of them altogether as improbable as they. Here are no stories told you of what is to be seen at the other end of the world, but of things at home, in your own Native Countrey, at your own doors, easily examinable with little travel, less cost, and very little hazard. This book doth not shew you a Telescope, but a Mirror, it goes not about to put a delightful cheat upon you, with objects at a great distance, but shews you yourselves.[72]

For Childrey, vulgar opinion was not to be a yardstick for deciding what did or did not have a natural explanation. Rejecting the truth of all seemingly strange and improbable facts would prove one's ignorance about nature's operations rather than a good policy of investigation. But in endorsing the project of unnatural natural history, Childrey made a point with wider implications. The world may abound in the grandiose "sports of Nature," he claimed, but our land is no less than a "compendium" of all the countries, having the distant marvels copied on to its native soil. "I would have those that know other Countreys so well, not to be strangers to their own."[73] The Commonwealth of Learning should use the book as a contribution to the "several Histories in his Lordships Catalogue, at the end of his *Novum Organum*." But not until the young "relatours" compile the rest of the "circumstantial histories" of their county rarities would there be a credible *Aetiology* of those phenomena.[74] The testimony of witnesses also ought to be scrutinized: the reader, Childrey recommended, as well as the prospective compiler, should act with interpretive moderation when presented with written documents, and with a caution when on site. Whatever a "relatour" discovered in this manner he should communicate, "and in particular, if he be of Dorsetshire, that he would bestow upon us a punctual account of that raining of blood at Pool with all its circumstances."[75] For Childrey, describing rarities and strange events was a valid historical method. The natural historian should not reject as untrue everything that seemed strange. The *Britannia Baconica* was a proper contribution toward a natural history of pretergenerations, even though it allowed for a more liberal treatment of extraordinary facts.[76]

Rehabilitating the unusual

In 1666, Childrey's work found support in Thomas Sprat (1635–1713). Sprat's *History of the Royal Society* (1666) is in many respects a manifesto of the principles of the new science and a programmatic exposition of the Baconian historical method. In particular, Sprat's discussion on the reports of marvels and prodigies contributes significantly to his examination of the character of historical evidence. Following Bacon, Sprat dismissed the ancient natural histories of Pliny, Aristotle, and Solinus on the basis that, instead of matters of fact, they contained only "pretty Tales and fine monstrous Stories." They were "[r]omances, in respect of True History," which multiplied extraordinary events and uncommon appearances.[77] The new natural history, in contrast, ought to be based on an unbiased and systematic gathering of facts, because, as Joseph Glanvill put it, without "a considerable inlargement [sic] of the History of Nature, [our] Hypotheses are but Dreams and Romances, and our Science meer conjecture and opinion."[78] As a consequence, the "considerable enlargement of the history of nature" ought to cover not only the non-controversial realms of nature, but also those characterized by strange things. The following "renovated rationale for the unnatural natural history"[79] represents Sprat's rendition of the premises of Bacon's "superstitious history of marvels" as it was envisioned in light of the above methodological criticism:

> When my *Reader* shall behold this large number of *Relations*; perhaps he will think, that too many of them seem to be incredulous stories, and that if the *Royal Society* shal much busie themselves, about such wonderful, and uncertain *events*, they will fall into that mistake, of which I have already accused some of the *Antients*, of framing *Romances*, instead of solid *Histories* of Nature. But it is certain that many things, which now seem *miraculous*, would not be so, if once we come to be fully acquainted with their *compositions*, and *operations* . . . To

make [the *Natural History*] to consist of strange, and delightful Tales, is to render it nothing else but *vain*, and ridiculous *Knight-Errantry*. Yet we may avoid that extreme, and still leave room, to consider the singular, and irregular *effects*, and to imitate the unexpected, and monstrous *excesses*, which *Nature* does sometimes practice in her *works*. The first may be only compared to the fable of *Amadis*, and the Seven Champions: the other to the real *Histories* of *Alexander, Hannibal, Scipio*, or *Caesar*: yet there is nothing that exceeds the *Truth* of *Life*, and that may not serve for our *instruction*, or *imitation*.[80]

Seemingly "miraculous things" should be reported because future naturalists may prove them true. "Irregular effects" of nature should be recorded because they were as natural as regular effects. Sprat used the analogy between natural and civil history to stress two different kinds of improbability. The first was characterized by narratives of straightforward falsity, such as appeared in the "romantic" genre, i.e., the "old wives' stories." In contrast, the second was like that of the "real histories" of antiquity whose integrity could have been challenged only by lesser men, unable by nature to sympathize with great human achievements. Two years after Sprat's publication, Glanvill suggested an analogy in which he equated the opinions of these lesser men – vulgar opinions – with those of the ancients: Archimedes's science would sound no better to common people, wrote Glanvill, "then those of *Amadis de Gaule*, and the Knight of the Sun." Correlatively, the certainty of modern accounts would have been subject to the same historical myopia had they been known to bygone ages, because "to have talked of a new Earth to have been discovered," wrote Glanvill, "had been a Romance to Antiquity."[81]

It is important to note here that Childrey and Sprat emphasized that their new historical venture should be both a collaborative and socially agreeable practice. In the *Britannia*, Childrey desired that prospective naturalists should be young and leisured gentlemen. Their work should be pleasant, inexpensive, and edifying. During the 1670s, Childrey began to recruit correspondents. On several occasions he voluntarily acted on behalf of the Royal Society in personally distributing questionnaires concerning agriculture and the countryside in general.[82] For Sprat, collaboration had a strong epistemological advantage: "the Histories of Nature we have hitherto had, have been but an heap and amassment of Truth and Falshood, vulgar Tales and Romantick Accounts; and 'tis not in the power of particular unassociated Endeavours to afford us better." Sprat was also adamant about the importance of provincially-based correspondents who might bring with them "a good assurance of likelihood, by the integrity of Relatours; and withal they furnish a judicious reader with admirable hints to direct his Observations."[83] Within this feedback program, the "admirable hints" multiplied at a steep rate. Sprat summarized some of the early Royal Society's mail as:

Relations of *Parelii*, and other appearances seen in *France*: of the effects of *Thunder* and *Lightning*: of *Hurricanes*, and *Spouts*: of the bigness, figure, and effects of *Hailstones*: of *Fish*, and *Frogs* said to be rain'd: of the raining of Dust out of the *Air*; . . . of changes of Weather, and a way of predicting them: of the vermination of the *Air*: of the suppos'd raining of *Wheat* in *Glocestershire*, which being sown was found to be nothing but *Ivy Berries*.[84]

In cataloging local phenomena of such a kind, Childrey and early correspondents of the Royal Society consistently blurred the distinction between marvels understood as prerogative instances of induction (Bacon's view) and marvels as titillating items of the non-scholarly press. They also blurred the distinction between the "superstitious history of marvels" – which Bacon considered as necessary as any other natural history – and "old wives' stories," which he rejected as false.

These tendencies were expressed in a particularly precise fashion in the early *Philosophical Transactions* of the Royal Society. During the seventeenth and eighteenth centuries, the "unofficial" journal of the Royal Society became a major repository of letters on unusual meteors. These letters were written by both prominent natural philosophers from London and Oxford and less prominent country correspondents, mainly clergymen and physicians. For example, in 1666, Robert Moray wrote about "extraordinary tides" in the West Isles of Scotland, Thomas Neale on "a sad effect" of thunder and lightning, and in the same year there was a report about four suns and "two uncommon Rainbows."[85] During the following three decades, such "extraordinary" letters became familiar. A correspondent from Somersetshire described a "strange frost" at Bristol, John Wallis reported from Oxford about "an unusual meteor" known as *Draco Volans*, and Dr Nathaniel Fairfax wrote on hail stones of "unusual size" that fell in Suffolk.[86] Other phenomena included a "very terrible whirlwind," a "mischiveous Waterspout," a very "extraordinary thunder," a "surprising clap of thunder," an "eruption of fire" out of the earth, an "unusual *Parhelion*," and a "strange effect" of the storm in Sussex.[87] As BOX 4 illustrates, the key adjectives of meteoric reportage are "remarkable," uncommon," and "extraordinary."

These meteorological communications witness a general disposition of the seventeenth- and eighteenth-century naturalists to report exceptional weather. The "empiricisms" of these reports reflected both public fascination with marvelous meteors and the theological scrutiny that aimed to prove them to be *true* signs of God. These forms of empiricism were then contrasted with Bacon's inductivist philosophy which granted marvelous meteors a place of special importance, but declined to accept as true all the contemporary reports about these phenomena. As long as the reports about the preternatural were considered false the "inductive philosophy" of meteors was restricted to examples, illustrations, and compendia of phenomena recovered from canonical texts. Bacon's own history of winds was one such catalog.

From the mid-seventeenth century, however, naturalists like Joshua Childrey, Thomas Sprat, and Joseph Glanvill began to reclaim marvelous narratives into the program of natural history. Such narratives, they claimed, could be true even if improbable. They also could prove useful in the future study of pretergenerations. Therefore, the rehabilitation of existing forms of *non-meteorological*

BOX 4   Uncommon meteors in the *Philosophical Transactions* of the Royal Society of London, volume (year) page, 1661–1771

Robert Moray, "Extraordinary Tides in the West Isles of Scotland," 4 (1666): 53

"Of Four Suns and Two Uncommon Rainbows Observed in France," 13 (1666): 219

Throughout Thomas Neale, "A Sad Effect of Thunder and Lightning," 14 (1667): 247

Nathaniel Fairfax, "Hail Stones of Unusual Size," 26 (1667): 481

"A Narrative of Several Odd Effects of a Dreadful Thunderclap," 65 (1670): 2084

"Strange Frost about Bristol," 90 (1673): 5138

Signore Sarotti, "Of Red Snow Seen in Genoa," 136 (1677): 863

William Cole, "On the Grains Resembling Wheat which Fell Lately in Wiltshire," 186 (1687): 281

George Garden, "Effects of a Very Extraordinary Thunder," 222 (1696): 311

Robert Mawgridge, "Surprising Effects of a Terrible Clap of Thunder," 235 (1697): 782

Edmund Halley, "Account of an Extraordinary Iris," 240 (1698): 193

Stephen Gray, "An Unusual Parhelion and Halo," 262 (1700): 535

John Fuller, "A Strange Effect of the Late Great Storm in Sussex," 289 (1704): 1530

Edmund Halley, "An Account of the Late Surprising Appearance of the Lights Seen in the Air," 347 (1716): 406

Samuel Molyneux, "An Account of the Strange Effects of Thunder and Lightning," 313 (1708): 36

Richard Richardson, "An Account of a Wonderful Fall of Water from a Spout in Lancashire," 363 (1719): 1097

Ralph Thoresby, "The Effects of a Violent Shower of Rain in Yorkshire," 372 (1722): 101

William Derham, "Uncommon Appearances Observed in an Aurora Borealis," 410 (1729): 137

Robert James, Lord Petre, "A Letter Concerning Some Extraordinary Effects of Lightning," 464 (1742): 136

Peter Collinson, "An Observation of an Uncommon Gleam of Light," 483 (1747): 456

Peter Daval, "Of an Extraordinary Rainbow," 493 (1749): 193

Sir Andrew Mitchell, "Of an Extraordinary Shower of Black Dust," 50 (1757): 297

Thomas Barker, "Of a Remarkable Halo," 52 (1761): 3

John Stephens, "Of an Uncommon Phenomenon on Dorsetshire," 52 (1762): 108

James Stirling, "Of a Remarkable Darkness at Detroit, in America," 53 (1763): 63

Alexander Wilson, "Of a Remarkable Cold Observed at Glasgow," 61 (1771): 326

discourses within a new *meteorological* discourse occurred as a consequence of discussion on the nature of historical evidence. In other words, meteorological reporting in the late Stuart era appeared to have been a sort of re-contextualization of the narratives about meteorological prodigies from public discourse, religious argumentation, and an informal literary treatment to a mere epistemological milieu. The universal interest in exceptional meteorological events, intrinsic to the wonder and millenarian traditions, served as a topical model in the Baconian history of meteors. The meteoric tradition constituted a methodological rehabilitation of public and religious uses of prodigious meteors.

It should be noted, however, that Childrey and Sprat did not suggest any *specific* stipulations about the content of the new natural history of meteors. They defined a *general* character of the phenomena which they desired to include in the Baconian program. They thus referred to a *class* of meteorological events – marvelous, strange, and severe – not to specific meteors – halos, fireballs, mock-suns, or earthquakes. In principle, the specific content of this material was a matter of negotiation. But as we have seen, early meteoric reports had an extraordinary topical resemblance to the popular wonder-tradition. But this resemblance was not the result of an influence which meteorological pamphlets exerted on the rise of meteoric reportage. Rather, both the meteoric genre and the popular treatment of meteors developed in response to a growing public awareness of the overwhelming significance of unusual weather occurrences. Meteoric reportage and the popular culture of meteors were chronicles of public weather. They reflected universal interest and concern about strange and destructive atmospheric phenomena. Storms, fireballs, and mock-suns of the seventeenth century are analogous to the twentieth century's tornadoes, hurricanes, and *el Niño*. These phenomena attracted *both* the general public and scientists. *Important* and *conspicuous* phenomena – portents, disasters, marvels, strange appearances – were therefore explored both in the public sphere and in academic discourse. To illustrate the public dimension of prodigious atmospheric events, the following chapter will examine the early eighteenth-century debates that surrounded the social, political, religious, and scientific uses of uncommon meteors.

3

Public meteors

The next surprising scene this year,
did in the northern heavens appear,
where, after sun-set, did arise
strange coruscations in the skies.

From sun-set to the break of day,
did these celestial fireworks play,
whilst crowds and mortals stood below,
beholding the tremendous show.

Some harden'd sinners seem'd to gaze,
with pleasure on the scatter'd rays,
as if the wonder was no more
portentous than a rainy show'r.[1]

Edward Ward, *British Wonders*

In a superstitious age we cannot wonder that these phenomena [meteors] have all been attributed to a supernatural agency; it is one of the nobler purposes of philosophy to release the mind from the bondage of imaginary terrors.

G. Gregory, *The Economy of Nature.*[2]

Early modern understanding of meteors amounted to describing violent storms, unusual displays of the northern lights, oddly-shaped hailstones, prodigious colds, snows, heats, floods, or droughts: the phenomena of both the sublunary and public spheres, nature and society, people and God. These meteors ruined property, disrupted life, conveyed God's warnings and executed "His judgment". They transmitted astral influences and portended future destructions. During the seventeenth century, however, some writers developed arguments to combat the public and symbolic dimensions of meteors. They rejected "superstitious" readings as the socially subversive products of "vulgar enthusiasts." "Prodigious" events of nature were to be accommodated within a different framework, less emblematic and more tangible.

This chapter looks at this process and examines the arguments used in denouncing religious interpretations of extraordinary meteors by describing the social and cultural effects of the two major meteorological events of the early eighteenth century: the storm of 1703 and the Northern Lights of 1716. It is useful first, to observe the tangible effects of the homiletic uses of meteors on the political arguments surrounding the storm of 1703 and the ways in which these arguments were substantiated and second, to describe the methods used by some naturalists to condemn polemical readings of meteors, and outline the ways in which they were able to consolidate a reaction against the cognitive competence of the "vulgar" examining their social interests and political motives along the way.

Providential visitations

Moral and ideological instruction lay at the heart of the early modern literature on unusual meteoric occurrences. In sixteenth- and seventeenth-century sermons, uncommon meteors were represented as signs of divine will. Whenever disaster struck, religious writers could choose to connect it with the delinquency of its victims and so provide moral guidance.[3] After the London earthquake of April 6, 1580, John Aylmer, Bishop of London wrote to Lord Burleigh that he thought "it were requisit, without further delaye to geve some ordre and direccion to stirre up the people to devocion, and to turne awaye Gods wrath threatened by the late terrible earthquake."[4] When in 1588 the westerly gale – the famous "Protestant Wind" – caused great damage to Phillip II's Spanish fleet, a commemorative medal issued in England had the inscription "God blew and they were scattered."[5]

Unusual meteors and weather disasters lent themselves to didactic purposes. Since they were local, it seemed natural to consider them as intended for specific communities, parishes, towns, houses, or individuals. Storms, minor earthquakes, and deaths by lightning were especially convincing in this respect.[6] Ingeniously, even when a disaster bypassed a place, it could be used as a reminder of "what could have been if": "Nothing is here presented to thine eyes, to fright thee," explained a pamphlet from 1608, "but to fill thee with joy that this Storme fell so farre off, and not upon thyne owne Head. Yet Beware, for the same Hand holds a Rod to strike every one that deserves punishment."[7] From a common-sense perspective, such arguments capitalized on the perception of weather as a daily component of travel, agricultural and commercial activities. Popular religious belief required homely tokens – storms, earthquakes, and meteoric showers – which could be transferred from communal memory into Reformation morality (FIGURE 5).

From a theological perspective, such reasoning could be grounded in the doctrine of Divine steering (*gubernatio*) that presented God as the ultimate moral coordinator of natural, human, and collective affairs. Significantly, the doctrine entailed a *causal* link between nature and human affairs so that nature occasionally served as a direct Divine agency in history and thus fell in the domain of God's special providence. On the other hand, God's continuous preservation of, and cooperation with, the Creation (*conservatio*) was designated as general providence. For John Wilkins, general providence was God's government over nature via secondary (natural) causes, whereas special providence acted in the domain of humanity through an immediate Divine action. The appeal to special providence made sense when either natural mechanisms failed to produce the observed effects or at times when humanity failed to be grateful for the benefits of general providence.[8]

For some seventeenth-century theologians, special providence was a bulwark against the rise of natural theology. Natural theology shifted the emphasis in both Anglican and Dissenting thought from interpretation of the sacred text to reason as a primary if not ultimate arbitrator in matters of theodicy.[9] John Ray, William Whiston, and William Derham summarized this program stating that natural theology dispensed with the necessity of God's direct intervention, because the world's observed perfection was sufficient proof of a benevolent Creator. Opponents of this view perceived this as an untenable diminution of God's immediate sovereignty because it relegated the necessity of intermittent intervention necessary for the moral sustenance of the Creation. If all that remained was the *a posteriori* argument of natural religion, no one would be bound to any other moral principle than one's own view of the laws of nature. Natural theology would entail "casting away the whole basis of

FIGURE 5   *Wonderful signs of strange times* (1680). A broadsheet with an engraving of prodigious natural events, apparitions and omens.

Christian sanity, throwing aside rewards, punishments, hell, damnation, terror, sin, etc."[10]

It was a matter of urgency to counteract so dire a consequence. During the last three decades of the seventeenth century, a number of sermons, pamphlets, and broadsheets sought to interpret meteorological phenomena as uncontestable instances of special providence. Compilations of historical and contemporary storms, floods, earthquakes, and various lesser disasters attracted large and varied audiences.[11] Meteors were considered "Torches and Voices of God" from which something was to "be further learned."[12] One writer juxtaposed hermeneutics and prognostics of seasons, writing:

> You are weather-wise, it seems, and can make prognostication of the seasons from your observing the sky and the clouds. But why do you not make remarks likewise of another nature? Why do you not lift up your eyes and behold the black and thick clouds which hang over your heads, and will in short time disburden themselves upon you in storms and tempests, in amazing thunder and lightning. Are you not sensible of the turbulent weather that hovers over you? Do you not perceive the tempestuous season which is near?[13]

The reprimand was occasioned by an earthquake and written by an anonymous "Reverend Divine." His *Practical Discourse* in theology portrayed God as the voice of prophets, prodigies, and extraordinary weather: "the strange and amazing tempests, storms and thunders, earthquakes, eruptions of fire, inundations of water, alterations in the heavens, strange appearance of the sun and moon, comets, or as vulgarly called, Blazing Stars."[14]

Outside theological scrutiny, the invocation of divine providence in seventeenth-century parlance was so repetitive that it almost escaped historical attention; "[c]onventional providentialism belongs to conventional piety, and conventional piety, the bread and butter of so much seventeenth-century thinking, can easily be mistaken for mere literary decoration."[15] That the common people reduced the notion of special providence to a colloquial trope, ("God's will"), testified to the fact that Protestants of all shades shared a common linguistic, psychological and spiritual world when they spoke of God's "mercies" and "visitations." During the Civil Wars and after the Glorious Revolution, the notion became a catchphrase which seemed to dominate social and natural phenomena.[16] It was said that even William of Orange's succession was assisted by a meteorologically benevolent Protestant God. When, in the autumn of 1688, the Dutch fleet remained fully prepared for the invasion, a persistent, westerly wind kept it in port as if to fulfill prayers of the Dutch Catholics for James II's preservation. This "popish wind" continued until early November, during which time the phrase became so notorious that "the magistrates of the Holland towns were obliged to forbid anyone to say in public that 'the wind is papist' on pain of a heavy fine."[17] In early November the wind veered round to a strong easterly, the famous "Protestant Wind," which now pinned the English fleet to the anchorage in the Gunfleet while the invaders swiftly reached their debarkation point in Devon. "I never found," said Gilbert Burnet, Bishop of Salisbury, "a disposition to superstition in my temper," but he confessed that "the strange ordering of the winds and seasons, just to change as our affairs required it, could not but make deep impression on me, as well as on all that observed it."[18]

Providentialism was not uncommon in the *Philosophical Transactions* of the Royal Society of London. John Eames, F. R. S., theological tutor to the Fund Academy and an editor of the early *Transactions'* abridgement, received in 1730 a letter from a certain Evan Davies from Pencarreg, Carmarthenshire, with a narrative of the temporary "death" of a woman and her children struck by "Thunder-Bolt, Thunder-Ball, Lightning &c." In the section on "Wounds and Bruises," Davies wrote that he did not draw any conclusion from the case, but that many had said that "it was God's judgment upon them. But as the Ways of Divine Providence are unreachable, who causes it to rain upon the just and the unjust, I think People ought not dive too far into those inscrutable Arcana's [sic]; but only pray to God to preserve us evermore."[19]

Similarly, Dr George Garden, Provost of the King's College in Aberdeen and a Scottish Episcopalian, received in 1696 a letter from an Aberdeen school teacher on the effects of "very tremendous claps of thunder" which fatally hit his school. "All the Children that were killed were in different places," wrote the teacher, "and as it were even picked out. I cannot stay to tell you the many other Circumstances wherein Providence over-ruled for they who deny it are grossly mistaken, and if they had seen what I saw, they would be Confuted and Silent for ever."[20] The same year, Robert Mawgridge, The Royal Kettle-Drummer reported about the "unusual Clap of Thunder that fell upon the Trumbull-Galley," whose flash of fire reached the Gun-Room, and set three bundles of matches on fire, without igniting the gunpowder. "This, with Thanks to God for his Deliverance of us from so eminent a Danger, is the best account that can be given thereof."[21] Even without explicit reference to providence, the choice of details described would be hard to explain without recourse to the ubiquity of the providential idiom. In the popular sermon tradition, if the accounts were increasingly "secularized," it was because readers were now left "on their own to draw conclusion."[22] Because providential "reading" was often taken for granted, the "plain" information on storm damage, lightning-wounds, mock-suns and other meteoric prose of the eighteenth century would remain unintelligible without reference to the wider "semiotics" of the natural world.

The "Great Storm"

English politics from 1689 to 1714 revolved around religious issues. The fear of atheism, deism, popery, and Puritanism was inflamed by issues such as the ambiguous interpretation of the Toleration Act of 1689, the ejection of Non-Jurors, and the occasional conformity of a number of the Godly ministry. By 1701, "the whole imagery of battle and siege became normal verbal currency in the sermons and writings of High Church controversialists,"[23] the imagery that will permeate the following discussions of the Great Storm of 1703 and the aurora borealis of 1716. Interest in both events was enhanced by the religious issues; the effects of the storm were connected to the rights of Dissenters, and the aurora borealis of 1716 to the Jacobite rebellion of 1715.

The burning issues at the turn of the century were those of a Catholic monarchy, the increasing recruitment of Jacobites among the High Anglicans and Non-Jurors,[24] and the perceived threat of Puritan resurgences. There was a strong Tory/Whig polarization on issues such as the war with France, dynastic succession, and legislation regarding religious minorities. The Toleration Act of 1689 and the growth of Nonconformist congregations gave High Churchmen and Tories some grounds for fear that the wave of Dissent might irreparably damage the Anglican Establishment. The fear was exacerbated by the spreading of "Occasional Conformity," through which the Nonconforming ministers could occupy a civil office if prepared to take the Sacrament in an Anglican Church just once in the year before they could stand for local election. Worried about the resulting increase in the number of the municipal corporations controlled by Dissenters, the Tories launched a legislative campaign (1702–04) against occasional conformity. The tensions were further intensified by divergent political analyses of the War of Spanish Succession, the outcome of which had the potential to decide England's religious future.

During this agitation, the summer of 1703 turned unusually wet. In May, the amount of rain that fell in London set the ten-year record and the rest of summer remained exceedingly humid. European newspapers gave accounts of great rains during July. October and November were very warm. On November 12, the wind picked up and gradually increased in strength during the following two weeks; some chimneys broke in London and several ships were lost off the coast. Then, on Thursday, November 25, when everyone already believed that the weather had become unbearable, the wind greatly intensified and the Great Storm began.

"[N]o storm since the Universal Deluge was like this, either in its violence or its duration," Daniel Defoe estimated.[25] Its passage was marked by a 300 mile wide belt of destruction across southern England and Wales, northern Germany, Denmark, parts of France, Sweden, and even Finland.[26] In England, it was first felt in the West, and some conjectured that it originated in the area between Virginia and Florida where the confluence of vapours from the lakes and marshes supplied sufficient amounts of the stormy "matter." William Derham at Upminster identified the peak of the storm's intensity to be in early on Saturday morning, November 27: throughout that Friday, "the wind was S.S.W. and high all day, and so continu'd till I was in Bed and asleep. About 12 that Night the Storm awaken'd me, which gradually increased till 3 that Morning. And from thence till near 7 it continued in the greatest excess: and then began slowly to abate, and the Mercury to rise swiftly."[27]

Hardly anyone in London slept. Many expected their houses to fall but nobody ventured outdoors as bricks, tiles, and stones flew through the streets with deadly force. Many alleged that they had felt earthquakes, heard thunder, and "in the country the air was seen full of meteors and vaporous fires: and in some places both thunderings and unusual flashes of lightning, to the great terror of the inhabitants."[28] There was evidence of whirlwinds in a few places: at Whitstable in Kent, a ship was lifted out of the water and deposited 270 yards from the water's edge. According to Defoe, when the people eventually crept out of doors the next morning, with the "distraction and fury of the night visible on their faces," their first concern was the fate of friends and relatives. Scenes of destruction had not been experienced since the Great Fire, and the great majority of people spent the next two days wandering around and gazing at the devastated city.

Estimates of the effects of the storm listed 8,000–9,000 people dead; 300,000 trees broken or uprooted, 800–900 houses knocked down, 400 windmills destroyed, 100 churches stripped of their covering lead and numerous steeples shattered, innumerable barns and stables demolished, 15,000 sheep drowned in the Severn at high tide. The *London Gazette* published a report on losses sustained by the Royal Navy: in Portsmouth, Her Majesty's Ships *Newcastle* and *Lichfield's Prize* were lost, the Fireship *Vesuvius* was stranded; in Bristol, the *Suffolk* and *Canterbury* storeships were destroyed. The total damage was assessed between £1–4 million.[29]

The storm produced a strong impression on the government. With the ongoing war of the Spanish Succession, the loss of ships was perceived to be particularly serious. "What Amazing Strokes!" exclaimed a pamphleteer, "what numerous spoils to so many at Sea and Land, requiring much Time and Treasure to heal our Wounds, to make up our Private and Publick losses; which at this Juncture of War when we must open our Veins, to give Life and Strength to others, must need be a very formidable blow to the whole Kingdom."[30] The Queen announced a day of fasting and humiliation and declared the storm a "token of divine displeasure." The Dissenters, claimed a nineteenth-century author, "notwithstanding their objection to the interference of the civil magistrate in matters of religion, deeming this to be an occasion wherein they might unite with their countrymen in openly bewailing general calamity, rendered the supplication universal, by opening their places of worship, and every church and meeting-house was crowded."[31] It was evident, however, that the crowded meeting-houses and chapels harbored less than peaceful sentiments. Capitalizing on the Toleration issues, the ministry speedily exploited the storm's effects to stress conflict rather than harmony. Sermons and pamphlets flooded the market.

"When the Sins of a Nation are very great and provoking," one of the "storm-preachers" asserted, "it is God's usual Method to pronounce destruction against the Nation."[32] In the words of Thomas Bradbury, an influential Dissenting minister, the age was corrupt and "it's a National Stroke that we are now lamenting, and I know no greater piece of National Service [than reformation of manners], that we are capable of."[33] Preachers and pamphleteers singled out the theatrical plays performed after the storm as evocative of the moral decay of the nation. The Lord Bishop of Oxford thought the representation of a mock-tempest an unprecedented piece of prophanity. Preachers exploited the concept of divine management of all "Publick Calamities." Snow, vapour, and stormy wind, claimed Bradbury, "fulfill His Word [and] the Tempest flies upon its Master's Errand." Other authors seized the opportunity to address the issue of interpretation itself. *The Terrible Stormy Wind*, an anonymous sermon from 1704, mocked the pre-posterous "philosophical" claim that the storm "was nothing but an Eruption of Epicurus's Atoms; a Spring-Tide of Matter and Motion; a Blind Sally of Chance: so throwing *Providence* out of the Scheme."[34] Instead, the storm could not possibly be a natural phenomenon, but the Almighty's punishment of a corrupted nation.[35]

Such arguments fed on a pronounced discrepancy between the storm perceived as a scar on the face of the country – an overwhelmingly significant public event – and, on the other hand, the storm seen as merely natural and thus normal occurrence. The latter perspective could have easily been taken as a cynical posture of those who put their philosophies before the suffering of their fellow countrymen: an implicit accusation which could be readily substantiated by the biblical iconography of the wind's inscrutability and the limits of knowledge. The wind was well-suited to those arguing for a less than syllogistic proof of God's existence. On the one hand, it stood for the inscrutability of a natural world in its resistance to the process of reasoning. In his analysis of the storm, Defoe appealed to the Scriptures to argue that, unlike rationally explicable things, God "holds the wind in his hand, and has concealed it from the search of the most diligent and piercing understanding."[36] Gifford thought that whether winds arose from the motion of Earth, from reciprocal action between this and other orbs of the planetary system, or from the air and vapors, "all remain uncertain."[37] On the other hand, wind was a paragon of divine volition whose patterns eluded human understanding. Assuming the divine origin of the Great Storm, Defoe surveyed the theories of wind in Aristotle, Seneca and the Stoics, arguing that "no one hypothesis . . . has yet been able to resolve all the incident phenomena of Winds." This allowed him to move from an unsatisfactory natural explanation to a theological explanation proper. But in arguing for a providential interpretation, Defoe also played down natural theology. In his opinion, no ordinary natural phenomena, creatures and processes – however well suited to their functions – would suffice in formulating a consistent and true theodicy of the Creation. Instead, one ought to turn to Baconian defects and excesses of nature, and observe their true "causa finalis":

> [W]e never inquire after God in those works of nature which depending upon the course of things are plain and demonstrative; but where we find nature defective in her discovery, where we see effects but cannot reach their causes; there it is most just, and nature herself seems to direct us to it, to end rational inquiry, and resolve it into speculation: nature plainly refers us beyond herself, to the mighty hand of infinite power, the author of nature, and original of all causes.[38]

On December 6, 1703, a London newspaper ran Defoe's advertisement asking readers in the provinces to submit accounts of the storm.[39] In this undertaking, as Ilse Vickers has recently argued, Defoe embraced Baconian notions as well as the method of collecting information through parochial enquiries.[40] But where Childrey, Plot, Lhwyd, Hooke, and others had to rely on a scattered record of local wonders, Defoe's advertisement was very effective – the response from the countryside was overwhelming. Even when only a portion of the material was published, it amounted to over seventy signed descriptions of damage, deliverance, and unusual phenomena written by clerics, yeomen, merchants, peasants, both women and men. The collected information was intended for circumstantial evidence of divine justice which proved that God sided with the Dissent. Presbyterian Scotland evaded the calamity; Anglican England suffered terribly. "They say this was a high church storm,/Sent out the nation to reform;/But th' emblem left the moral in the lurch,/For 't blew the steeple down upon the church."[41] In Defoe's undertaking, Baconian natural history became the method of dissenting providentialism.

But circumstanial evidence also helped the Anglicans. In 1708, a lengthy pamphlet argued against Puritan enthusiasm by refuting the interepretation put forward in *Mirabilis Annus*, a Puritan collection of special providences.[42] *Mirabilis Annus* claimed that the destructive storm of 1661 was an act of a Puritan God, thus showing his displeasure with the Restoration of the monarchy. Because some pamphlets about the Great Storm asserted the same, the *Harvest* sought to "overmatch [the Dissenter polemicists] in giving them a most particular and authentic account of the stupendous accidents that happened" in 1703. This counter-interpretation centered on the issue of authenticity. It alleged that the Puritans failed to identify witnesses of uncompromised integrity. For example, it asserted that as *Mirabilis Annus* involved the work of several compilers, it was unlikely that they shared the same degree of sincerity; indeed, the second instalment of *Mirabilis Annus* (1662) confessed to seven false reports. Could there be more? The *Harvest*, on the other hand, asserted the authenticity of its information and its authors claimed to be in a better position to judge the meaning of evidence. Indeed, the facts unambiguously supported the position of the established church: "God Almighty did upon this Great Occasion, give us more signal proofs of His Justice of his Mercy."[43]

Defoe's and the *Harvest's* reliance on evidence transcends issues of epistemology. As was the case with the empirical methods used in the authentication of miraculous meteors, Defoe's search for "matter of factness" was neither ideologically neutral nor politically innocuous. In the case of extraordinary events and bizarre feats of nature, the procedure for establishing a definitive "matter of fact" was central to interpretive purposes and, by extension, to the religious and political discourse at large.[44] The focus on detail in these accounts was neither ornamental nor informative but religious and political. In this sense, the witnessing and accreditation of sources in early modern meteorological discourse cannot be seen as a prerogative of early modern science by which the experimental philosophers sought to map the audience, dominate "the literary technology of experimental reporting, and [establish] matters of fact as worthwhile products of experimental work."[45] Because circumstantial witnessing served theological (and political) discourses, its relevance was derived in a significant part from those extra-epistemic concerns.

Aerial appearances and astounding apparitions

To move from informed theological tracts to sermons, and from sermons to popular ephemera and pamphlets, is to move from relatively dynamic doctrinal debate to the world of everyday providentialism and fascination with the extra-ordinary. Seventeenth-century England witnessed an immense production of broadsheets and catchpenny quarto pamphlets describing military skirmishes in the sky, warships, castles, men with flaming swords, crosses, and other images among clouds.[46] What agencies lay behind these phenomena? Was there a social dimension which governed these appearances? In which way did these phenomena differ from meteors?

On the night of September 3, 1654, the day of the Dunbar and Worcester victories of the English over the Scots, three soldiers of the Hull Garrison were in a state of shock when, between nine and ten o'clock, the sky became illumin-ated by fiery-colored streams. The streams disappeared very quickly, and to the amazement of the soldiers, a great battle "of horse and foot appeared in the air." A body of pikemen on the eastern sky faced a northwestern army, and after skirmishing, the bodies engaged in battle, charging each other with pikes, "break-ing through one the other backwards and forwards." Drawing upon the reserves, the fight continued for a quarter of an hour. Then, when the western army seemed to be winning, a dark army entered the sky from the North. Its soldiers instantly fired muskets and cannons and the battle ended with the red army in flight. The author stated:

> Reader, what Interpretation thou will make of this Apparition, I know not; neither shall I adde anything of mine own to the Relation: onely take notice (and believe it) it is no Fiction nor Scar-Crow, but a thing real, and far beyond what is here reported: for the Spectators (such was their astonishment) could not recollect so much as they saw, afterwards to make a true report of.[47]

Similar reports can be found in other seventeenth-century sources (FIG-URE 6). In March 1621 two armies were seen in Austria fiercely fighting at mid-day; in 1622, two swords standing against each other were seen above the city of Lintz while in the midst of them two armies were engaged in a "pitch'd Fight." In 1624, armies waged wars over Bohemia while in 1631, above Asherleben in Saxony, a northern army won over a southern one. "The coming in and total defeating of Duke Hambleton's Army, anno 1648," reported *Mirabilis Annus*, "was clearly portended by the appearance in the heavens of a Southern and Northern Armies in Yorkshire, and the N. armies being beaten by the other." In 1716, people observed a flaming sword over Chipping Norton while Londoners saw two armies fiercely engaging each other for several hours.[48]

Understanding the status, timing, and meaning of these phenomena is a complex problem. First, it is wrong to treat them in the same way as the northern lights folklore because such treatment would beg the question on the epistemological status of "folklore."[49] Second, it would be inadequate to think of sky wars as either pamphleteers' allegorical constructs or perceptual errors of "unenlightened crowds." This would entail the very "objectivity" which invites allegorical or perceptual approaches. Third, to contend that what people observed was merely a natural event would be to side with skeptics, and would not explain the *persistence* of the claims which supported the supernatural character of these occurrences. Indeed, "superstition," "deception," or "error" were labels used by those with high (ideological) stakes in proving the opposite.[50] These and similar interpretations are premised on what Stuart Clark has recently identified as a "realist" model for the studies of early modern witchcraft and liminal phenomena in general. The realist model requires a distinction between true and false utterances by assuming that the former refer to the phenomena in a "real world" whereas the latter do not. For example, because in late modern societies "witchcraft" and sky wars do not refer to real activities, historians have tended to see them as either irrational or "epiphenomenal reflexes of other [genuinely real] things."[51] In other words, the realist approach in historiography has so far failed to interpret these old-fashioned beliefs as beliefs. We should instead entirely disregard the issue of reality by abandoning the need for reference to the Authorities and consider reality as an attribute given to things by historical actors. Only with such an approach would the historian be able to make sense of the beliefs and events which have no direct parallel to modern experiences.

Given the critical historical junctures at which "strange news" flourished, it may be suggested that the reasons which its authors and readers had for asserting

FIGURE 6   A broadsheet engraving representing the appearance of ships fighting over the town of Goeree in Holland. The eyewitnesses in the figure are meant to substantiate the claims made about an improbable event by associating it with a public consensus. *A true and perfect Relation, of a strange and wonderful Apparition in the Air*, 1664.

its reality were embedded in a context in which such reality made sense of the events which either preceded or were expected after such strange appearances. More specifically, meteorological apparition narratives could be seen as a platform from which those, who perceived themselves in a precarious social, confessional, or existential condition, could voice their discontent and even threaten to change it. These heavenly images, it may be said, were hieroglyphics of social struggle. But, as we will see below, they also enabled an assertion of collective identity which did not discriminate between exaggeration, allegory, belief, gossip, providentialism, and prevarication.

This identity may be put in the context of the fact that early modern societies perceived life as inherently uncertain. The laboring poor used Christianity and folk beliefs to seek "assurance, guidance, protection, and some explanation of the world" which they experienced as precarious and largely unknown.[52] The poor lacked a "predictive notation of time" in the sense that "opportunity is grabbed as occasion arises, with little thought of the consequences, just as the crowd imposes its power in moments of insurgent direct action."[53] Semiology of the apparition narratives could be viewed as a major resource for using those opportunities, refracting in this way the discontent, fear, expectation, and physical stress. In this sense such narratives could even be used as an extra-legal form of sanction, a "species of popular disapprobation," or, on a more general level, a subversive "expressive symbolism" deliberately imposed upon or incited within the lower orders.[54] In short, the sky-war narratives will be presented here as enduring not because they described how things happened in the world, but how the world could become.

Their cultural endurance is especially visible in that the accounts of heavenly conflicts obeyed rigid stylistic norms: they described the weather before the event – an unusual heat or a strangely shaped cloud – and described the encounter with regard to the insignia, arms, and movements of the troops. Armies either won or lost, were separated, or disappeared. Political interpretation featured in a small number of these materials, but it was generally believed that the events in question foreshadowed future disturbances. Writers drew up lists of witnesses or produced numbers of reliable observers. Particularly important is the fact that psychological and public reactions to these sky battles played a crucial role in the larger purpose of the narratives. Observers' astonishment was not to be understood solely as a response to the strangeness of these phenomena, but primarily as a response to their divine origins. God, not nature, was the object of fear.[55]

Thundering heavens, as an author put it in 1677, extend their "aery throats . . . to alarm [the] Inhabitants," and to display how "Factious Meteors Muster all their Forces, and contend with Warring Elements, as if they would unhing [sic] the Globe, or reduce the Universe into its primal Chaos, and in an undigested Heap of all Confusion."[56] Dismayed, people would leave their work and gape up toward the sky, fall on their knees, and say that "verily the day of Judgment was come." The "wonder [of those observing the cross in the Moon] was by and by forced up to the highest pitch of amazement and terror," while the eerie streamings of light in 1716 elicited "a most dreadful Cry among my Neighbours; all thinking to be consumed by it immediately; or that the last Day was come." Some authers invoked etymology to propose that "Heaven" comes from an Hebrew Root, which signifies to amaze and astonish.[57] During the 1665 epidemics of the plague, Defoe wrote, Londoners were addicted to prophesies to the point at which it seemed a "wonder the whole body of the people did not rise as one man and abandon their dwellings." It was unclear, however, whether this attitude had been raised by those "who got money by it," or by the "phlegmatic hypochondriacs"; at any rate, people were "terrified to the last degree." As their imagination in consequence turned "wayward and possessed," some of those who constantly gazed at the clouds eventually happened to see "shapes and figures, representations and appearances," and were so positive of what they claimed to have seen that "there was no contradicting them."[58]

Whatever their reasons, the "lower" orders were declared to be especially prone to observe aerial battles. The falsity of such reports was evident because they came from the "unthinking" and that what truth there was in an event, was distorted by word of mouth: "[r]eport in such distractions as these, hath thousands eyes, and sees more than it can understand; and as many tongues, which once set a going, they speake anything."[59] Among philosophers, belief in marvels, prodigies, and miracles was correlated with the stratification of society. Spinoza claimed that the common people frequently invoked the miraculous because "partly from piety, partly for the sake of opposing those who cultivate the natural sciences, they prefer to remain in ignorance of natural causes."[60] Descartes and Sprat assigned the sky-war narratives to the socially inferior and by the early eighteenth century both religious and secular writers tended to label such stories as figments of the "vulgar" imagination. In 1712, the chorographer John Morton, commenting on reports on wheat- and frog-rains, said that they reflected "the easiness of the vulgar and their proneness to believe any strange thing, that they are not very nice in distinguishing upon such subjects. That which looks like truth is often as passable with them as which really is so."[61] The learned thus abstained from "common" opinion and not only on doctrinal grounds. Henry More argued that human proneness to suspect the presence of God in things "great or vehement" derived from a physiological distemper. The "fits of Zeal" with which a religious enthusiast expresses his devotion to God and which effected the "belief that there is something Supernatural in the business" were but the flatulence of melancholic minds."[62] Decades later, when the anti-clerical *Freethinker* characterized popular predictions as the "phantoms of a distempered imagination," it also suggested that humankind's great frailty, in addition to its propensity to vice, was its strong inclination "towards everything, that is mysterious, Dark, and incomprehensible."[63]

Arguments like these did not affect the production of apparition accounts, however. As if entirely immune to these objections, pamphleteers sometimes proceeded with an attack on natural explanations and "men [who] deride the

Wonders of the Middle Region, as if they were only the effects of Heat and Moisture."[64] The author of an account on aerial ships observed by a man in Cirencester noted that "men will object it was only the fancy of a whimsical Brain; but how could 6 people be seized with the same impediment at the same time."[65] Early in the nineteenth century, John Howie provided a historical account of sky battles which were seen, not once and by individuals only, but many times and by the whole families of ordinary villagers with "credit and reputed integrity." For Howie, as for his seventeenth-century predecessors, the sheer numbers of witnesses represented the central criteria in establishing the authenticity of the battle reportage. If the appearances had been seen only by a few persons "– and these being perchance of a timid nature, or given to drink and exaggeration, or subject to trances and diseases – then it might have been said that the darkness of their own nature, or the passion of fear, had so wrought upon their fancy or imagination, and so beclouded and overheated it, as to make them unable to judge aright of what they had seen." Yet none of this was true: the armies were observed in broad daylight by "a good number of witnesses," free of the pretense of divine inspiration. Howie did allow that some if not all of the sights were explicable by secondary causes but went on to assert the precedence of the "primary cause" in the production of these phenomena, i.e., God's special providence.[66] One of the most cogent maneuvers in this regard was the Reverend Divine's counterclaim aimed at Descartes:

> There are some who imagine the Clouds may by chance fall into the shape of Horses and Men, and the Winds ruffling the Clouds and beating them backward and forward may make them seem to encounter one another: and upon Thunder and Lightning there maybe a resemblance of great guns going off: thus they impute all to the Natural Position and structure of the Clouds. But to think that this is the true solving of this Phenomenon is so fond and idle that I cannot believe it will be the Sentiment of any sober and considerable Person. Spectacles of so composed a frame are not works of mere chance. Besides the experience both of the wise and the vulgar attests these to be presages of approaching Evils and Calamities which shews that they are not Casual and Fortuitous.[67]

"Lord Derwentwater's lights"

Many contemporaries recognized socio-political subversion at the heart of sky apparitions. The *Harvest* of 1708 explicitly argued from such a position.[68] John Flamsteed complained about the popularity of prognostications with "the superstitious Vulgar: of what ill consequences their predictions have been and how made use of in all commotions of the people against lawful and established sovereignty . . . our own sad experience, in the late Wars, will abundantly shew the considerate."[69] Astrologers considered apparition reports as a *deliberate* formula of the discontented "to raise, spread, and promote stories of a strange tendencie, on set purpose to amaze and amuse our late-distracted and yet not well-quieted Kingdom."[70] But the most notorious controversy about the meaning of sky battles arose in the wake of the mysterious illumination of England in March 1716. Where politicial polemicists wished to see the phenomenon as another aerial battle, naturalists sought to construct it as a natural event. In their bid to normalize the sublunary sphere, philosophers counteracted the heterodoxy inherent in the symbolic readings of aerial wars.

In its March issue, the *Historical Register* wrote on a luminous appearance seen in London on Tuesday evening, 6 March 1716: "the display appeared at first like a huge body of light, compact within itself, but without motion; but in a little time it began to move and separate, extruding towards the west, when it seemed to dispose itself into columns or pillars of flame. From thence it darted south east, with amazing swiftness, and after many undulatory motions and vibrations, there appeared to be a continual fulguration, interspersed with green, red, blue, and yellow."[71] Even though the writer used the terms "vibrations," "undulations," and "fulguration" to describe the affair, he failed to suggest its probable causes. Neither could others. The *Flying Post* wrote about "a pale sort of a Light [which] broke out in the NW of our Horizon, which looked like the Dawn of Day, or rather like the Moon breaking through Clouds. It darted many Streams towards all parts of the Sky, which look'd like Smoak. It proceeded towards the SE, and continued by several intervals till Midnight, when it finally disappeared."[72]

Witnesses from all parts of the country groped for adequate terms and analogies; A gentleman from Edinburgh declared he knew not "how to describe it . . . though in the Night gown and Slippers I walk'd up to the Castle Hill, to have a more distinct and open Prospect." Another wrote that he could "not possibly find Words to give you a perfect Description of it."[73] To some it appeared as a "theater of light," a "seeming dawn," the "Changing and Shakings of the Light" or an "explosion [of light] from horizon." William Derham compared it with luminous "lances, cones, spires, whatever they may be called;"[74] Others likened it to the representation of "Glory wherewith our Painters in Churches surround the Holy Name of God," or to the radiating "Starrs wherewith the Breasts of the Knights of the most Noble Order of the Garter are adorned." The astronomer Edmund Halley was afraid that "in delivering the *Etiology* of a Matter so uncommon, never before seen by my self, nor fully described by any either of the Ancients or Moderns, I fail to answer their Expectation or my own Desires."[75]

Notwithstanding widespread agnosticism, the phenomenon appeared to a great number of people far less "surprizing" and less difficult to describe. A clergymen at Watford, Hertfordshire, was awakened by his servant, who in "great fright begged of [him] to come out immediately, for there were Two Armies fighting in the Sky." At Oxford, several people "saw Swords drawn, and Armies fighting in the Air." In Upton-upon-Severn, Worcestershire, many people observed the armies with weapons in full use; the smell of the gunpowder and sulfur filled the atmosphere. Sights of this nature were so frequent, a correspondent from Edinburgh reckoned that "by this Post you [Whiston] will hear of very dreadful

Stories of Apparitions in the Air, seen by all People of this city."[76] Even those failing to observe battles and armies reported on the "*Contentions* betwixt these Meteors, being all in Confusion, and darting one against another, with an incredible force and swiftness [yet] thro' all that Region of the Air where these *Confusion* and *Strife* (for I can term it nothing else) was, the Stars appeared Clear."[77]

The nation was appalled. On Edinburgh's Castle Hill, "Young and Old, Poor and Rich, in short, the whole Body of People gather'd together." In Elston, "it was observed by a thousand People, not without a Greatest Wonder and with strange Apprehensions." The people of Wakefield, Huntington, and Hatford spent a sleepless night.[78] As for London, it was aghast: "All the People were drawn out in the Streets, which were so full one could hardly pass, and all frightened to death,"[79] because:

> Then appeared in the sky [over Lincoln's Inn Fields] two great Armies which contained thousands of Men and Horses, these seemed fiercely to Encounter Each Other, and the Battle seemed long and doubtful, as if Fortune was in debate with herself, not knowing to whom she would give victory. [After 11 o'clock], Armies seemed to be separated, there appeared a continual stream of fire from the NE to SW to the Astonishment of all that beheld it. For it was no Meteor or Vapour raised from a Natural Cause, but Something Supernatural, Strange, Frightful, Astonishing and Amazing, as is testified by Thousands of People.[80]

As in the case of the Great Storm of 1703, it is the political rather than epistemological context which provides a more adequate approach to under-standing the meaning of these incidents. In this period of intense electoral politics, no fewer than ten general elections occurred between 1695 and 1715.[81] In 1695 there were riots at Oxford, Exeter, Westminster and elsewhere; the election of 1705 provoked a disturbance at Coventry where 600 or 700 people took over the town hall and held it throughout the election. Following Sacheverell's prosecution by the Whig administration in 1710 and the series of riots induced by his sentence, the autumn elections were held amid intense party feeling and disturbances broke out in York, Canterbury, Norwich, Nottingham and other towns.

In the spring of 1715 conflicts between Jacobite and Hanoverian factions intensified; Tories and High Churchmen feared that the dynasty's Whig bias heralded an attack on the Church's privileges. Attacks on Presbyterian meeting houses took place in London and in the Northwest; in Staffordshire, "scarcely a Whig or Dissenter could escape insult or more serious injury."[82] The autumn of 1715 also marked the date of the most serious mainland uprising of the Jacobites. It was associated with the Earl of Mar and Thomas Forster. Arrested earlier that year by the Whig Ministry on charges of conspiracy against the King, they both escaped and began to organize the uprising. But the Earl of Mar suffered early defeats and despite the Pretender's arrival in Scotland in December 1715, the rebellion there was suppressed. In England, Forster and James Radcliffe, Earl of Derwentwater, raised about 300 men and proclaimed the Pretender (son of

James II) as "King James III". By the end of the year they victoriously marched into Lancashire where, after being attacked by the regular Army, the Jacobite leaders surrendered. The Earl of Derwentwater, William Gordon, and others were arrested, transported to London, and tried for high treason. The trial reached its end on 24 February 1716, when the Earl of Derwentwater was executed on Tower Hill.[83] Several days later, the hearse with his remains started its journey towards Dilston in Northumberland, where the burial was to take place. At about 6 o'clock on the evening of 6 March, Francis Dunn, an old family servant of Derwentwater, met the hearse and wrote the following in his diary:

> A most Beautiful glory appeard over ye hearse, wch all saw, sending forth resplendant streams of all sorts of colours to ye east & west, the finest yt ever I saw in my Life. It hung like a delicate rich curtain & continued a quarter & half of an hour over ye hearse. There was a great light seen at night in several places & people flockt all night from durham to see the corpse. Its remark't yt att ye sameday & hour ye glory appear'd over my lord's hearse, ye most dreadfull signs appeard over London.[84]

The entry links the coffin and auroral displays. The people who observed the lights near Durham were the same who congregated around the nobleman's hearse. The "Northern Lights" of that night were known thereafter as "Lord Derwentwater's Lights."[85] In Newcastle, a Dissenter's journal affirmed that the lights were "said to have been interpreted by the Jacobite Party as an omen of God's displeasure for beheading the rebel lords." In London it was reported that some had seen two headless men fighting with flaming swords. Lady Cowper observed that both political parties manipulated the imagery of the Lights. "The Whigs," she noted, "said it was God's Judgment on the horrid Rebellion, and the Tories said it came for the Whigs taking off the two Lords that were executed."[86]

The Flying Post featured an article on "the late unnatural Rebellion," deploring the practice of some city clergymen who, on the eve of Derwentwater's execution, "prayed in a more particular Manner for the Lords condemned." But when the surprizing phenomenon appeared, "[t]he Disaffected Party have work'd this up to a Prodigy and interpret it in favour of their Cause."[87] Anti-Jacobite interpretations proliferated in the following months. The representation of a great dark cloud visible during the phenomenon, a pamphlet argued, was said to be darkness of the consciences of those seeking to destroy both Church and State. "The Man that appeared therein is a certain great Man" whose actions are dark notwithstanding his upright position. "As he carried in his Right Hand a Flaming Sword, it signified that his Heart is full of Fury, seeking the Destruction of all Love and Charity."[88] As the interest in the phenomenon grew daily, an advertisement appeared in the *Daily Courant* about the intention of Mr William Whiston to shed light on the mystery at the Marine Coffee House in Birchin Lane and in the large lecturing room in Villard Street. Lecture tickets were sold to several coffee houses in central London; "All good accounts of this Appearance will be acceptable either before or after" Mr Whiston's performance. The lights had become a public commodity.[89]

At this point in time, the northern lights became part of eighteenth-century "public" science about which recent historians have had much to say. Historians have shown that from the early eighteenth century, an ever-increasing group of philosophical entrepreneurs worked to present natural knowledge as public good. Itinerant lecturers and demonstrators captured the attention of genteel audiences to whom they marketed skills, instruments and demonstrations as commodities in an enlightened age. In doing so, they not only made money and levered themselves into positions of power, but also promoted the ideals of a utilitarian and commercial culture and, by extension, acted as protectors of the existing social hierarchy.[90]

Simon Schaffer has used the notion of philosophical "audience" to suggest that experimental philosophy in eighteenth-century England can be usefully analyzed in terms of "a practice of public display" and the "love of the marvelous," to be found among the eighteenth-century practitioners of magnetism, electricity, geology, and meteorology. Atmospheric phenomena in particular, symbolized for natural philosophy "a wider and grander theatre of power and *also* a space in which the new economy of understanding and control might operate."[91] The London earthquake of 1750 quickly became a commodity used by competing groups (natural philosophers, clerics, millenarians, journalists). Something of this nature had also happened after the luminous display of 1716: the phenomenon provided an occasion for antagonistic political and ecclesiastical commentators to cross swords on the issue of the political implications of the phenomenon. It was also appropriated by two other groups: by the less radical writers whose goal was to defend the Church in times of ecclesiastic crisis, and second, by natural philosophers. These appropriations epitomized the contentious nature of meteorological marvels but they also served as evidence for early Hanoverian views on relationships between religion and science.[92]

Thomas Sprat, Joseph Glanvill, or John Wilkins, for instance, presented Restoration experimental philosophy as an antidote to religious "fanaticism" of all kinds, leading some recent historians to propose that the rise of the new science was a force in a battle against Interregnum sectarians. In contrast to the "fury of enthusiasm," experimental science sought to secure the intellectual foundations of the restored monarchy.[93] Sprat criticized those who mistook figments of imagination as miracles, as well as those who "make general events to have a private aspect, or particular accidents to have some universal signification."[94] He recommended a strict scrutiny of extraordinary phenomena before they could be categorized as supernatural. If, in this process natural philosophers removed phenomena from the realm of marvels, they then classified them as the ordinary works of the same Creator.[95]

In opposition to such arguments, the more conservative Anglican clergymen rejected the adequacy of merely intellectual arguments for God's existence.[96] Science threatened the doctrine of revelation in particular: if theology were detached from revealed truths, revelation would be pronounced "superstitious"; if, on the other hand, reason sufficed in demonstrating religious truths, revelation would become superfluous. This thinking led several early eighteenth-century natural philosophers to reassert the weight of the miraculous and to offset the hegemony of reason by stressing prodigies, divine intervention, and omens, even if that path ran the risk of improbability.[97] The Reverend Divine argued staunchly for the predictive aspect of sublunary apparitions, writing that "the delineations and effigies of persons and things in the clouds were ever presumed to presage some strange events," and that the doctrine of prodigies had been vouched for by the long tradition of historians, and all other researchers into the works of providence.[98]

Whiston's lectures on the 1716 lights exemplify these divergent views. An impoverished former Boyle Lecturer with Arian leanings, Whiston homed in on the area of doctrinal ground beyond the arguments against revelation – those of superstition and redundancy – and argued for a non-superstitious, reasonable approach to revelation.[99] This was to be done in a form of an extended biblical exegesis in which Whiston relied on two scriptural passages. The first, Jeremiah 10:2, argues for a skeptical position with respect to prodigies: "Thus saith the Lord, Learn not the way of the heathen, and be not dismayed at the signs of heaven; for the heathen are dismayed at them." The second, Matthew 16:3: "O ye hypocrites, ye can discern the face of the sky; but can ye discern the signs of the times", asks Christians not to forsake prophetic revelations, but on the other hand, they should not be confused so as to forget the true meanings of such occurrences. In other words, Whiston's position requires a proof for the rational relevance of prodigies in the theology of revelation and such proof must be based on a separation of the public/sensational/psychological impact of unusual events and their spiritual/doctrinal dimensions. One should observe prodigies for their message, not for their spectacle. In Whiston's interpretative *via media*, the 1716 appearance of the lights, while not truly miraculous, represented never-theless a warning "to raise the Affection of Reverence, Fear, Trust, Admiration, Worship, and Praise, to the World's great Creator and Governer."[100] The position required Whiston to repudiate the premises of both vulgar *and* philosophic knowledge by rejecting both the subversive power of marvel-mongering and interpretation in terms of the ordinary course of nature. In the first section of the essay, the phenomenon was reconstructed from the letters of Whiston's correspondents. Whiston analyzed the principal features of the phenomenon and proposed a solution based on the secondary causes. Having accomplished this, in the following section, entitled "Reasons why our Solutions are so imperfect," Whiston reasserted the providential reading.

Some aspects of this lengthy account are particularly worth noting. First, Whiston produced evidence about the famous historical occurrences of similar lights: the list of chronicles and eye-witnesses included chorographers like Camden and Stow, philosophers like Gassendi, Seidel and Kirchner, and royals like King James II who, on Boyle's word, saw a display on a visit to Edinburgh. Second, the letters which Whiston prefixed to the account exemplify the genre of "strange but true" news: graphic narratives which used authenticating techniques, described the crowds of frightened spectators gazing at these extraordinary phenomena. Most of them had a religious implication. These "perlocutionary" properties were strengthened by the belief that the phenomenon had a power to cause, attract or intensify other anomalies. Thus some reported to have seen a comet just before the streaks of light began and the people of Yorkshire observed parhelia and peculiarly colored rainbows.

Whiston's account eschewed colloquialism as much as possible and focused on theoretical issues such as duration, size, colors, movement of the lights, and the strong nitrous smell which "affected and disordered spectators to a great degree." Regarding the cause, Whiston supposed that large quantities of sulf-nitrous exhalations ascended into the atmosphere, where they were the subject of a vigorous fermentation. The lights were merely a faint fermentation of the cold regions of air, like thunder and lightning were ripened fermentation of the warm regions. Whiston's intention, however, was neither the reporting of philosophical truth nor philosophical polemic. The scientific section of the account served instead to establish two mutually supporting claims: first, that the phenomena were "certainly agreeable to the Natural Course of Things," and second, that they were not signs of "a Properly Miraculous or Supernatural Power."[101]

These claims were designed to forestall the first line of attack on the doctrine of special revelation, the charge of superstition. Rather than finding a *particular* natural explanation, his concern was instead to secure the possibility of *natural* explanation *per se*. Whiston's project was no less than that of supplying theoretical grounding for natural philosophy itself. He perhaps realized that adopting any *particular* natural explanation of the Lights would automatically presume the validity of natural explanation *per se*: a *petitio principii*. But if the content of natural philosophy could not provide justification of natural philosophy *per se*, where was one to look for an independent source of legitimacy?

The solution was in a social rather than philosophical strategy, that is to say in departing from vulgar knowledge. It is worth emphasizing that in an important sense, the "philosophical" grounding of natural knowledge owed much to the power of claims that presented new science as rational because it was "non-vulgar" and "non-superstitious." Natural philosophy was to be disassociated from the perception of the ordinary people. In this realm of disciplinary and social control, individuals like Whiston, Morton, Halley, and Pointer, extolled virtues of science *at the expense* of the incompetence of the masses.[102] Whiston thus wrote that "*seeming* Clashings and Conflicts" observed during the night of March 6 were brought about by the agitation of vapors; the same cause produced those "strange Mutations and Disorder" in the heavens which "the unthinking, affrighted, or superstitious called Armies, Spears and Battel in the Air, which yet to a sober and judicious Person they did no way properly resemble."[103]

Edmund Halley differentiated on the same grounds in his own account of the lights, published in the *Philosophical Transactions*. He apologized to readers for not being able to offer any conclusive explanation of the phenomenon, partly because he had missed the beginning of the event, which "however it might seem to the vulgar Beholder, would have been to me a most agreeable and wish'd for Spectacle." Writing about a grand ball of fire two years later he made it clear that even though the astonished vulgar spectators saw in this phenomenon not more than a mere spectacle, it still afforded "no less subject of Enquiry and Entertainment to the speculative and curious in Physical things." Such attitudes were shared among earlier writers who, like John Morton from Northampton-shire, said that the reports on the appearances of "Aerial Warriours" had been found even among "better Ranks than the Vulgar. Those in the Fen-Countrey who were spectators of this strange Prodigy, peradventure saw a great many small clouds of uncouth Shapes . . . and all the rest [was] the Product of their own Superstitious Imaginations."[104] (FIGURE 7).

Natural philosophy and natural theology could both be used as antidotes to the presumed radical implications of vulgar knowledge. In this context, Whiston's project, rather than a wholesale naturalistic deconstruction, was in reality an

FIGURE 7   A representation of the northern lights. Benjamin Langwith, "An Account of the Aurora Borealis that appeared October 8, 1726", *Philosophical Transactions* 34 (1726): 132.

attempt to define a stance in which the 1716 lights – and, by extension, other marvels in the sublunary sphere – could be perceived as having less volatile political effects. The legitimacy of natural philosophy was in this way linked to an ameliorative social policy. Thus when naturalists sought to expose the fallibility of the ordinary people, they also wished to banish subversive meteorological hermeneutics.

Such a strategy was embraced by the political writers. The Whig *Flying Post*, – seeking to repress the Jacobite reading of the phenomenon by accusing "some ignorant people" (Jacobites) of distorted vision – wrote that heavenly armies could have been observed only by those whose ideas overpowered their senses. Indeed, there was nothing there that could not be explained from "Natural Causes, the sun having being hot for two days past, by which Vapours were exhaled both from East and West etc."[105]

In rationalizing discourse, the symbolism of sky prodigies became too suspect to be given public credibility and political weight, especially if it could be of help to the rebellious.[106] Following Whiston's idea for the sufficiency of secondary causes, John Pointer distinguished between false and true interpretations by showing the former to be morally flawed. He feared that the resort to "prodigious" readings would render special providence unnecessarily "exposed for Our Pleasure and Conveniency," concealing "our Own Ignorance, [and saving] us the Trouble of Enquiring into Natural Causes." Those who turned the normal and natural into the marvellous and ominous were guilty not only on religious, but also on civil grounds. Pronouncements about ominous marvels, Pointer claimed, would lead to widespread astonishment, panic, and eventually to the "desperation" of civil subjects: "Insomuch that should we suppose a Foreign Prince or Potentate to re-invade us, How flushed (hereupon) with Hopes of Success might the Invader be? And how dismayed, and already Half-overcome with Fear, the Invaded?" Prophesies were, in Pointer's view, by their nature "very impolitick, because oftentimes of dangerous Consequence to the State."[107]

The various theoretical analyses of the aurora borealis end here in a politically motivated censure of the socially irresponsible "plebeian public." If the "auroral" tradition revealed the popular political anxiety of the 1710s, the rebuke which naturalists mounted against this tradition can be viewed as a maneuver aimed at annulling the volatile imagery which pamphleteers sold to discontented audiences. The epistemic stigma attached to catchphrases such as "vulgar ignorance," "superstition," "fanaticism" or "enthusiasm," was in learned discourse transformed into a suppression of the undesirable social developments that radical political hermeneutics of meteors threatened to incite. The denial of supernatural, millenarian, or providential readings in the context of philosophical and theological enquiries served as an antidote to a disorderly state. It was one of the strategies by which eighteenth-century elites endeavored to empower themselves against the threat "from below." Indeed, recent historians of the eighteenth century have claimed that the locus of the social evolution during the *ancien regime* was to be sought in *culturally* defined groups, rather than in *economically* defined classes. During the early decades of the eighteenth-century, natural philosophy and low-church ratiocination emerged as formidable forces in counteracting unorthodox intellectual, religious, and social opinions.[108]

This process of collective self-fashioning was symmetrical; whereas philosophers labelled as superstitions the news of wonderful, unnatural, and extraordinary weather phenomena, so the lower classes resisted this practice by continually emphasizing the importance and frequency of such phenomena. As Patrick Curry has shown, eighteenth-century popular astrology persisted among the laboring poor precisely because it promised to offer resistance to the labor-oriented organization of time which threatened the traditional rhythm of "natural time." The vitality of the many "un-canonical" forms of knowledge – particularly at times when such forms were under severe criticism by coteries of philosophers, reformers, and naturalists – was no more than a "measure of resistance, whose passivity, far from being merely chthonic and inchoate inertia, was precisely its own best and most appropriate kind of defense."[109] In this sense, the early Hanoverian period witnessed the "*self-generating* aspects of plebeian political culture" emerging in spontaneous protests and snubbing the Whig Establishment by "drawing on a well-established repertory of political symbolism."[110]

Apparition narratives – and sky-battle reports in particular – were one of the items in this repertory. They were perceived as being increasingly unwelcome in a period of widening cultural division between the laboring and other strata of English society. In this process, the gentry became detached from the rest of society, leaving symbolic culture relatively independent from the body of learned knowledge.[111] As the continuing supernatural readings of meteorological phenomena testify, such culture remained remarkably robust and increasingly detached from the rational culture of the English Enlightenment.[112] When in 1798 thousands of "prejudiced people" observed a heavenly crown "announcing general peace" in Europe, a scientific author present at the site thought it necessary to persuade "the majority that these signs were the effect of the vapours and exhalations occasioned by heat," admitting that "ten years ago, I should not have been so successful."[113]

The eighteenth-century persistence of portentous, providential and political meteorology – even as it was continuously challenged – suggests that large sections of society continued to live in an emblematic environment and an anthropocentric culture of weather. While the attributes and different social implications of philosophical and symbolic understandings of nature might have been widely acknowledged, it was far from evident that the majority would find it either reasonable or opportune to accept the rationalized, normalized, and secularized world of philosophical reconstruction. The public, psychological and existential presence of meteorological phenomena made such an acceptance especially unlikely.

4

"Memorials of uncommon accidents"

I have known a gentleman in Ireland deny their climate being moister than
England's; but if they have eyes let them open, and see the verdure that
cloathes their rocks.

Arthur Young, *Arthur Young's Tour in Ireland 1776–9*[1]

The mixed nature of prodigious meteors – their public visibility and their epistemic
value as prerogative instances – was especially exemplified in seventeenth- and
eighteenth-century British local geographical study, chorography. In chorography
– an early modern genre executed on a county basis and devoted to local history,
genealogy, antiquities, and natural productions – prodigious meteors were
important because they were *local*, because they were *place-specific*. Local curiosities of nature reinforced regional identity and highlighted the region on
the geographic and cultural maps of the country. Regions acquired identity through their weather and the weather was characterized by its regions:
Cornish storms, Yorkshire waterspouts, and the salubrity of Norfolk's air were
not only phenomena of nature, but also historically recognized peculiarities of
regions.[2]

This chapter shows that the detailed description of these unique and rare
meteorological phenomena – i.e. the meteoric tradition – served as a means of
providing British provincial culture with individual character and regional flavour.
The chapter will illustrate the ways in which the meteoric tradition was used for
this purpose by discussing the cultural function of Stuart chorography and by
examining the writings of its most influential late-Stuart representatives, Robert
Plot, John Aubrey, and their successors. Moreover, a survey of the *Philosophical Transactions* published during the editorships of Hans Sloane, Cromwell
Mortimer, and Martin Folkes, suggests that a majority of weather narratives of
the period came from the provincial gentry and clergy with strong antiquarian
interests. The epistemological dimensions of this are discussed in the context
of meteorological matters of fact: led by analogy to historical method, provincial naturalists made meteorology a study of the physical evidence of weather's
severity. By adopting this approach, they replaced a natural history of meteoric
"impressions" with a reconstruction of the "remains" these phenomena made on
material objects: houses, churches, trees, and so on. Meteors, hitherto eluding
analysis and description, became in some of these narratives the "inscriptions"
which the weather imprinted on property and natural objects. The interpretation of these inscriptions became an objective of early meteorological science.

The most important aspect of this reconstruction was the chorographic apotheosis of personal, first-hand inspection of evidence.

Historians of British regional geography – chorography – associate its early prominence with the works of William Lambarde and William Camden. These Tudor authors combined "a medieval chronicle tradition with the Italian Renaissance study of local description," and provided a topical, stylistic and geographical framework for a variety of subdisciplines that previously had been treated separately: heraldry, numismatics, iconography, genealogy, description of antiquities and monuments of nature.[3] Chorography of the Tudor-Stuart period thus included surveys of estates, genealogies of local families, and descriptions of natural and architectural attractions. Much of the incentive behind these enterprises lay in the economic ambitions of the middling gentry who needed surveys of their recently acquired monastic lands. But these individuals also searched for their ancient roots which they thought would admit them into the ranks of the gentry. Hence the development of geneological research as a part of an attempt to substantiate one's position through documentation: "[the new men] were hungry for information about their patch, its antiquity and importance, and by extension their own place and importance."[4]

The ideological strength of chorography lay in its connection with the meaning of landed property. Richard Helgerson has argued that early Stuart chorography was *the* crucial force in the formation of both the English sense of geographical place and political identity. Looking at the chorographic works of Camden, William Carew, and John Stow, Helgerson argues that in their emphasis on locality, they have transferred national identity from loyalty to the monarch to loyalty to the land, upholding in this way the values of local prerogative, accustomed privilege, and resistance to royal encroachment. In enterprises like these, the authorial self gave "the dumb and inanimate land voice and life, in exchange for which the land grants the self an impersonal and historically transcendent authority."[5]

The land thus becomes a means of representation, and, according to Helgerson, the essence of all representation is differentiation. Chorography's appeal in this sense lay in its ability to *distinguish* – rather than seek similarities – between places and create, in the words of E. W. Gilbert, "the *personalities* of regions."[6] On this reading, for instance, it is Cornwall's differences that make Cornwall worth describing and Richard Carew's chorography of this county worth reading. Because these differences make the county meaningful on a national map, so do they make meaningful and identifiable the lives of Cornishmen: "[t]heir individual identity and authority depend on their participation in a system of local differences that chorographers, Carew among them, make it their self-justifying business to describe."[7]

In seventeenth-century chorographies, these forms of individuation became enmeshed with the geographical differentiation of nature. Curiosity about local natural history and topography became an aspect of a proprietorial culture of representation. It was argued that weather, soil, waters, and the landscape determined one's place and one's character as much as the civic transactions. Investigating nature indigenous to a district, county, or a parish, could express a commitment to the locality. This was prominent among a number of Stuart chorographers and local historians, who after their studies in Oxford or London, returned to their native county to assume rectories, practice law, or serve as justices of the peace.[8] Vicar White Kennet wrote in 1695 that, next to his official duties, he knew "not how in any course of studies I could better serve my patron, my people and my successors than by preserving the memoirs of this parish."[9] In his study of Middlesex, John Norder glorified the fertility and fruitfulness of the whole of Britain, William Burton in the description of Leicestershire "adventured to restore [Leicestershire] to her own worth and dignity," and Camden confessed that it was "the glory of my country that encouraged me to undertake" the work on *Britannia*.[10]

This chorographic differentiation was reflected in the treatment of local climates and "airs." As the weather can be understood as an interaction of regional and global movements of air, so can a nation be defined as a negotiation between an overlying national character and its regional idiosyncrasies. Stereotypically, the discussion about the "air" or "temperature" of the region – containing a list of the weather's peculiarities – praised its salubrity and clemency in comparison to other regions. John Norden found Hertfordshire air "most salutarie," while John Speed considered the air in Cheshire of a quality far exceeding that of its neighbors. Robert Reyce, William Burton, and Tristram Risdon made similar claims regarding the air of Suffolk, Leicestershire and Devon.[11] Lambarde wrote that "the Aire in Kent, by reason that the Countrye is of sundry partes bordered with water, is somewhat thicke: for which cause . . . it is temperate, not so colde by a great deal as Northumberlande, and yet in manner as warme as Cornwall."[12] Speaking of Cornwall, Carew observed that the "ayre thereof is cleansed, as with bellowes, by the billowes, and flowing and ebbing of the Sea, and therethrough becometh pure, and subtill, and, by consequence, healthfull." He thought this could be supported by the fact that upon the return of the English fleet from Portugal, the state of diseased soldiers worsened as they progressed further inland. In Plymouth, where they landed, the diseases they brought, though infectious, were not as contagious, "yet not the verie pesstilence, as afterwards it proved in other parts." Salubrity notwithstanding, Cornwall was in Carew's assessment particularly exposed to storms, "which fetching a large course from in the open Sea, doe from thence violently assault the dwellers at land, and leave them uncovered houses, pared hedges, and dwarfe grown trees."[13]

From the mid-seventeenth century, however, a relatively marginal interest in local nature changed into one similar to Baconian natural history. Particularly after the work of Gerard Boate (1652),[14] the concept of county chorography implied a more practical and a more critical study increasingly focusing on nature rather than on human history. And while traditional chorographies continued to be published, the Restoration marked the emergence of the regional *natural* histories, coinciding with the introduction of inductive methods into chorography during the 1650s, and the foundation of the Royal Society of London.[15] Regional natural histories reflected the fact that both chorography and the Baconian program of natural history investigated non-textual materials. Where chorography looked at local inscriptions, ruins, topographic and hydrographic features, natural history of the inventorial type depicted the natural world as a repository of animals, plants, stones, and meteors. Both kinds of studies were conceived as comprehensive registers of animate and inanimate productions of nature – even as some specialized treatises also appeared in print[16] – and shared a tendency to become registers of miscellaneous particulars, as was exemplified in Childrey's work. This comprehensiveness is especially well-represented in the work of Robert Plot whose chorographic investigations exerted considerable influence on the practice of regional natural history, and, more importantly for us, on the development of eighteenth-century meteorology.

Curious plotting

The popularity of Baconian natural history in the late seventeenth century and the methodological interest in the marvelous created an unprecedented interest in the research of Robert Plot (1640–96), who made an impact – theoretical, literary, and ideological – on the chorographically-shaped meteoric tradition.[17] Plot matriculated at Magdalen Hall in 1658, received his MA in 1664, and eventually became Vice-Principal and Tutor of the college. In 1676 he left Magdalen for the commoner's position at University College, where he resided until 1690. In 1670 he issued a questionnaire entitled "Enquires to be Propounded . . . in My Travels Through England," which included headings on heavens and air, waters, earths, stones, metals, plants and husbandry. Plot was prepared personally to "perambulate" the whole of England after the fashion of Camden and John Leland; a rather grandiose scheme of an enthusiastic man in his twenties.[18]

Such projects had already been institutionalized in the thirty-two-member Georgical Committee, appointed by the Royal Society in 1664, which drew up the "Heads of Enquiries" in the format of a questionnaire to be dispatched to farmers throughout Britain and Ireland for the purpose of improvement of agriculture. Before the Georgical Committee, the physician Arnold Boate had used this form of information gathering in the *Interrogatory* or "Alphabet" of queries in relation to the natural history of Ireland. This was appended to the second edition of Hartlib's *Legacy* (1652). William Petty in Ireland and John Aubrey in Wiltshire were made responsible for digesting the reports, but the returns were meager, just as Childrey had experienced.[19] In 1666, Robert Boyle proposed the new "General Heads for Natural History of a Countrey, Great or Small," part of which referred to observation of the weather: the temperature of the air (a general disposition of air) and its qualities – weight, clarity, refractive power, subtlety, and density. The prospective observers were asked to take notice of duration, place and the order of production of meteorological phenomena.

Several Restoration writers followed Boyle's idea: Robert Hooke wrote a "Method for Making a History of the Weather," while Martin Lister, Sir John Hoskyns and Edward Lhwyd circulated their own surveys. Lhwyd's was entitled "Parochial Quiries" and intended for a projected survey and natural history of Wales.[20]

In 1683, Plot was one of the founders of the Oxford Philosophical Society. He took an interest in instrumental weather recording, and suggested that an account of weather should be kept at all times, "either according to Mr. Lister's, or some such compendious method and that at the end of this, and other years, the account of the weather of preceding years, one or more, should be printed with an almanac for the year to come."[21] Weather records accordingly arrived in London during the following spring. One was from Whitely Hall, near Wakefield, Yorkshire, another from Keighley in Craven, dealing in particular with the extraordinary February thaw. Samuel Molyneux's observations, based on Lister's scheme, were dispatched from Dublin in the summer of 1684. And in January of 1685, Plot drew up his own table for keeping a "Diary or Weather-Journall, by observing the station of the mercury in the baroscope, the place of spirits in the thermometer, the points of the Wind, and the state of Weather."[22]

Again, the response to the questionnaire was scant. Despite occasional references to instrumental observations and standardization, the information gathered with the aid of such questionnaires was defined by a tacit, yet universal, interest in what was rare, peculiar, unique, and perhaps unrepeatable. This is amply illustrated in Plot's *The Natural History of Oxfordshire* (1677), a work so erudite it was said to have inspired Elias Ashmole to donate his collection to Oxford University (FIGURE 8). Indeed, in 1683 Plot became the first Keeper of the Ashmolean Museum and, as a consequence of the immediate popularity of the work, Secretary of the Royal Society, and the editor of the *Philosophical Transactions*.[23]

In 1684, Plot took up the professorship of chemistry at Oxford while planning a survey of Kent. When John Aubrey, then somewhat unsuccessfully carrying on his research on Wiltshire, asked Plot to use his material on Surrey, Plot declined saying: "the next county I go upon in this kind, will be Staffordshire, and if any other, it shall be my native County of Kent, *which is a great County*."[24] Like those planned for a natural history of London and Middlesex, the account was never produced.

Plot's *Natural History of Oxfordshire* can be seen as a well-researched and a better illustrated version of Childrey's compendium. It was begun in Plot's spare hours, gradually to encompass no less than a description of "animals, plants and the universal furniture of the world;" Nature's "extravagances and defects, occasioned either by the exuberancy of matter, or obstinacy of impediments, as in Monsters;" and natural phenomena restrained by "Artificial Operations." In this unmistakably Baconian program, Plot claimed to have based his work only on authentic information from past writers and on what his informants could inspect personally.[25]

One of the main characteristics of Plot's enterprise – and of those of his followers – is that the criteria in the selection of first-hand information favored those events that were observable unaided, and on the spot. For this reason, the first chapter, "Of the Heavens and Air" ignores astronomical information on the celestial bodies, on the basis that they were so distant as to need the "help of Art" for their observation. Rather, the chapter opens with the description of *parhelia* (the optical phenomenon also known as "mock-suns") in Ensham on

FIGURE 8   The title page to Robert Plot's *Natural History of Oxfordshire* (1677).

29 May 1673, "local" celestial phenomena observable with the unaided eye. Plot's insistence on first-hand inspection led him to visit the banks of the River Cherwell which, in the wake of a storm, had flooded over the bridge at Magdalene College, a scene which "I had then the curiosity to go see myself, which otherwise, perhaps, I should have as hardly credited as some other persons now may do."[26]

Importantly, however, such emphasis did not exclude textual evidence. Hence a detail from a diary in an account of the rare appearance of lunar rainbow – "even now while I am writing this, at Oxon: on the 23d of November 1675, about 7 at night, [I] behold the Moon set her Bow in the clouds" – is immediately followed by the description of the phenomenon as reported in Pliny, Aristotle, Seneca, Cardan, and others.[27] Combining in this way the information from textual sources with facts observed first-hand, Plot revived the procedure which Ann Blair found in early modern works of natural philosophy and termed the "method of commonplaces."[28] Originally intended for notes for reading, the commonplace book accommodated non-textual materials and became a means for storing natural knowledge. The commonplace book thus may record "the origin of a fact (whether bookish or reported by a witness or an artisan), but [it] treats each entry independently of its source, as potentially useful knowledge equivalent to every other entry."[29]

The heterogenous quality of information gathered according to the commonplace method is especially evident in Plot's second major work, the *Natural History of Staffordshire* (1686). The text contains a collection of information gleaned from the canonical authors (Aristotle, Seneca, Agricola, and Gassendi) with that obtained from people who Plot credited with the same degree of credibility – a Thomas Rudyerd, Esq, Thomas Gent, Robert Conny, and numerous other provincial "relatours." Plot looked at information from all available sources dealing with the county and the result was a thoroughgoing compilation of local legends, stories, and authenticated observations about the natural curiosities of Staffordshire. Plot was especially inquisitive about bizzare accidents which he described with the zeal of an antiquarian (FIGURE 9):

> [An] unusual accident (perhaps not to be met again in many Ages) was shewed me at *Statford*, by the formention'd worthy Gentleman the Worshipful *Francis Wolferstan* Esquire, who having built a new *Gate* before his house *Anno* 1675, and placed fair *Globes* of the finest and firmest stone over the *Peers* of it (whereon He depicted with his own hand two *Globe Dials* in oyl colours, and on the *terrestrial* the several *Empires* and *Kingdoms* of the *World*, that he might see how day and night succeeded in each of them) in *January* 1677 had them both struck with *lightening* in the same *point* (where the *great Meridian* of the *World* and the *North* part of the *Polar circle* meet). Yet the *Globe* on the left hand the *gate* going forth, seemed to have been stricken first, and the *other* not till the day following, which is so, the *accident* was so much the more strange, as they should be stricken in the same *point* at different *times*.[30]

This sort of antiquarian meteorology became a staple of early modern weather reports. One should however not be surprised to find such curiosities in a

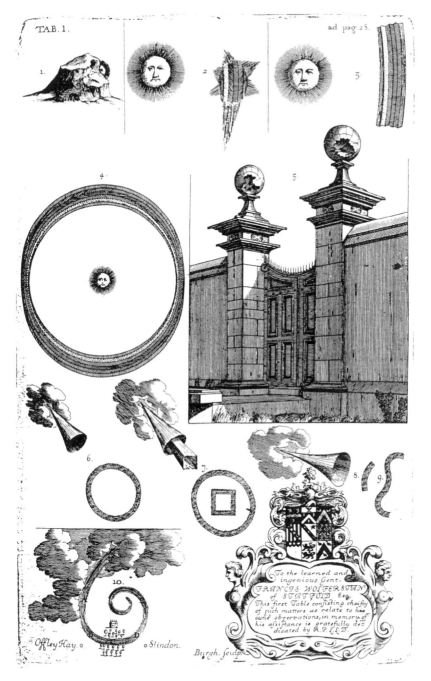

FIGURE 9   A dedicated plate showing the gate globes damaged by lightning. Two insets are illustrations to Plot's discussion on the nature of "hurricanes" (whirlwinds) and thunder. Robert Plot, *Natural History of Staffordshire* (1686).

chorographic study. By definition these gave as much, if not more, value to the place of natural events and its, as it were, proprietal implications than to the event itself and its philosophical analysis. For instance, such emphasis informed the eighteenth-century rural manuscript maps in which rustic draughtsmen reinforced proprietal claims with a kind of attention to detail that entailed the representation of, for instance, landmark trees cleft by lightning, stones with facial features, and other landmarks with individual traits.[31]

With a place-oriented curiosity as its guiding principle, Plot's *History* features narratives about "more uncommon Meteors" which fully satisfy the chorographic, epistemological and sensational premises of the genre. Thus John Nash, the Vicar of Ranton Abbey, is reported as seeing at Michaelmas 1676 a great fire of a globular figure, "moving by jerks and making short rests, at every one of them letting fall drops of fire." A music of invisible winged creatures was heard by a Francis Aldridge of Hammerwich, agonizing shrieks were common at Frodley, and the barking of "Gabriel's Hounds" was reported from the coal pits around Wednesbury. The frog-rain at Bowling Green was so heavy "it has been found difficult not to *tread* on [frogs] in *walking*." In November 1672, Mr Miller, Vicar of Wednesbury Church, noticed a bright ball against the door of the church, where it continued for about seven minutes, and after which a storm of hail and rain followed immediately. Plot connects this to the story from Seneca in which a fiery star-shaped meteor stuck to the spear of a Roman soldier marching to Syracuse; this is in turn followed by a digression on a similar "sticking fire" accident from Fromondus.[32]

This new idiom of the titillating "wonder tradition" was certainly justifiable by invoking the *raison d'être* of the regional scholarship – a complete register of a county's curiosities – but there can be no doubt that it offered some satisfaction to those in search of the latest "strange news". In a more relevant sense, however, Plot's work was eagerly anticipated by his aristocratic and gentlemanly subscribers. These had seen their names go public in the subscription lists and on the dedication plaques on the engraved plates of such publications. In local history, this mechanism was used in James Wright's *History of Rutland* (1684) and onward, when it became usual to include figures of the houses of local nobility and gentry. From the author's point of view, this had an economic rationale as most of those whose houses were represented felt obliged to accept the dedication of the plate offered by the author or publisher, and to subscribe for at least one copy of the book. With Plot this practice was continued into the parts dealing with natural history so that the local families switched from a rather literal self-promotion to a more elegant way of patronizing nature (FIGURE 10).

The work of Plot's colleague John Aubrey exemplifies this inclination in the same, spatially defined framework of chorographic curiosity. A son of an estate owner from Wiltshire, Aubrey was educated at pre-war Oxford and briefly studied law at the Middle Temple until financially ruined by several lawsuits concerning the estate after his father's death. In the 1650s he entered the Oxford circle, where he met Samuel Hartlib and Christopher Wren. From 1656 Aubrey

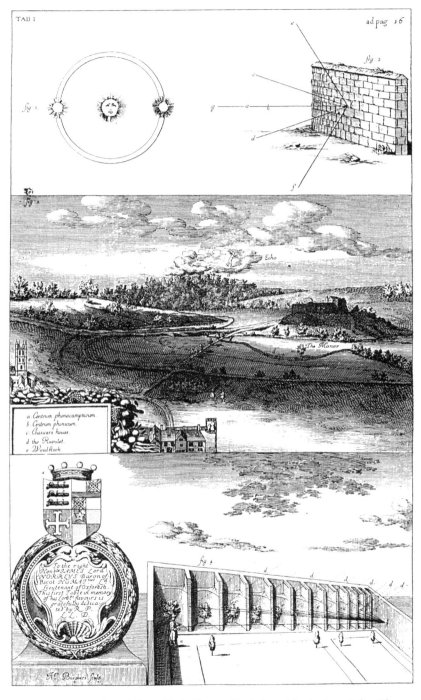

FIGURE 10   A dedicated plate with the figures of parhelia and local echoes. Robert Plot,
Natural History of Oxfordshire (1677).

was engaged with the surveys of Wiltshire and Surrey, the product of which was the "Essay towards the Description of the North Division of Wiltshire," an antiquarian study followed by the "Perambulation of Surrey" (1673). On Plot's suggestion, Aubrey compiled "The Natural History of Wiltshire" in 1685, but its contents displeased John Ray, to whom Aubrey presented an early manuscript. Ray wrote:

> I think that you are a little too inclinable to credit strange relations. I have found men that are not skillful in ye history of nature, very credulous, and apt to impose upon themselves and others, and therefore dare not give a firm assent to anything they report upon their own authority; but are ever suspicious that they either may be deceived themselves, or delight to teratologize (pardon ye word) and to make a shew of knowing strange things.[33]

This important reference to the inferior status of 'teratologizing' – best translated as a search for wonders – makes it quite probable that many seventeenth-century authors who indulged in such explorations, perhaps even Childrey himself, could choose to resort to the methodological principles of Bacon's history of pretergenerations without fully realizing – or perhaps conveniently avoiding to realize – that such history was meant as a complement to, not a replacement of "ordinary" histories. Whether describing the inexplicable workings of local *nature* or the monuments of *local* nature, natural history and antiquarian research of this kind were bound more by the tastes of their readership than by the abstract norms of methodical inquiry promoted by the circle of metropolitan *cognoscenti*. In Aubrey's "Wiltshire," for instance, one would have difficulty in finding much that was not out-of-the-ordinary. Meteorological topics ranged from strange whirlwinds and mists, to sounds and meteors, and to extraordinary snowfalls, hurricanes, and storms, including the then famous "Oliver's Wind," said to have been providentially caused by the death of Cromwell. There are stories on the lightning damage of church-steeples, a report about a ball of fire, "big as a bushell," seen by the author while riding in the north lane of Broad Chalke in Wiltshire.[34]

As we have noted in chapter 2, sublunary accidents of a similar nature comprised a large part of the early *Philosophical Transactions*. But such reports continued to figure in the eighteenth-century chorographic works, and they are particularly worth noting in Charles Leigh's *Natural History of Lancashire, Cheshire, and the Peak in Derbyshire* (1700) and John Morton, *The Natural History of Northamptonshire* (1712). Leigh was an Oxford graduate, and since 1685 a Fellow of the Royal Society. During the 1680s, he had heard Plot's lectures in chemistry, and became acquainted with Plot's histories of which, however, he did not think much from a literary point of view. Ironically, Leigh's own work was declared too vague and too selective by some, and of little value by others.[35] Admittedly, *Lancashire* was intended as a compendium of rarities and a collection of mere "Matter of Fact," because Leigh had seen "the Misfortunes of many, in swelling their Books with digressive Quotations, and Chimerical

Hypotheses." He hoped the work would add to the honor and the natural history of the counties which afforded "so great a variety of Mines, Minerals, Metals with other choice Products, and the most *surprising* phenomena of nature."[36]

The significance of Leigh's work lay in the fact that it validated premises of the tradition that had matured in Plot's county studies. In those, as we have seen, natural and human "matters of fact" deserved only so much attention as they were judged rare and hence "worthy" of knowing because they underscored regional variety and uniqueness. Echoing the patriotic sentiment of Stuart chorography, Leigh wrote of the air of Lancashire as "for the most part mild, serene and healthful" going on to note the occurrences of hollow murmuring noises heard by the coastal inhabitants, during which the surface of the sea was elevated by "eruptions [of exhalations] from the Bowels of the Earth," that struggled to pass through the body of the ocean and into the atmosphere.[37] But as such murmurs occurred every year, Leigh noted, they were less likely to arouse the curiosity of the reader, at least not as much as the description of a "terrible Tempest" with "amazing Claps of Thunder" and lightning which killed several people in a local village and where Leigh went to make observations on the state of the health of survivors. Leigh adds a discourse on remarkable echoes and other exceptional qualities of the local air and begins a new chapter on waters "remarkable for their Levity, Subterraneous Eruptions and Remarkable Rivers and Ponds."[38]

John Morton's is a similar collection for the county of Northamptonshire where he had been "a Searcher after nature already many years, and the Gentlemen of the Country are pleased to Naturalize me."[39] The selection of the material is a result of Morton's refusal to deal with a Mr Gibbon's observations of halos as having "nothing extraordinary or uncommon" about them. This is the reason why space is given to a hurricane in 1702, a whirlwind of 1694, the fiery meteor of 1697, and the Hon. William Craven's report on the globe of fire in 1701. A long discussion on the nature of the "star-shot" (a gelatinous body the common people believe to come from the stars), closes with the first-hand demystification: "In the month of Sept 1708, I saw a [bird] shot down to the ground that on her fall upon the ground when almost half dead, disgorged a heap of half-digested earth-worms, much resembling the gelly [sic] called star shot."[40]

As these examples were intended to show, Leigh and Morton employed the form of reportage that Plot had rendered almost equivalent to the "wonder genre". For these writers, natural rarities deserve attention because they define the "physiognomy" of a region more than ordinary phenomena. Properties *common* to regions or those which occur on a regular basis are not chorographically interesting; in fact, they are contradictory to chorography as an enterprise of cultural identification. This local-patriotic empiricism organized early eighteenth-century chorographies into compendia of (preter)natural, antiquarian, and historical rarities which sought to emphasize and even define county identities. On a different level, chorographic compendia catered for the cultivated audiences whose patriotic sentiments advised them against the Grand Tour. Thus eighteenth-century guidebooks of the English countryside portrayed domestic nature as a theater of sublimity and a microcosm of the wider world. In chorographic writings, moving from a historical knowledge of "sports of nature" to a guidebook mentality of touristic curiosities required only a slight change of emphasis.[41]

Collectable weather

We may at this point extend the relevance of the chorographic method to other early modern genres devoted to the investigation of nature. The affinity between chorographic meteorology and the history of unusual weather described in the *Philosophical Transactions* can be seen as based on the common focus of both enterprises to identify curiosities on a regional/local basis. Given the miscellaneous content of the early *Philosophical Transactions*, its natural history and antiquarian subjects can be seen as parts – distinguishable only in theory – of a grand chorographic project "atomized" into pieces of information supplied by provincial information gatherers. This view can be justified by the historical analyses demonstrating that the antiquarian and topographical recreations of the provincial correspondents of the Royal Society comprised a large number of articles published in the early *Philosophical Transactions*. The next section looks at meteorological contributions of these provincial naturalists as part of the chorographic enterprise and focuses on epistemological principles involved in such reportage.[42]

The previous section showed the chorographic dimension of unusual meteorological phenomena. This section explores some epistemological consequences of such practices. Chorographic methods applied to identifying cultural and historical individuality were also applicable to the study of regional nature. Early eighteenth-century chorographers and naturalists approached meteoric occurrences using methods that they normally applied to dealing with the human past. They sometimes defined the history of meteors as a collection of specimens such as thunderbolts. At other times they equated weather reportage with the reconstruction of meteors from their physical effects by describing victims of and the damage caused by severe weather. They occasionally wrote about meteors as if these were public events, i.e., using the literary idiom which emphasized the spectacular aspects of the atmosphere. In practices like these, however, meteorologists always sought to maintain high standards of trustworthiness. They credited only first-hand observation and, in the instances in which such observation was unavailable, they resorted to alternative means of authentication for example, collecting either physical remains of meteoric activity or information supplied by the people at large.

The similarity of historian's and naturalists' methods in the seventeenth century has been discussed by Barbara Shapiro and Joseph Levine. They suggested that the resemblance could be explained by a set of epistemic concerns shared by natural historians and historians proper. Both "historical" and "scientific" observations were considered "matters of fact," created by specified "literary technologies," yielding probable rather than certain knowledge. Beyond this common theoretical dimension, the two discourses converged through an overlapping membership: "It would be difficult to decide," wrote Shapiro, "whether Aubrey, Plot, Ray, Charleton, and Thomas Gay belonged among the historians or the naturalists."[43] She points out that the terms usually used to name seventeenth-century scholars obscure the fact that so-called humanistic, scientific, historical, antiquarian, or philological investigations were "neither specialized nor professionalized and that intellectuals moved with relative ease from one kind of intellectual enterprise to another."[44] In Plot's and Aubrey's chorographic studies, eclecticism of this kind tended to homogenize facts of disparate origins and present them according to the criteria of visibility, strangeness, or rarity, rather than their ontological provenance, that is to say, as human, artificial, or natural. Thus Aubrey, recognizing difficulties associated with an omnivorous "Commonplace book" mentality, apologized for laying down the information about Wiltshire "as if tumbled out of a sack; . . . Antiquities and Natural Things [mixed] together."[45]

The methodological similarity between humanistic and natural studies can be especially observed in seventeenth-century chorography. Through its hybridization of civil and natural history, chorography served as a conduit through which "the legally derived concepts of witnessing and evaluation were transmitted" from human to natural domains.[46] Antiquarianism – an important part of chorographic projects – remained prominent in the science of the late seventeenth century as "cognate with natural history" at least with regard to the type of activity involved (field work, collection, display, classification)" and the stress on "firsthand knowledge and reports from reliable witnesses and records."[47]

In ascertaining authenticity of evidence, personal inspection, scrutiny, and observation played a critical role. Personal examination had already been recognized as a part of both scholarly and vernacular repertoires of persuasion and an appeal to seeing things "with one's own eyes" was used as a traditional, even if rhetorical, "antidote to falsehood and credulity" surrounding popular knowledge of events, places, and things in early modern culture.[48] Stories about the effects of violent weather enticed local clergymen to visit the "disaster areas," verify what they had heard, and interpret the meaning of events. But with the beginning of topographic surveying, personal observation of places and monuments emerged as the central vehicle in chorography. There was a recurring awareness of the power invested in a personal tour of places and a direct assessment of local findings. In the preface to his *Northamptonshire*, Morton wrote:

> the Exact Description of Things, however small and seemingly contemptible, and faithful Accounts of what is observable in them, will always be of Use to those who study Nature, to what End soever that be; . . . I have spared no Pains, or Cost, to make [this book] as compleat as I could; I have visited almost every Quarry, Wood, Spring: and, to be short, every Thing, that, I could think, merited Remark and Observation.[49]

The first-hand and unaided access to information has been addressed within the medical and physiological contexts. Andrew Wear has shown how Andreas

Laurentius, the early-seventeenth-century anatomist, explained in his *Historia anatomica* that anatomy could be established in two ways: by inspection (*autopsia*) and instruction (*doctrina*). By *autopsia* the anatomist observed the internal motion of the parts of animals, while with the *doctrina* he might try to explain final causes and purposes of these parts. *Autopsia*, being historical and descriptive, was more certain than *doctrina*, which was speculative and prescriptive. Following Laurentius, Harvey adopted this distinction and in his *De Motu Cordis* (1627) increased emphasis on the superiority of *autopsia* over theory. The ocular observation in particular became for Harvey "a form of knowledge in its own right . . . by contrast to the Aristotelian position that knowledge of causes was only knowledge."[50]

In a more general sense, the requirement to use the personal inspection method meant that the sensory examination of material evidence – casual objects, epitaphs, coins, landscape features, storm damage, experimental findings – warranted the accuracy, reliability, and value of such descriptions. Not only were these advantages of *autopsia* constructed in opposition to the scholastic *doctrina*, they were increasingly recognized by some as the only path to knowledge in natural history.[51]

In which ways did the methods of autopsy determine the accuracy of meteorological reporting? We may approach it through an analysis of the reports in which naturalists aimed to reconstruct meteorological phenomena through a minute inspection of its effects on material objects. This form of reporting was modelled on the antiquarian procedure of using physical evidence to resurrect the past and the application of this method to uncommon meteors was sanctioned by the early modern belief (addressed in chapter 1) that meteors had a radically different existence than did other objects described within natural history. Unlike minerals, plants, fossils, and the air-pump observations, meteors were brief and unforeseeable occurrences. Their development was literally "meteoric."

Plot, for instance, was aware of the predicament that such characteristics caused for his method of chorographic work. He divided the objects of natural history into two kinds. Objects such as plants, shells, fossils, medals, or shields belonged to the category of *portable* curiosities. Monastery ruins, ancient monuments, and topography were curiosities that were *inseparable* from their locale, but observable either through Plot's description or by a direct inspection at the site.[52] But on this scheme meteors did not fall into either of the categories: they were neither portable, nor inseparable from a locality. They appeared as non-collectibles, yet collected they had to be.

There were some collectible meteors, however. By the early eighteenth century, Plot, Morton, and Edward Lhywd had information about the "Elf's arrows," i.e., flint-stones presumed to be shot at cattle by Highland elves. Hans Sloane acquired some for his collection, describing one as "an ancient gray stone hatchet . . . called by some thunder stones," and another as "an Irish hatchet made of green spleen stone found after a shoure [sic] & thunder by a ditcher who thought it hot."[53] Others referred to these as thunderbolts, i.e., the stones believed to be produced by the coagulation of sulfurous vapors during violent thunderstorms. Thus an eighteenth-century naturalist wrote about a farmer, who after a storm "found a beautiful yellow Ball lying on the Turf, which he gladly took up, in hopes it would well reward him for stooping." Because the ball smelled badly, the man took it to the nearest scholar who admired its "[e]fflorescence of fine, shining, yellowish Crystals," and conjectured that it might have been intended for one of those explosions of atmospheric "bituminous Matter . . . but by some Accident miss'd firing."[54] In 1733, a man in Dorset saw a flying object resembling a kite; his friend informed him that it fell on "the Ground somewhere about the Kennel-garden, whither I accompanied him in Expectation of finding some of those *Jellies* which are supposed to owe their Beings to such Meteors."[55]

However, not all meteors could be collected in this manner. If they could not be "collected," they could be identified by their effects. The Yorkshire natural historian, Ralph Thoresby, was among those who gave prominence to this peculiarly "antiquarian" treatment of meteorological accidents. In his letters to the *Philosophical Transactions*, Thoresby revealed how the meteoric tradition became a study of people and objects affected by severe weather (FIGURE 11).[56] Thoresby's first meteorological communication to the *Philosophical Transactions* was "An Account of a young Man slain with Thunder and Lightning, Dec. 22.

FIGURE 11   An early representation of the path and damage of a waterspout. Zachary Mayne, "Concerning a Spout of Water that Happened at Topsham," *Philosophical Transactions* 17–18 (1693–4): 28–31.

1698."[57] It is a page-long description of the death by lightning of a Jeremiah Skelton, a laborer from Halifax, Yorkshire. The account gives information about the time and place of the accident and about the damage which lightning made on the barn in which Skelton sought shelter. Upon coming to the barn, writes Thoresby, Skelton's relatives found a young man laying face-down, naked, "save a small part of his Shirt about his neck, and a very little of Stocking upon one foot, and so much of a Coat-sleeve as covered the Wrist of one Arm, his Clogs driven from his Feet, one not to be found, and the other Cloven."[58] His hair and beard were singed and a little hole was found below his left eye. The report contains neither theoretical speculation nor Thoresby's opinion; it is closer to a post-mortem medical examination of a lightning victim.

In another piece sent to the Royal Society, Thoresby concentrated on the physical impact of lightning that hit the cottage of a Henry Parker, went down the chimney and through the floor to the lower room, melted two pewter dishes and a candlestick, and filled the room with a bituminous smell. Thoresby took the candlestick as a "memorial of so uncommon an accident," and proceeded to investigate the case of a Thomas Lambert, the victim of a thunderbolt of September 1672.[59] On another occasion, Thoresby wrote of a gardener caught in a storm during which lightning set fire to his wooden stick. Having arrived at Leeds, the gardener gave the stick to the Mayor of the town who in turn presented it to Thoresby's museum of curiosities. Thoresby wrote to Sloane that: "[i]t yet retains part of the blackness, tho' the man had beat off much of the end of the Rod (*little minding it as Curiosity*) by forcing the Horse forward, to get the sooner out of the fiery Incalescence."[60]

As the episode illustrates, in changing owners' hands, the stick moved from the realm of utility into the realm of curiosity, or, in words of Krzystof Pomian, it became a *semiophore*, an object of absolutely no use except its ability to communicate with the invisible (i.e. not directly observable).[61] Things representing the invisible might vary – from scrolls and medals to dishes and candlesticks – but they invariably brought the invisible within the horizon of semiophores' possessors. Through Thoresby's "memorials of uncommon accidents," such accidents were removed from the past into the present, and from a fleeting existence into the domain of tangible articles. Through the latter, meteors became Plot's "portable" curiosities that could be stored in antiquarian cabinets.

Reports similar to Thoresby's were frequent in the eighteenth-century *Philosophical Transactions*. In 1725, the Northamptonshire rector Joseph Wasse, a classicist and a natural historian, wrote a letter to the court physician, Richard Mead, with a description of a death by lightning.[62] Wasse gives information on the victim's position on the ground, his wounds and his clothes and belongings: the man's belt was thrown more than forty yards away, "and the knife in the right Side Pocket of his Breeches was broken in Pieces, not melted."[63] On a later occasion, Wasse asked his nephew – "a Student of Merton a pretty good philosopher" – to examine the holes made by a ball of fire in the ground near the local church, but the owner of the ground insisted his conscience could not allow the inspection. Prevailing upon him with "Money, Ale, and other rural arguments," the Merton philosopher eventually explored the holes and found in them several hard-glazed stones which he unsuccessfully tried to pierce.[64]

In insisting on the physical impact of unusual meteorological situations, eighteenth-century naturalists were consistently blurring the distinction between the relevant and the perfunctory, and between their tastes and methodological principles of 'high' scientific culture. The interest in, as it were, the meteorology of domestic utensils shows that the naturalists' attitudes derived from a domain of immediate human concern with the weather and property, not from a theory of exhalation and *antiperistasis*. Outside this existentially driven curiosity, it would be difficult to make sense of the essays which reported on the shapes of deformed domestic utensils found in lightning-struck houses or packages sent to the Royal Society containing half-melted knives and forks.[65] Inspired by similar interests were the reports of Sir John Clark on an unusual effect of lightning on an oak in Whinfield Park, Cumberland,[66] and Stephen Fuller's communication about the state of churches, houses and barns in Cambridgeshire after a "hurricane."[67] Some of these communications included illustrations of the houses which suffered destruction (FIGURE 12).

Most of these were accounts of phenomena which had not been directly witnessed. The author of such correspondence would be typically informed of an event in the neighboring area where he would then go to discover further details by investigating whatever evidence was locally available.

The larger group of meteorological articles in the *Philosophical Transactions* deals with the events witnessed by naturalists. Such correspondence combined information about the progress of meteors – for example, the trajectory of a fireball, movement of a storm, position of a halo, tremors of an earthquake – and the information about their effects on people, property and landscape. Typical in this regard was the piece by Sir William Hamilton, British envoy to the Court of Naples and Fellow of the Society of Antiquaries. On March 20, 1773, Hamilton wrote to the Secretary of the Royal Society, Matthew Maty, on the effects of a thunderstorm on the house of Lord Tylney: "On Monday last, about half past ten at night, I had the satisfaction of being one, of many witnesses, to several curious phenomena, occasioned by the lightning having fallen on Lord Tylney's house, in this city."[68]

The event occurred during an assembly night at the Tylney's house; ministers, Englishmen of distinction, and Italian nobility (250 invited and 250 servants) were having after-dinner conversation when lightning hit the mansion, causing utter confusion. The doors were instantly jammed with dignitaries, several serv-ants were burnt and "a Polish prince, who was playing cards, hearing the report (as he thought of a pistol) and feeling himself struck, jumped up, and, clapping his hands to his sword, put himself in a posture of defense." Hamilton himself had been conversing with Monsieur de Saussure, arguably the most influential European meteorologist of the day, and after the initial turmoil was over, they walked about the mansion to note the damage and discuss the intensity of the "report:" Hamilton thought the explosion was like that of a pistol, Saussure thought of an Indian fire-cracker. They observed the presence of a sulfurous smell while examining blackened gilding of the cornice and furniture. The next day they returned to inspect the rest of the house. Hamilton's letter includes pictures of three rooms showing the path of the lightning (FIGURE 13).

In articles similar to Thoresby's, Wasse's and Hamilton's, the first-hand inspection of things of an enquiring *virtuoso* became a practice in itself. As could be concluded from the early Augustan satires of science, accumulating minute particulars about the natural world could be easily perceived as an aimless pursuit of its own. Unlike investigation conducted under the explicit aegis of inductivism, the methods of autopsy did not require a complementary relationship with other ways of knowing (deduction, generalization, hypotheses) with which it could constitute a full philosophical knowledge. Because autopsic practices were in this sense less an introduction to a philosophy of nature than an independent mode of inquiry, they provided a justification for the meteoric tradition itself. Even if the description of isolated meteoric activity could be perceived as methodologically insufficient – if for no other reason than because it implied no causal correlation between locally observed prodigious meteors

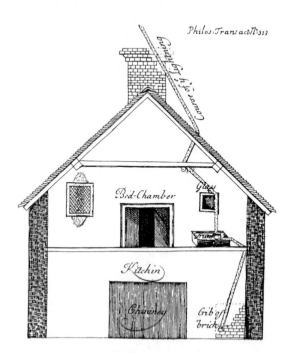

FIGURE 12   Reports reconstructing lightning paths were perhaps the most typical among eighteenth-century meteorological reports. Samuel Molyneux's "Account of the Strange Effects of Thunder and Lightning," *Philosophical Transactions* 26 (1708–9): 36.

– it still could be perceived as a self-justifying knowledge of true particulars of nature. Indeed, autopsy can be retrospectively identified as the central methodological feature of the eighteenth-century meteoric tradition.[69]

The common voice

Unusual weather was always a shock. Taking notice of it was as much a meteorological as a psychological exercise. Meteorologists, being literally inside the object of their research, must make an effort to render their 'philosophical' accounts objective. Thus eighteenth-century meteorological writers presented their findings not only as narratives of unusual atmospheric facts, but also as their personal *experience* of such facts. This practice resembles that used by the early English experimentalists, in whose reports "[a]n 'experience' was now an event of which the observer was a part."[70] The event-narratives were instrumental in making an eminently *modern* form of philosophical authority by consciously opposing those narratives to the "statics" of textual scholasticism. In the new science, the natural world was to be grounded in discrete events, to be described in detail and in an active voice: "the literary form itself represented a credential formerly supplied by references to an appropriate ancient, or by a different literary form."[71]

FIGURE 13   A highly detailed engraving of the interior with lightning marks in Lord Tylney's mansion in Naples. William Hamilton, "An Account of the Effect of a Thunder-Storm," *Philosophical Transactions* 63 (1773): 327.

The circumstantial nature of such reports introduced the so-called "virtual witnessing": "if one wrote experimental reports in the correct way [i.e. as a description of discrete events] the reader could take on trust that these things happened."[72] This technology was central to the definition of new natural science because it excluded both "'secretists' and the 'vulgar' from the community of experimental philosophy."[73] In other words, the new matters of fact were constructed as circumstantial stories of laboratory events and presented as public items of indubitable truth. This process guaranteed an exclusion of the "common," "vulgar" or "superstitious" parlance of popular culture.

Early meteorological reports were dense in circumstantial detail. They were not laboratory narratives, however. Because meteorologists wrote about events which had already occurred in public – as opposed to those *constructed for* the public in the laboratory – their accounts were more likely to be permeated with a "vulgar" commentary about the cause, extent, or intensity of these phenomena. Sometimes, these reports described the reporters' emotional response to them. This broadening of the meaning of "empirical" was facilitated by the approval of information supplied by ordinary people. To a certain extent, personal presence and personal inspection of evidence – again, unlike laboratory constructions of fact – were agencies of inclusion rather than the exclusion of unphilosophic opinions. For if the personal autopsy inaugurated a more fruitful approach to meteoric events, it also transmitted public chat into "high science." In fact, when late eighteenth-century meteorologists pointed out the inadequacy of observing unique and unconnected meteorological events, they especially criticized their predecessors' exclusive interest in meteorological "news" about spectacular meteors, rather than the "objective" approach to all matters of fact. These reformers demanded that meteorological observation be de-publicized to become scientific.[74] The next section analyzes this inclusion of popular public meteorology into scientific discourse, and illustrates it by examining the use of collective memory and hyperbolic speech in giving evidence about the intensity, esthetics, and emotional dimension of extreme meteors.

For the writers within the meteoric tradition, personal presence mattered more than the social rank of the observer. In many instances, the naturalist would acknowledge the information received from the local people and consider it as reliable as his own. The authenticity of such information was thus not exclusively linked to the codes of gentlemanly civility.[75] Martin Folkes ordered his watchman to monitor auroral lights; Philip Percival reported his servant's observations on a luminous appearance above Dublin; John Pringle's description of a fireball came from a farmer, while John Dorby received an account of a whirlwind from several "judicious workmen." Similarly, Stephen Fuller interviewed "common people" for his account of a fiery meteor, William Gostling's report of a fireball was that of a coachman and Stephen Hales questioned soldiers on duty in St James's Park. Lord Lovell's letter on a fiery meteor referred to the observations of local "ploughmen" and "turnip-houghers,"[76] and, when in 1736 His Grace Charles Duke of Richmond and Dr Edward Bayley wrote about a minor earthquake, their account was "backed by the united Testimonies" of other persons of veracity from the town of Chichester: eleven men, six women, and three female children.[77]

Lay reports usually supplemented naturalists' own observations. Provincial naturalists thus visited places where people were said to have seen or heard the phenomena; these reporters examined surrounding objects, and re-examined the observers' positions and movements. They used azimuths and theodolites to convert ocular information into numerical data and encouraged eyewitnesses to amend their original testimonies or to recall additional details. They drew up figures, showed them to informants, and asked them to attest to their accuracy. When there were disagreements, evidence was given different degrees of likelihood. The most interesting meteorological phenomena of the eighteenth century – the fiery meteors of 1718, 1741, 1759, 1783, the hailstorm of 1699, and numerous eighteenth-century appearances of aurorae boreales – were reconstructed by this mosaic-like technique, grounded in ocular autopsy, instrumental measurements, and testimonies from the neighborhood. Direct observation, informants' written accounts, newspaper reports, re-inspection of sites of observation, oral communication with witnesses, re-evaluation of information, quantification of narratives, and comparison of the collected information – all of these methods constituted scientific testimony whose value always originated from the actual presence of eyewitnesses at the place and time of events.

Despite these interventions, however, many meteorological reports preserved popular idiom; they carried the information which could be heard in people's conversation or read in personal journals. This penetration of common into learned discourse was known in the sixteenth and seventeenth centuries, when use of the testimony of commoners and information based on popular traditions were part of the historian's and antiquarian's method. D. R. Woolf has drawn attention to the topographers' and antiquarians' use of the "common voice," or "common fame," that is to say, what most of the local inhabitants believed to have happened in the past.[78] The common voice was transmitted orally, but antiquarians and chorographers had little reason to question the authority which derived from a consensus of local inhabitants about the events in their area: historic battles, crimes, apparitions, landslides, and other items peculiar to the place. John Leland called such consensus *in hominum memoria* which referred to the information warranted by the memory of living men. Leland wrote of a bridge in Lostwithiel that had in "tyme of memorie of men lyving" sunk deeper and deeper into the sand.[79]

The common voice could also assimilate the long-standing tradition of local customs, beliefs, and natural folklore. Writing about the weather of Ireland, the chorographer Gerard Boate explained that if the morning dew did not occur during the months in which it was normally the most copious, it was a "certain sign to the inhabitants, that great rain is to fall suddenly."[80] John Aubrey commented that in South Wiltshire "the constant observation" of the locals was that the year would have a good yield of peas if the dew hung from the hedges on Candlemas Day: "it is generally agreed on to be matter of fact." Writing about the extraordinary snowfall of 1578 he used information from "some ancient people of the parish that did remember it."[81] During the 1650s, Joshua Childrey collected material about British rarities from the "odd stories" related by "common people" who measured the strangeness of a phenomenon by the standards of common memory. Childrey held that earthquakes had rarely occurred in Ireland because they "hardly come one in an age: and it was so long ago since the last of all was, that it is as much as the most aged persons now alive can ever remember."[82]

Idiom of this nature was customary among early modern diarists. As private records, diaries were unhindered by both the standards of empirical method and the precision of speech. The common voice of diaries often became a descriptive trope used to highlight the uniqueness of narrated events. Mark Browell, the seventeenth-century Newcastle attorney recorded that since April 1, 1688, the weather was so showery, "as had not been within rememberance of man." Jacob Bee's diary mentions three floods in Durham in 1689, which "exceeded all the floods that had been these many years," and Samuel Pepys writes about the warmth of the winter of 1662 "as never . . . known in this world before here." The antiquarian Abraham de la Pryme spoke of "the dreadfullest thunder and lightning that ever was seen," John Hobson about the "forwardest spring that ever was known in the memory of man," and John Evelyn about the "severest winter that man alive had known in England."[83]

Such idiom had its origins in the common voice, not in the comparison of long series of numerical "chronicles" of the weather. Even if these individuals did keep instrumental weather diaries, the lack of pre-instrumental record would render the superlative speech equally inaccurate. The common meteorological voice was so widespread that there were some fears that it would contaminate even scientific discourse itself. Robert Boyle advised a strict separation of the scientific and "common voice" meteorologies:

> And truly when I consider, that things of the greatest consequence do oft-times depend upon the most common observations; and that matters of the highest improvement do receive their beginning from mean, small, ordinary experiments; I would have no man, who has leisure, opportunity, and time, to think it a slight thing to busy himself in collecting observations of [meteorological] nature. It being much more commendable for a man to preserve the history of his own time, . . . , than to say, upon every occasion that offers itself, this is the hottest, or this is the coldest, or this is the rainiest, or this is the most seasonable or unseasonable weather, that ever he felt; whereas it may perhaps be nothing so.[84]

Boyle's advice was partially heeded. However accurately naturalists wished to present their observation, narratives about rare meteors drew upon sources which Boyle would have preferred not to use. Indeed, the meteoric tradition can be portrayed as a chronicle of the events reverberating in the public's memory. This aspect can be stressed by the fact that not even the editors of the

Philosophical Transactions had qualms about the common voice found in the journal's articles. In 1698 a Yorkshire vicar witnessed an eruption of water which "hath not been heard of within memory." And an apothecary from Hertfordshire, assessing the damage of a hailstorm to amount to the sum of £4000, believed such enormous cost to be "never read nor heard of."[85] A generation later, Stephen Fuller wrote about "the most violent hurricane of wind in these parts, that ever was known since the memory of man," and Lord Petre thought that the "most terrible thunder ever heard" in Thorndon happened one Tuesday in June 1742. In a communication to Stukeley, Mr Perry went so far as to judge that the earthquake in Lisbon of the year 1755 "was certainly the most awful tremendous calamity, that has ever happened in the world."[86] Until very late in the eighteenth century, the *Philosophical Transactions* published correspondence that abundantly confirmed the fact that the purging of the common voice and its hyperbolic idiom was a largely unaccomplished task and that a vernacular perception of severe weather was a major element of the meteoric tradition.

Emotions also pervaded the genre. The features of meteoric reportage showed strong congruity with contemporary literary expression. Like voyage and pilgrimage narratives, the natural history of Hans Sloane's time was, despite numerous disclaimers, perceived as a category of fiction, "dealing as it did with medical oddities, monsters and extraordinary occurrences, places and phenomena. Some wrote 'The History of Tom Jones,' others 'A History of Bees.' "[87] Apart from the topical similarities, integration of the language of natural history was, according to Soupel, particularly strong in moments when naturalistic description took on a "personal dramatic function." The quasi-medical descriptions of poisonings, deaths by lightning, or the ostensibly factual accounts of weather-induced damage all reverberate with horror, and in these cases, "the sentiment practically invades whatever is scientific."[88]

In meteoric reportage the sentiment ranged from admiration to fright. The great botanist Dr Richard Richardson described the panic of those caught in a waterspout on the Lancashire Moors, saying that "on approaching the place, I was struck with unspeakable horror," as the ground was torn up to the very rock. Earlier in the century, a diarist had written that "at evening most eager storm of all, near upon full W: or at least S.W. Veering and hovering that way, best part of day. About 8 there was a dreadful, fierce, and violent vehement tempestuous wind and storm as shook house and blew down part of our Pous-Rick."[89]

In 1719, William Maunder wrote to Samuel Cruwys, about an *aurora borealis*: "you have doubtless been *surprised* again afresh by the wonderful Lights."[90] An anonymous Dubliner declared to have been "*agreeably surprised* with a kind of Coruscation" in the south West, followed by a "*beautiful* Tremour or Undulation in that Subtile Vapour." "This *beautiful* Spectrum might be likened to the Star worn by the most Noble Order of the Garter," judged Dr John Huxham at Plymouth of something his colleague from Exeter found indescribable: "It is impossible to express the *beauty* of this *glorious* umbrella." De la Pryme beheld with "*great satisfaction* this extraordinary phenomenon [waterspout] and found it proceeded from a gyration of the clouds, by contrary winds meeting in a point of centre." And the antiquarian John Swinton was in his parlor in Oxford when on September 9, 1769, he observed "with no small degree of astonishment, through the glass, such a redness in the sky as proceeds from the reflection of a great fire."[91] And the explosion of a meteor in 1761 above Whitby, Yorkshire, "made so beautiful an appearance in the heavens and was gazed at with wonder and delight by the connoisseurs."[92] In these accounts, the enlistment of emotional response was a norm which did not detract from the "objectivity" of reports. In fact, it underscored the observer's presence at the site of an event and guaranteed that the observation had indeed taken place. Rather than curbing them, the method of autopsy invited emotional categories as an additional vehicle in achieving the descriptive completeness of the Baconian natural history.

In concluding these analyses, we may note that the eighteenth-century meteoric tradition, its common voice and the idiom of emotion demonstrate the extent to which the public dimension of extreme weather informed the structure and content of meteorological reports. These features exhibit a hitherto neglected aspect of eighteenth-century natural science in that they show us that Hanoverian naturalists often violated – rather than enforced – the literary and epistemological requirements put forward by the seventeenth-century reformers of natural knowledge. Instead of *excluding* "vulgar" understanding by policing the linguistic boundaries of proper locution, meteoric reporters *included* vestiges of such understanding because they could not – or were not willing – to distance themselves from the public dimension of the natural world.

The emphasis in this chapter has been exclusively on *observation* in meteorology; the meteoric tradition has been taken to refer to the brief descriptive reports on individual meteors, rather than theoretical analysis of their causes. This separation is based on the internal evidence, as many eighteenth-century meteorologists thought of their accounts as inductive instances toward a natural philosophy of meteors. In a more extreme case discussed here, this empiricism took on a form of genteel curiosity which introduced a wholly new domain of concerns about the medical, architectural, financial, and esthetic implications of strange and severe meteoric activity. As later critics of these interests pointed out, no amount of such isolated and extraordinary facts could suffice for a most elementary law of atmospheric behavior. But, as it has been stressed, such isolated instances served a preeminently cultural purpose in the tradition of geographical and, as we will shortly see, social differentiation in the English countryside. In this regard, the chorographically informed study of meteors both epitomized and defined eighteenth-century natural history as a mimetic and "museological-diagnostic" enterprise with a tenuous link to natural philosophy proper.[93]

5

Provincial weather

Nor be the Parsonage by the Muse forgot,
The partial bard admires his native spot;
Smit' with its beauties, lov'd as yet a child,
Unconscious why, its 'scapes grotesque, and wild.
Gilbert White[1]

Sunday, April 13th. Wind westerly, pretty high. Lectured on Psalm 68 v. 22.
Preached on 1 Cor. 15, vv 56 and 57.
George Ridpath, *Diary*[2]

If meteorological reportage had meaning in an early modern quest for regional identity, it is not surprising that it was a quintessentially provincial endeavor. But if it appears that the eighteenth-century provincials participated in empirical meteorology as miniature human observatories, it is because data-collection has usually been seen as an acquisition of knowledge rather than a matter of esthetics, the constraints of "stationary residence" or local attachment. This chapter turns to these grass roots of natural history and argues that the appreciation of eighteenth-century "field sciences" should take into account more than its organizational, theoretical and methodological aspects and look in addition to the confluence of the personal, scholarly, and political motivations of the provincial natural historians.

In 1690, Ralph Thoresby, after meeting with Mr Nicolson, the Archdeacon of Leeds and an eminent antiquarian, set out collecting topographical and historical information about the Parish of Leeds. The results appeared in 1715 in the form of a folio volume entitled *Ducatus Leodiensis*, a work dedicated to the Right Honourable Peregrine-Hyde Lord Marquise of Caerwarthen to whom Thoresby wrote the following:

A Natural Propension to the Study of Antiquities inclining my Thoughts that Way, an innate Affection to the Place of my Nativity, did more particularly fix upon the Present Subject. The inclination was the more excited, even when a School-Boy, by an Expression of our Learned Vicar Mr. Milner, in a sermon, that the Town was of great Antiquity, being expressly mentioned by venerable Bede, who flourished near a Thousand years ago.[3]

Thoresby explained his naturalistic beginnings with a reference to his historical roots in his native county. For Thoresby, one's geographical situation was not an unfortunate circumstance to be cured by ersatz cosmopolitanism, but an opportunity to study local history and nature. In this regard, *provincial* learning and learning about one's *province* were equivalent.

Recent scholarship on eighteenth-century provincial science has noted that the Augustan naturalists used science to reject their cultural parochialism and the backwardness which it symbolized in the city-oriented society: as Roy Porter argued in an influential paper, "[t]o provincial eyes, enlightenment values offered a leg-up from rusticity, associated with barbarity and riot, towards metropolitan – indeed, cosmopolitan – urbanity."[4] Due to the presence of the notion of *emergence* of cosmopolitan values, provincial learning was often seen as an emulation of city-made rationality. On this reading, provincials were attuned to the trends of contemporary social excellence when they followed city dwellers in attending philosophical lectures, buying mathematical instruments, and reading the London scientific periodicals. Catering to this demand, itinerant lecturers and demonstrators moved out of London, circulating libraries lent popular works in natural philosophy, and the production and sales of scientific instruments soared. Politically, these promotional strategies reflected the interests of Whig oligarchy in that the scientific ideology of order and harmony, in the words of Margaret Jacob, "complemented the political stability over which [the Whigs] sought after 1689 to preside so comfortably." But in asserting that natural sciences acted as a focal point of the "rational culture within a broad Enlightenment civility," this view leaves several problems unaddressed.[5]

While the idea of cultural emulation may reflect some aspects of economic development in the countryside and some elements of provincial culture, those which Michael Reed has called "distractions of the *villetes*" – provincial newspapers, lectures, debating societies, music, and theater, – it does not make allowances for an outlook which promoted and sustained specifically *local* scholarship.[6] Why did provincial naturalists acquire such cohesion in *this* rather than in some other time? Why has the principle of emulation never been invoked to explain Henry Oldenburg's network of scientific informants? Why did the clerical naturalists acquire such prominence in the eighteenth-century culture of natural history?

To answer these questions, it may be advisable to abandon the social- and class-oriented historiography, and pay attention to *where* scientific knowledge was made. Can the present, merely *descriptive,* distinction between the provincial and the metropolitan, become analytical? Is there anything to learn from the fact that a large number of contributions to the eighteenth-century *Philosophical Transactions* came from the provinces? Did the "provincial" origin of such information shape their content, method, and purpose? These questions are pertinent because they are asked with regard to the fact that eighteenth-century county historians not only continued but even expanded the scope of antiquarianism, archeology, and natural history as it was defined by Restoration scholars.[7] In the first half of the century, few counties were left without a multi-volume history and almost all were "perambulated" by naturalists. In agricultural theory, the emphasis shifted from the philosophical to the practical, from savants to rustics. The mid-century physician Thomas Short even thought that the savants "deserve[d] to be ridiculed" for setting aside observations of rural countrymen "as not flowing from any natural connection that [the savant] knows of; and because he knows them not, they cannot be."[8]

While we have already noted that chorographic research acted as the prime organizer in the quest for spatial and social physiognomies of regions, it must now be mentioned that the eighteenth century saw provincial cohesion strengthened by such forms of assembly as geological and botanical walks, the flower-shows of clubs like Norwich's "Sons of Flora" and Lichfield's "Society of Gardeners," private antiquarian collections, and the formal gentleman's clubs devoted to local natural and humanistic study.[9]

Thoresby's "innate Affection to the place of Nativity" was a key word for those who, in this period, tried to secure the relevance of local history and defend provincial investigation in the face of looming national culture. Moreover, Thoresby's affection was not only a source of scholarly enthusiasm, but his intellectual cachet. It vouched for his authority in local ethnology, etymology, numismatics, or meteorology because, as a native and being well-connected, he could elicit respect and access local patronage. Personal involvement in the history of a region could thus provide epistemological grounds for an enquiry into regional curiosities. In 1712, Sir Robert Atkyns the younger, a Tory lawyer and MP for Cirencester, wrote in the Preface to his study of "Glostershire" that it was the "true and hearty love for his Country that excited him to [the] Performance."[10] Similarly, the Brasen Nose Society of Spalding, founded by the antiquarian William Stukeley, cultivated natural philosophy but also research devoted to "preserving the memorials of persons and things fit to be transmitted to posterity."[11] To local naturalists, local history was as relevant a pursuit as natural philosophy.

The scholarly implication of these cultural factors has been largely shunned by historical analysis. Social historians of science have overlooked the role of "domicentricity," – the home-centered interest – in the development of British natural history during the eighteenth century.[12] In geographical studies, however, the concept of place and locality is addressed in connection with the concept of *topophilia*, the love of place, referring to "all of the human being's affective ties with the material environment."[13] In an extended sense, topophilia could be a substratum of national sentiment: John Mulso, for instance, thought that Gilbert White's notes from his tours over the country would "make an agreeable Pocket volume [that] might tempt Gentleman to examine their own country before they went abroad and brought home a genteel Disgust at ye Thoughts of England."[14] Less ambitiously, some naturalists considered their studies justified by the loyalty of a county readership and simultaneously used them as the master-key to the gates of cosmopolitan learning. The career of William Borlase epitomizes such symbiotic maneuvers.

The Rev. Borlase and his network of correspondents

The case of the Cornish clergyman William Borlase (1696–1772), a naturalist whose pursuits are considered to be as an important contribution to eighteenth-century meteorology, illustrates the relationship between national and local society.[15] Borlase's vision of the natural historical subject was inclusive, yet still scholarly when compared to specialized studies: by the mid-century Borlase was considered one of the most respected of meteorologists. In 1713, Borlase matriculated at Exeter College, Oxford and took his MA in 1719. After ordination in 1720, he briefly stayed in London, and returned to Cornwall to assume the Rectorate at Ludgvan. During the 1720s he "frolicked" with the gentry, read Horace and Dryden, worked in the garden and began collecting minerals. He consulted gardening manuals written for the provincial clergy.[16] Remoteness from London made "matters of importance lose half their impression by the time they come down to us," but within the span of two decades, his philosophical friends encouraged him to begin a study of monuments and of natural history in general. By the late 1730s, he was acting as a tutor for the sons of Cornish aristocrats and was, with most landowners absent from their estates, the chief man of the Parish.[17]

Borlase's visitors during this time included the squire John St Aubyn and the Reverend Edward Collins with whom he debated, among other things, over "[h]ow worthy Cornwall (though hid as it were in the extreme angle of Britain) was of further enquiry." Borlase's first opportunity for positive action was on the occasion of receiving a letter in 1735 from John Andrew, a student of medicine under Boerhaave at Leyden. He wrote on behalf of the university and asked Borlase for samples of Cornish minerals for the university's mineral cabinet. The recipient of Borlase's specimens was the chemistry professor John Frederick Gronovius, who immediately reported back regarding the good impression they made on the faculty and on Linnaeus himself, who happened to be at Leyden at the time. Seeing more opportunities for Borlase, Andrew expressed his surprise "that a person of your taste and abilities should not think it worth his while to enquire into . . . the Natural Curiosities of his country, which I am ashamed to say are almost wholly unknown."[18]

Borlase responded with diffidence and asked Andrew to oversee his work, which gradually encompassed a variety of pursuits, from fossil and mineral collecting and marine botany, to landscape drawing, linguistics, heraldry and numismatics. This work clearly fell within the tradition informed by chorographers such as Robert Plot, Ralph Thoresby, and Charles Leigh. And although Borlase's clerical career spanned the period of the rise in the popularity and specialization of natural history, his work embodied the cultivated curiosity, symptomatic of all previous Plotian projects: "[M]y endeavours shall all tend to enquiries in the natural world, to view every curiosity of that kind, to draw faithfully what is too large to be procured, to collect minerals and fossils of all kinds etc."[19] Antiquarianism and natural history "travel sociably together," he wrote to Andrew, using at the time several fieldwork manuals and John Woodward's *Brief Instructions For Making Observations*.[20] In the 1740s, Borlase discovered advantages to his isolation, saying that if "the little spot I live in could furnish enough to give pleasure to a set of learned men who have seen the Sweden, Hungary, and German mines, I too easily entertained some thoughts of pursuing this part of knowledge."[21]

Borlase had a series of contacts during the following decade. In 1746, he hosted the young naturalist and traveler Thomas Pennant, who later credited Borlase with having inspired his own passion for natural history.[22] Emanuel Da Costa, mineralogist and a Fellow of the Royal Society, visited Cornwall in 1749, undertook field-work, and helped establish communication between Borlase and William Stukeley relating to matters of ancient history. Upon his return to London, da Costa proposed Borlase for election to the Society. Borlase was elected after publication of his treatise on Cornish diamonds in early 1750. From 1750 he established a friendship and correspondence with the clerical naturalists Charles Lyttelton and Jeremiah Milles of Exeter, and John Hutchins, Rector of Wareham, who was working on the topography and history of Dorset. His London acquaintances were the microscopist Henry Baker and the marine zoologist John Ellis.[23]

In January 1752 Borlase started a weather journal, first in narrative form and then, from 1754, in tabular arrangement. Following Milles's suggestion, he began to circulate a set of printed queries on parochial history and topography, based on the model used by Lhwyd in Wales and Yorkshire.[24] This locally-oriented search – originating with the seventeenth-century work of Gerard Boate – had numerous champions in the eighteenth century. Francis Peck, for instance, prepared in 1729 "Queries for the Natural History and Antiquities" and circulated it in Leicestershire and Rutland. In 1731, Anthony Hammond, Esq., published "Inquisitio Parochialis." On a larger scale, a Mr Blomefield circulated questionnaires in Norfolk in 1736, John Hutchins in Dorset in 1739, and Richard Rawlison in Oxfordshire during the 1740s. The Royal Dublin Society prefixed a set of similar queries to their history of county Down (1744), which were taken up by a Dr Burton in 1758 who accompanied them with proposals for illustrating the county of York. In 1755, Edward Cave published the queries drawn up by the Society of Antiquaries. George Allan circulated his "Address and Queries for the Palatinate of Durham, 1774."[25]

Using his own observation and the scant information received through the questionnaires, Borlase prepared several manuscripts for print. They were used as the basis of his two chief works, *Antiquities of Cornwall* (1754), followed by *The Natural History of Cornwall* (1758) (FIGURE 14).[26] These works are particularly relevant for understanding the intellectual and spatial placement of natural knowledge. The natural history of a particular district, argued Borlase, ought to appeal to local inhabitants to whom its content should by necessity be the "most interesting disquisition": even if observations relative to Cornwall have no importance to the world, they should claim "the most serious thoughts from the inhabitants of the County."[27] These claims must be read in light of the fact that Borlase's work, like the work of other chorographers in the eighteenth century, was the result of a munificent subscription. Over 400 hundred people subscribed to Borlase's *Cornwall*. These included the Princess Dowager of Wales, the Archbishop of Canterbury, Horace Walpole, Thomas Birch, and, most importantly, a large proportion of the local nobility. Over a hundred subscribers came from Cornwall.[28] The book was dedicated to Sir John St Aubyn, the landowner and opposition Whig, who subscribed for fifty copies. John Prideaux Basset subscribed for twenty and paid for a plate of antiquities found on his property. Borlase's correspondent Hutchins captured the importance of the landowner's subscription when he remarked in 1770s that "had it not been for the financial

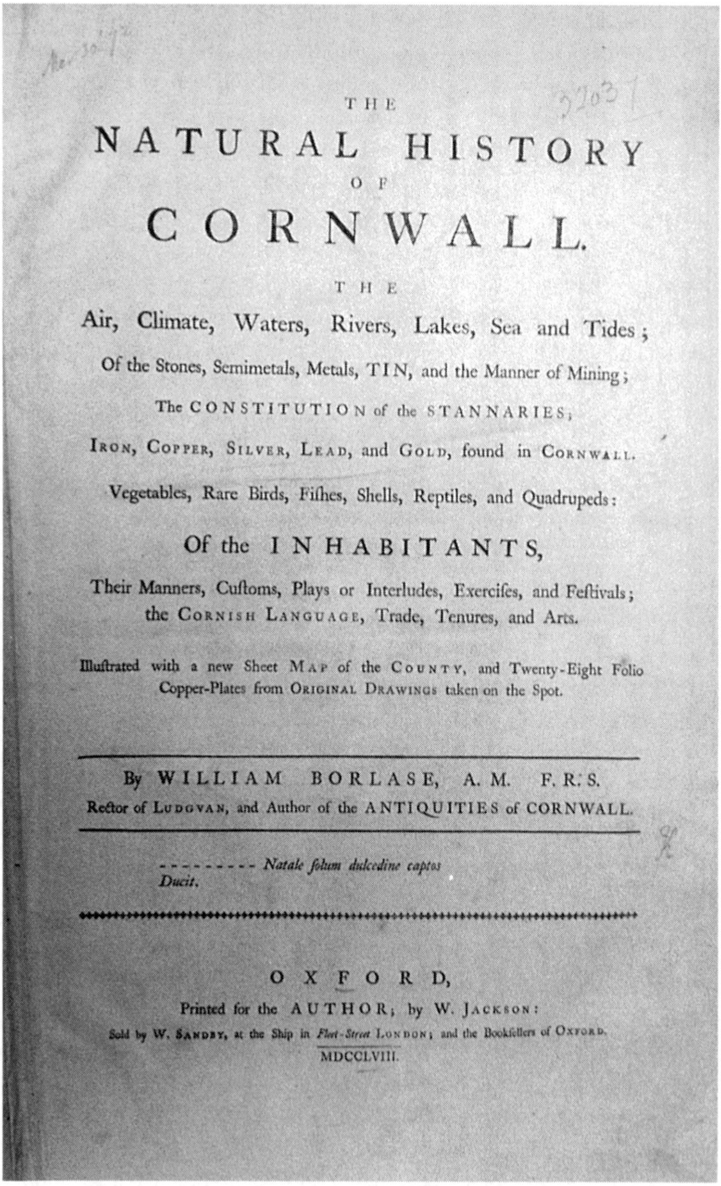

THE

NATURAL HISTORY

OF

CORNWALL.

THE

Air, Climate, Waters, Rivers, Lakes, Sea and Tides;

Of the Stones, Semimetals, Metals, TIN, and the Manner of Mining;

The CONSTITUTION of the STANNARIES;

IRON, COPPER, SILVER, LEAD, and GOLD, found in CORNWALL.

Vegetables, Rare Birds, Fifhes, Shells, Reptiles, and Quadrupeds:

Of the INHABITANTS,

Their Manners, Cuftoms, Plays or Interludes, Exercifes, and Feftivals; the CORNISH LANGUAGE, Trade, Tenures, and Arts.

Illuftrated with a new Sheet MAP of the COUNTY, and Twenty-Eight Folio Copper-Plates from ORIGINAL DRAWINGS taken on the Spot.

By WILLIAM BORLASE, A.M. F.R.S.

Rector of LUDGVAN, and Author of the ANTIQUITIES of CORNWALL.

--------- Natale folum dulcedine captos
Ducit.

OXFORD,

Printed for the AUTHOR; by W. JACKSON:

Sold by W. SANDBY, at the Ship in Fleet-Street LONDON; and the Bookfellers of OXFORD.

MDCCLVIII.

FIGURE 14   The title page of William Borlase, *The Natural History of Cornwall* (1758).

assistance from the gentry," his topography and history of Dorset "would not be completed."[29] Chorographies epitomized the mutual exchange of benefits between the financially insecure authors, and the local nobility, who saw such projects as a form of family genealogy writ large.

In the *Natural History of Cornwall*, Borlase recommended the study of nature as the handmaid of religion, a guide to piety and public benefit, and a source of entertainment: "if the mind thirsts after variety, and a fresh succession of objects where can she find for contemplation so numerous and various a treasure?"[30] He refused to be bound to any predetermined plan, hypothesis, or method, in all probability alluding to Linnaeus, whom he held incomprehensible. His science was of "a most extensive" scope, "taking in all animate and inanimate substances which Land, Air, and Water contain . . . in short, giving a recital and detail of the whole visible Creation." His study of Cornish weather developed within this agenda and was stimulated by his interest in philosophic and topographical instruments: in 1749 he used a hydrostatic balance, a hydrometer, quadrants, two microscopes, a theodolite, both perpendicular and diagonal barometers, and several thermometers. Starting in 1752, when he commenced a daily record of barometer readings, temperature, wind, and the state of weather, Borlase began contributing to the *Philosophical Transactions.* The similarities between these and the articles discussed in the previous chapter is striking.[31]

His early account on the effects of lightning on a hill near Ludgvan was highly detailed, including a sketch of the shapes, and "geology" of the lightning-dug furrows as well as the information obtained from local women. Another sketch shows the floor plan of a farmer's house in which the dwellers, shocked by the lightning, remained senseless for quarter of an hour. Borlase visited the household, drew the positions of family members who, when asked of the incident, "kindly enter'd into a detail of particulars."[32] (FIGURE 15). In 1758, Borlase employed the autopsic procedure of on-site observation and consulted parishioners about the intensity of a minor earthquake. He also received observations from local gentry, but it is not clear whether he explicitly solicited them as part of his earlier attempt to establish a parochial information network (BOX 5).[33]

What is important is that in his explorations, Borlase engaged with the social and cultural spheres of both the national and local community. Such double engagement was characteristic of most local scholars since the later Stuart period. For example, in Essex, in the early decades of the eighteenth century, Samuel Dale and William Holman pursued their natural and historical investigation at the intersection of two correspondence networks. They communicated with both botanists and natural historians of national repute and the group engaged in the parochial history of the county.[34] Beyond these, however, Borlase also had access to the beliefs and daily life of his parishioners. When visiting the Isles of Scilly in 1752 for the purpose of archeological digging at the site called Giants' Graves, a violent storm followed the day after. Borlase remembered that a local woman asked him whether he thought that he had disturbed the Giants, adding that "many good people of the Islands were of opinion that the Giants were offended, and had really raised that storm."[35]

This straddling between the rural and national worlds is especially visible with regard to Borlase's meteorological activities. Some time after Borlase's earthquake paper was read to the Royal Society, Thomas Hornsby, the Savillian Professor of Astronomy at Oxford, visited him at Ludgvan and suggested an improvement of his meteorological pursuits by instituting the measurement of rainfall. Hornsby became Borlase's "meteorological mentor," urging his protégé to begin a more elaborate journal and acquire an ombrometer, a rain gauge.[36] Weather observation became such an important part of Borlase's work that in 1762 he and Charles Lyttelton discussed a plan for keeping systematic records at widely scattered locations and sending them to the Royal Society for comparison and preservation.[37] But, as with the parochial queries, volunteers were hard to find; Lyttelton encouraged a certain Dr Carlyle at Carlisle and Borlase advised Lancelot St Albyn, a clergyman at Bridgewater, but there the network ended.[38] Despite this, Borlase continued the series of observations for his own entertainment and for the pleasure of comparison. From 1768–72, his weather summaries and rainfall were printed in *Philosophical Transactions* and the *London Chronicle* as the "official" Cornwall data.

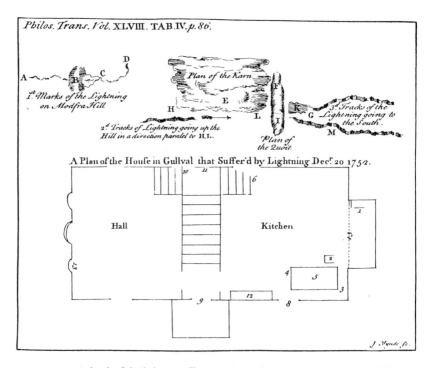

FIGURE 15 A sketch of the lightning effects in a house and on the nearby grounds. William Borlase, "An Account of a Storm of Thunder and Lightning," *Philosophical Transactions* 48 (1753–4): 86–98.

BOX 5   Networks of the Rev. William Borlase: a select list of correspondents[39]

LOCAL NETWORK

Francis Basset, Esq, at Tehidy (includes a letter with information about *his* informants)

Hon. J. Harris, Esq, at his seat, the summer house in Keneggy, near Penzance (his gardener questioned)

Mr J. Nantcarrow, "intelligent captain of the mine", from Cornwall

Thomas Olivey, a farmer of good substance, from Cornwall

James Tillie, Esq, with neighbors and servants, from Cornwall

John Trehawk, Esq, from Plymouth

William Veale, Esq, at Trevalier, two miles from Penzance

Rev. Anthony Williams, curate, from Cornwall

(INTER)NATIONAL NETWORK

Henry Baker, FRS, microscopy, London

Henry Ellis, governor of Georgia

Peter Collinson, FRS, electricity, natural history, London

Emanuel da Costa, FRS, geology, London

John Ellis, FRS, electricity, London

John Frederick Gronovius, Professor of Chemistry, Leyden

John Hutchins, topographer, Dorset

Charles Lyttelton, President of the Society of Antiquaries, London

Matthew Maty, President of the Royal Society, London

Jeremiah Milles, antiquarian, historian, Devon

Thomas Pennant, FRS, naturalist, London

Alexander Pope, poet, London

William Stukeley, antiquarian, naturalist, Cambridge

Furthermore, Borlase bridged a gap between the meteoric tradition – public meteors of the parish and county – and the instrumental weather record – physical parameters of the scientific atmosphere. He thus published narratives of Cornish storms as well as daily measurements of atmospheric variables and had no difficulty in seeing the importance of both the isolated reports and the synoptic monitoring. And while it is possible to see Borlase's instrumental record as part of the international effort to correlate weather and health, weather and crops, or winds and sea-travel, it is doubtful whether these interests were Borlase's personal preferences. Despite his duty-like fervor in the routine of daily observation, Borlase harbored doubts as to the utility of the theoretical results of such measurements. Writing to Henry Baker, his adviser in matters of instrumental accuracy, he noted that

[i]t would not be much amiss to observe, with greater than the present attention, the several degrees of wet and dry, heat and cold, storm and calm, in the

different parts of the Island, and transmit them to the Royal Society for their Transactions. This may in time either facilitate some more perfect theories of winds and weather in our climate, or, which is altogether as likely, show the uncertainty and vanity of all such attempts; in short, the atmosphere is such a various irritable mixture . . . and the action of the heavenly bodies so perpetually shifting, that nothing permanent and sure is to be expected; no apparently similar circumstances will always produce the like, nor is any thing to be foretold from analogy and review.[40]

The comment makes clear the fact that the mid-eighteenth century quantitative observations did not always reflect the inductivist ethos. It has been shown that the eighteenth-century uses of meteorological instruments could not be divorced from the consumer culture, short-term forecast, and notions of refinement and gentility.[41] But even if observation reflects inductivist ideals, what purpose would Borlase see in his own record if he considered it as a methodological blind alley? His writings suggest that if he had a view on these matters, it was a view that preferred the meteoric tradition over instrumental data, either because meteorology still meant the science of singular meteorological events, or because unusual meteors had more appeal to his parish informants, readers, curious correspondents in London and himself as a cleric with pious regard for the works of Providence.

This preference can be confirmed by other features of the eighteenth-century weather record. For instance, the issue of standardization of instruments, measuring units, and practical regulations (like the location of the instruments and the frequency of daily readings) concerned only a minority of diarists. The very idea of an indoor thermometer, for instance, the most common of instruments during the eighteenth century, makes the program of organized measurement suspect from the start. How was one to correlate room temperature with the outside changes of weather? Another issue is connected with the seasonal commuting of some landed gentry, which resulted in a number of diaries being kept only during the summer months, when the routine fitted into their conception of cultivated leisure.[42] On the scientific side, even when in the 1770s the diaries grew in number, their purpose remained enigmatic. Samuel Horsley observed that "[t]hough the practice of keeping meteorological journals is, of late years, become very general, no information of any importance has yet been derived from it." Additionally, as we will see in the next chapter, the common understanding of the purpose and efficacy of meteorological instruments was far from unanimous: there was no consensus as to how reliable the barometer was in recording and predicting the weather.[43]

However important these issues may look from the perspective of the development of modern meteorology, they were of secondary importance at the time. Borlase mentioned on several occasions that he kept a diary to compare it with those made at other locations. Some of his correspondents kept their own to compare them with Borlase's. For these individuals, the impulse to record, tabulate, compare, and speculate about the varieties and extravagance of local weather derived from an acute awareness of geographical place, rather than some "global" methodological perspective. The members of Borlase's network would perhaps approve of the schemes of James Jurin's ambition, but, for the time being, they found little personal incentive to engage in a project which valued the centralization of local data more highly than the study of local weather for its own sake. Servicing metropolitan "projectors" (those with grand ideas) – even when that could be praised as forward-looking – had far less appeal in the provinces than at the Royal Society quarters, because those asked to participate had little, if nothing, to gain from such enterprises. Without material compensation and with their data "anonymized" by metropolitan calculators, provincial enthusiasts could have lost the *last* motivation for the activity which they performed voluntarily anyway. Instead, they could have easily cited Robert Boyle's doubts about the use of collecting weather data, who thought that, observations of this nature would do most if they "conduce to compleat the natural history of *any place.*"[44] (FIGURE 16).

In this context, the range of Borlase's interests from gardening through occasional mineralogical tours to full-time natural study is hard to explain in terms of the utilitarianism of the European republic of letters. However enthusiastic Borlase, Lyttleton and Milles might have been about the prospects of global networking, the factors determining the scope of their research lay in the nature of their clerical vocation: their leisure, their secure living, and their settled lifestyle.[45] Borlase's history of weather was less of an achievement in empirical philosophy of science than the combined effects of geographical isolation, a sense of duty, and the esthetics of place.[46]

Calendars, diaries, narratives[47]

Borlase was a prototype of the "clerical naturalist," whose numbers were growing among the lower clergy during the mid eighteenth century. Local clergymen used parochial documents, birth records, genealogies, and inspected antiquarian and geological sites. Their miscellaneous work made sense within a chorographic framework, with many pursuing regional studies or a more intense study of nature as an aside to their nominal occupation. During the 1740s and 1750s, for instance, William Henry, from Gloucestershire, educated at the University of Dublin and rector in Urney, undertook a number of mineralogical and meteorological investigations which appeared in the *Philosophical Transactions.*[48] The Rev. William Cole a native of Cambridge, combined classical hermeneutics, philology, topography and study of local monuments. Like Thoresby he undertook several "sketchbook" visits to France and Flanders and, after holding several administrative posts in Cambridgeshire, began an ecclesiastic career in 1744 as Curate at Withersfield, Suffolk and Hornsey, Middlesex. In 1753, he moved to the rectory in Bletchley, Buckinghamshire where his antiquarian zeal produced more than a hundred folio volumes about "Parochial Antiquities of Cambridge-shire," "Parochial Antiquities for the County of Buck," and "Parochial Antiquities

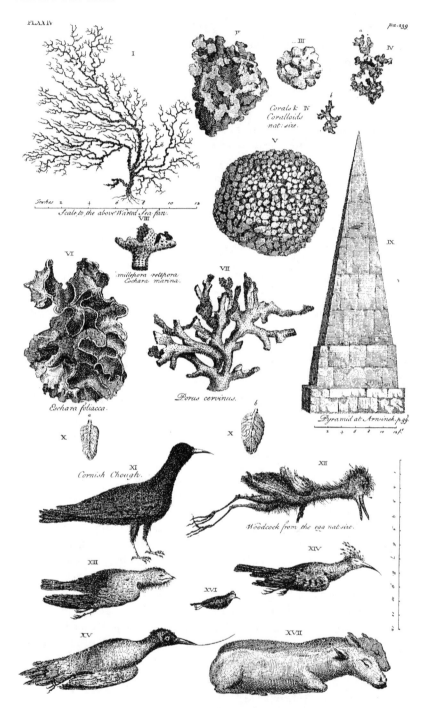

FIGURE 16 A plate representing Cornish corals, birds, a pyramidal monument, and a monstrous calf. The juxtaposition of natural and historical objects epitomizes the comprehensive eclecticism of eighteenth-century chorography. William Borlase, *The Natural History of Cornwall* (1758).

for the County of Huntingdon." A note on the nature of his weather diary testifies a clear lack of interest in communicateing his observations to any institution: "[F]or these thirty years [I have] kept a sort of Diary of the Weather and Journal of other things, being as someone justly enough calls it, The Importance of Man to his own Self."[49] Cole's self-routine testifies to a private meteorological attitude which in the case of the Rev. James Woodforde shaped his clerical duties. Estimated by Addison to have been "one of our best clerical naturalists," Woodforde's typical diary note has him baptizing a child on a February morning, preaching at Weston in the afternoon and heading home in the freezing air: "Excessively cold indeed, colder than ever, if any thing . . . The Thermometer down to 56, tho' in the Study and a fire constantly kept there – at eleven this Morning. The Barometer also down to 28–18 in the same Room."[50] An entirely descriptive diary of weather was kept by George Ridpat, a Presbyterian Minister of Stitchel, Roxburghshire, who from 1764 started to publish proposals for printing by subscription a work on the history and antiquities of Berwick, Roxburghshire, and Northumberland.[51] In 1784, the Rev. John Cullum from Suffolk, a botanical informant of Joseph Banks and Willam Cole, sent in an account of a remarkable frost to the *Philosophical Transactions*. Similarly to Ridpath, he worked on the local history of Hawsted and Hardwick in Suffolk. Such interests were shared by the Rev. Benjamin Hutchinson, who solicited Joseph Banks's patronage for a history of Huntington while also publishing *A Calendar of the Weather for the year 1781*.[52] In Tooting, Surrey, Henry Miles, a Dissenting minister, kept an instrumental weather record, publishing on microscopy, electricity, and meteorology.[53]

An important factor in the foray of the clergy into local studies was that the careers of most of them were financially uncertain especially in the first half of the century (BOX 6). For the "deplorable pool of unemployed or spasmodically employed clergy" a more secure living lay in promotion or patronage.[54] Making oneself known as a chorographer of local estates attracted subscribers whose benefactions could lead to promotion and reward. Nathaniel Salmon, a parson, towards the end of his life, wrote three county histories, of which the third, his

---

BOX 6 **Several prominent eighteenth-century clerical meteorologists**

William Cooper, Archdeacon of York (storm, 1784)
George Costard, Oxford (fiery meteors, 1745)
John Michell, Cambridge (meteor, 1755)
Mr Prince, Barnstable, Devon (agitation of the sea, 1755)
Benjamin Ray, Cowbit, Lincolnshire (waterspout, 1751)
Dr Shipley, Silchester, Hampshire (fiery meteor, 1755)
John Swinton, Oxford (optical phenomena in the sky, 1761, 1764, 1751)[55]
William Turnbull, Abbotrule, Cumberland (meteor, 1755)
Anthony Williams, St. Keverne, Cornwall (thunderstorm, 1771)

book on Essex (1740), was referred to by Richard Gough as "his last shift to live." One or two of these historians wrote their works under contract to patrons. The eleventh Duke of Norfolk paid John Duncumb two guineas a week for some twenty-five years to collect materials and compile a history of Hereford-shire; the same duke also employed James Dallaway on a history of Sussex. Both works were abandoned when the duke died in 1815.

On the other hand the incentive for pursuing regional studies could be mainly intellectual. In a position to use local sources of information, clerical naturalists – as well as those who were not clergymen – could find a respectable niche for themselves in the national republic of letters. Scholars in correspond-ence considered it permissible to seek assistance from each other, and, as Anne Goldgar has recently pointed out in this respect, "it was generally believed that if someone was *sur les lieux* – on the spot – he was fair game for requests in that place."[56] Isolation did not mean the end of education. As Borlase wrote to Stukeley: "it being my fortune to live at a great distance from places of publick resort, and my profession confining me to a small round, I found myself obliged to amuse myself with such remarkable as were within my reach, or utterly to abandon that share of curiosity which I have imbibed during the time of my education."[57]

During the middle decades of the century, a more organized method of meteorological study emerged with the popularization of "naturalists' calendars," which recorded practical agricultural and botanical information. These calendars documented moments in vegetative and climatological cycles, and thus resembled meteorological journals with respect to a sequential organization of facts. They also resembled another serial genre – "the diary of a country parson" – in which "the tenor of the regularly turning seasons and the profession of the Church's calendar imprint themselves" upon the content of these writings, in which "men with Oxford and Cambridge degrees combat the progress of vegetation with the exercise of words upon their environments."[58] But the keepers of these calendars committed themselves to recording *specific, important,* and *local* natural events in the annual cycle: the calendars in this sense lent themselves to the highly regarded comparative study of the seasons, places, and local vegetation.

Benjamin Stillingfleet of Trinity College, Cambridge, played a major role in these developments. As the leading English expositor of Linnaeus he published *The Calendar of Flora* in 1755, a programmatic calendar setting the standards for its future format and use. The calendar was to begin an international "bank" of information on the seasonal cycle of vegetables with emphasis on plants known to predict the change of weather. He wished to present the calendars to the public early in the year, so "that others might be induced to keep journals of the same kind . . . I think therefore it would be a pity, that an opportunity be lost of making so curious comparison between [the] different climates."[59] The division of months was made according to the phases of the annual vegetative cycle, following the model advanced by the Uppsala group, a group of Linneaus' Students at the University of Uppsala. Information was included on temperature observations and the general state of the weather.[60]

Stillingfleet dedicated his work to his benefactor Viscount Barrington, then Secretary of War, and eldest brother of the naturalist Daines Barrington. The latter started as a lawyer at the Inner Temple, held posts in the Admiralty and Greenwich Hospital and was at one time involved in lobbying for Arctic exploration. He ended as Commissary-General of stores at Gibraltar, beginning there the study of nature and antiquities. His *Naturalist's Calendar* of 1767, a diary designed on the Stillingfleet model induced the famous Gilbert White to systematize his parochial record.[61] When they eventually met in London in 1769, two years after White befriended Thomas Pennant, it was Barrington who envisioned White's observations in the form of a book. Much like Borlase, White responded that "it was no small undertaking for a man unsupported and alone to begin a natural history from his own autopsia!" Nevertheless, "[i]n obedience to your repeated injunctions I have begun to throw my thoughts into a little order that may reduce them into the form of an *annus-historico-naturalis* comprising the natural history of my native place."[62]

The meteorological pursuits of Gilbert White are suggestive of a pervasiveness of a chorographic ethos in eighteenth-century natural history. White is clearly a chorographer of Selborne and its environs and as such belongs to the tradition of local-oriented study. Indeed, White's biographers' emphasis on his originality in the history of literary popularization of natural history has overshadowed his identity as a clerical chorographer. His personal development, however, clearly reveals this role. As an adolescent, White was known as an acute observer of nature, but it was after he acquired Phillip Miller's *Gardening Dictionary*, and visited in 1735 his future brother-in-law Thomas Barker, that he began the kind of record which led to his chief work. In fact, his early contact with Barker was the main source of White's subsequent study.

Certainly the most conscientious observer in the eighteenth century – his journal spanned sixty years – Thomas Barker's study of the weather had an interesting history. His grandfather Augustin had been a patron of William Whiston, whose daughter had married Samuel Barker, Thomas's father. Lyndon Hall in Rutland, where the Barkers had owned land for several generations, became Whiston's second home; it was even mentioned in Whiston's article about two mock-suns he and Samuel Barker saw in October 1721.[63] Thomas recorded his first weather observations when eleven and continued a journal for the rest of his life (FIGURE 17). The journal began as a record of the seasons and unusual weather, the migration of birds, the leafing and flowering of trees, but it gradually branched into "The Memoranda Book," with information on crops, harvests, seeding-times, and so on; and the "Meteorological Journal," with an almost uninterrupted record of pressure, temperature, rainfall, and weather in the period from 1733 to 1795.[64] When in 1735, Gilbert White (15 years of age) visited Barker during the Easter holidays, they jointly observed weather and birds.[65] Gilbert's weather record dates from their later meeting on the occasion of Barker's marriage to Gilbert's sister Anne in 1751.

In 1759, Barker was appointed Deputy Lieutenant of Rutland and a local hospital foundation, but as the meetings at the foundation took place only twice a year, they did not affect his meteorological journal nor his writings on astronomy and meteorology. [66]

Journals similar to Barker's, in format and in purpose, were kept by others. Robert Marsham from Straton in Norfolk sent a summary of his naturalist's journal (1736–88) to the Royal Society.[67] Later, they were also kept by Gilbert White's brothers, Henry White of Fyfield Rectory and John White, Vicar of Blackburn in Lancashire.[68] In 1751, presumably taking the advice of Barker or his neighbor Stephen Hales, Rector of Farringdon, Gilbert White began his diary. Initially it consisted, like Borlase's and Barker's early records, of notes on daily events in the garden, and a chronicle of the weather. These were later accompanied by the notices of experiments and natural history observations, visits to friends, and family events. In 1765, White changed the name of the *Garden Calendar* to *Calendar of Flora and the Garden* and eventually, in 1767, to the *Naturalist's Calendar*.

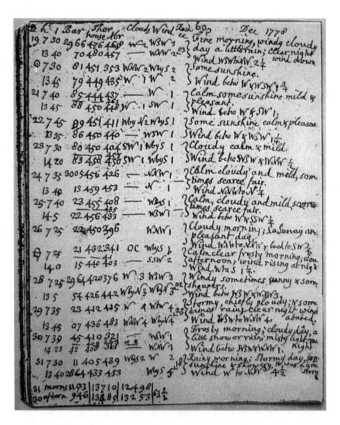

FIGURE 17   The weather diary of Thomas Barker.

White's major work, the *Natural History of Selborne* revolved around the parochial vegetation, birds, insects, customs, and weather, presented in personal letters to Pennant and Barrington.[69] The book was not, however, solely a transcription of White's calendars, diaries and notebooks; White perceived it as new and self-sufficient in respect to both content and method. In so creating a specimen of what many perceived as an "outdoor" sub-genre of the natural historic literature, White made a distinction between the genre of diarist non-fictional prose and a documentary-based quasi-fictional form by way of organizing his first-hand record around the axis of what he now termed parochial history. In the introduction to the first edition, he said:

> The Author of the following letters takes the liberty, with all proper deference, of laying before the public his idea of *parochial history*, which he thinks, ought to consist of natural productions and occurrences as well as antiquities. He is also of opinion that if stationary men would pay some attention to the districts on which they reside, and would publish their thoughts respecting the objects that surround them, from such materials might be drawn most complete county histories.
>
> Men that undertake only one district are much more likely to advance natural knowledge than those that grasp at more than they can possibly be acquainted with: every kingdom, every province, should have its own *monographer*.[70]
>
> [My undertaking] might induce more able naturalists to write the history of various districts, and might in time occasion the production of a work so much to be wished for, a full and compleat natural history of these kingdoms.[71]

Seeking to justify his meteorological pursuits within this quintessentially chorographic enterprise, White wrote that "the weather of a district is undoubtedly part of its *natural history*."[72] White's entries on the weather in his *Journal* and *Selborne* reflect this localist orientation by considering the weather within the immediate surroundings and by keeping track of both the unusual and regular changes in the atmosphere. He consulted the barometer and thermometer and thus – like Borlase, Barker, and Cole – combined measurement and the narrative meteoric tradition. An anecdote from White's diary, relating to the extraordinary frost of 1784, provides a good example of how quantification and narratives could thrive and reinforce each other. Between December 7 and December 9, about twelve inches of snow fell in Selborne, and during the day on December 9, the temperature fell so rapidly, White thought it would be "curious" to measure its progress during the night. To make the event more exciting, he hung two thermometers instead of one in the backyard. By ten o'clock, they fell to 21°F, at eleven o'clock to 4°F. The next morning, Dolland's thermometer read a half degree below zero, while the quicksilver in Martin's, "which was absurdly graduated only to four degrees above zero, sunk quite into the brass guard of the ball; *so when the weather became most interesting this was useless*."[73]

The weather was most interesting, according to White, when it was extreme. White's practice of *measuring* exciting weather and quantitative reports on uncommon high and low temperatures suggests that the meteoric tradition could even inspire systematic record. For the most part, however, White's letters describe the remarkable snowfall in London and the freezing of the Thames, the severe heat of the summers of 1781 and 1783, and the mischievous sulfur-smelling thunderstorm in June 1784. Letter LXV draws on the repertoire of the meteoric genre when it speaks of the events of the summer of 1783 – the summer of the eruption of Skaptar-Jokull in Iceland – as "an amazing and portentous one, and full of horrible phenomena; for besides the alarming meteors and tremendous thunder-storms that affrighted and distresssed the different counties of this king-dom, the peculiar haze, or smoky fog, . . . was a most extraordinary appearance, unlike anything known in the memory of man."[74]

We can add, in the memory of the inhabitants of Selborne, that if the numin-ous reading of odd phenomena was by this time beyond the pale for scientists, it was certainly as frequent as ever in the parochial common voice and its conven-tional belief in God's providence. For White, therefore, as for his other contem-poraries, inclusion of such descriptions was not driven by a mere curiosity, but by what they heard from their parishioners and what events they considered to be important in the local social and physical environment. Indeed, the Royal Society in the decades from 1750 to 1790 was flooded with accounts – stories, literally – of fiery meteors, halos, smoking cliffs, unusual agitations of the sea, remarkable darknesses of the sky, and the "agreeably surprizing" flames dancing on the ears of the horse that a mathematics teacher rode near Wakefield in Yorkshire.[75]

"Stationary intelligencers"

Clerical information-gatherers equipped with antiquarian skills mastered the genre in a delightfully rich manner. From Oxford, John Swinton wrote to Thomas Birch about a meteor that he claimed he had never seen before in his life: a dark semi-circular column of smoke extending itself "in a wonderful manner" over the evening sky which, during that time, "gave no disagreeable appearance."[76] (FIGURE 18). The antiquarian Hayman Rooke wrote about the hurricane in Thorsby Park near Mansfield Woodhouse in Nottinghamshire, which "com-menced its depredations on the elegant mansion, where it threw a sheet of lead on the sky-light in the dome, which threatened the destruction of the beautiful octagon salon."[77] Rooke kept an instrumental record and, importantly, realized that his interest in remarkable weather would not be approved of by meteoro-logical science. Yet he still believed that a tidy record would prove both a "source of useful information" and "a pleasant amusement," remarking that he did not "pretend to offer this as a Philosophical Register: such an Undertaking would require a stationary Residence, which I could not conveniently submit to."[78]

The meteorological engagements of the eighteenth-century could not be treated apart from the "stationary residence", but a "stationary residence" in-volved more than a residence in a place. In the provinces, such residence entailed an active participation in local life and sometimes an enquiry into its nat-ural setting. Contrary to W. J. Keith's suggestion that the interest in detailed examination of a limited geographic area was "a new idea," the sequence here appears unbroken between Plot, Leigh, Borlase, Barker, and White, the line along which one repeatedly discovers strong chorographic ethos in its validation of regional identity through a display of its historic peculiarities and natural curiosities. The execution of White's parochial history – as Plot had announced in his *Oxfordshire* a century earlier – depended on an eagerness of the provincial men of letters to engage in an exercise with high moral, social, and patriotic qualities. In this "tradition of the parish scholar" provincial naturalists made themselves and their subject "worthy of respect and greatly raised its standing in the eyes of the local public." These pursuits were hard to separate from the parochial role of the clergymen who – unlike the indoor classicist or experimenter – fulfilled their communal duty as they inspected fields, mines, shores or lightning victims, "affording ample opportunity for frequent intercoures with his parishioners. In this way their reciprocal acquaintance is cultivated, and the clergyman at last becomes an adviser and friend, as well as spiritual teacher."[79]

In view of these social roles, we may suggest that if Hanoverian clerics contributed to the program of large-scale weather research, they did it accidentally. If they approved of such a program, they were rarely motivated by the results it promised. If to us they appear to have participated in Baconian projects,

Philos.Trans. Vol. LII. TAB. II. p. 94

FIGURE 18   Representation of the optical phenomenon of anti-sun. John Swinton, "An Account of an Anthelion," *Philosophical Transactions* 51 (1761–2): 94.

it is because these programs had much in common with chorographic methods of inquiry. The chorographic natural study closely resembled the empiricism of new science, but the two had different origins. Where the empirical philosophy of nature required data-collection as a means toward a *cognitive* end, chorographers directed these means toward a *cultural* end. By means of a munificent subscription to their monographs, chorographers like Thoresby, Leigh, Morton, and Borlase – whose work considerably defined eighteenth-century meteorology – made it possible for themselves *and* for country gentlemen to become participants in a public sphere defined by the consumption of philosophical *belles-lettres*. These interests of the landed gentry and of provincial authors reinforced each other: the gentry paid for the display of regional society, authors obtained security and academic renown.

The vitality of regional scientific consciousness, therefore, was not simply political nostalgia of the landed society.[80] Alan Everitt has pointed out that even with expansion of population, occupational migration, and improvement of transport, eighteenth-century England did not experience any substantial loss of local loyalties or regional traditions. On the other hand, the eighteenth-century villagers were remarkably resistant to urban cultural influences. They preserved distinctly rustic patterns of settlement, sociability, consumption, and expression that persisted throughout the eighteenth-century. These patterns were maintained by direct social interaction, material and print culture and were highly suggestive of the abrupt topographical, environmental, and intellectual divide between the urban and rural worlds.[81] In fact, economic and demographic factors as well as the development of dynastic connections in provincial towns and villages "tended to anchor the mind of the local community [between the reign of Charles I and Queen Victoria] more firmly in its region."[82] A stable governing system by local gentry existed throughout the eighteenth century. [83] Relatively independent from central government, these elites maintained a paternalist social organization. The clergy in particular undertook a number of tasks of local administration and so offered a kind of "proxy paternalism" by which the priorities of landed society continued unchallenged. Administratively at the nexus of the local and the national, the provincial clergy held a unique status among their parishioners and their metropolitan brethren. This status would prove a critical precondition for their subsequent pursuit of natural knowledge.[84]

But if it managed to preserve its identity, provincial society was also under stress. Dror Wahrman's recent analysis shows a significant cultural-political divide emerging *within* the Hanoverian gentry between the "communal-provincial culture" and the alternative "national society." The "middling sorts" had a choice to either emulate the newly available London-centered culture, or to assert their distinct local culture inimical to the intrusion of metropolitan influences.[85] The dilemma of choosing between a national and a local culture was most forced upon the holders of official functions in the local communities: men of middling status, constables, church wardens, and Justices of the Peace. Norma Landau's study of eighteenth-century Justices of the Peace exemplifies this dilemma by identifying two groups of Justices: patriarchal and patrician paternalists. The former draw authority from the local community whereas the latter distance themselves from it and embrace norms of national culture.[86]

Just as these Justices found themselves on the two sides of the cultural divide, so did the provincial Augustan clergy. A group of Oxbridge-educated clerics and their correspondents found themselves exposed to both the parochial culture and to (inter)national science, philosophy, and *belles-lettres*. Due to the nature of their profession, the clergy was able to become a part of the national republic of letters by virtue of their privileged access to information of *local* origin – from church chronicles to parishioners' testimonies about extraordinary storms. Clergymen thus had an opportunity to integrate their occupational isolation and clerical duty into the scholarly cosmopolitanism of national society.

It is crucial to recognize that the privileges and disadvantages of provincial life are as equally conducive to the chorographic study of nature. With an emphasis on social accreditation and the geographical expansion of fact-gathering, eighteenth-century natural history had its paragon in the provincial clergyman. His isolated parish and his two-pronged authority – one among his lessers, the other among his fellow-clergymen – ensured that his scientific correspondence could be both trusted and circulated. In this way, the study of regional nature became an authentic form of participation in national society without becoming a distancing from – let alone a repudiation of – the clergy's parochialism.[87] Chorography and local natural histories emerged as genres and practices by which clergy entered the public sphere as authentic provincial scholars. During his tour of Scotland, Thomas Pennant observed "[m]any of the Parishes in North Britain are of such extent as to supply ample materials for history of each alone; so it is to be hoped some parochial *Geniuses* will arise and favour the Publick with what is much wanted, Local Histories."[88] The eighteenth-century natural history was sustained by just such individuals.

Although the scope of this book does not warrant any firm conclusion with respect to the politics of provincial natural history, it should be noted that by the time of George II's reign, the anti-Whig and anti-court rhetoric of political and social corruption had acquired conspicuous publicity.[89] The georgic writings, to which we turn in the next chapter, praised the virtues of rural labor and the agrarian retreat from city/court corruption. Interest in agricultural improvement and gardening, and education on the "physique" of the soil often became an issue of political allegiance with principles of dynastic and ecclesiastical legitimacy that were still associated with the Stuart cause. The rural environment was an obvious subject for sentimentalism of sorts and writers often associated its pathos to a "Rousseauan nostalgia for the simplicity and sturdiness of agricultural life, which embodie[d] in pristine form the essence and inner virtue of the community, uncontaminated by urban luxury and corruption."[90] Provincials looked to the pursuit of natural history as a means to uphold gentrified nostalgia by equating it with "domestic retreat from economic progress."[91]

If any of these impulses influenced scholarship, they did so through chorography. Chorographers' celebration of the countryside translated the political and cultural iconography of the landed society into a study of its land. Indeed, some local historians and naturalists – dependent on local patronage and bent on preserving the declining political significance of the land – exhibited a pro-country complexion, often representing the interests of Lord Bolingbroke's patriots.[92] Such was the case, for instance, of the Tory writer Sir Robert Atkyns, the historian of Gloucestershire, and of the Tory scholar John Hutchins, Rector of Wareham, Dorset. William Borlase, the Cornish chorographer and close friend of Hutchins, sympathized with the patriotic Whigs while openly criticizing Robert Walpole's ministry.[93] Whether a detailed prosopography would elicit more in the way of such political underpinnings remains to be seen.

But what can be inferred at this point, is that it was thus not only a fusion of consumption, politeness, and the urbane culture of reason, that prompted the rustics to polish their intellectual instincts. Their work continued a tradition of geographical investigation that could be traced back to Elizabethan times and they were engaged in an enterprise for which they had no need of an exclusively external warrant.[94] In this regard, *affirmation* more than *negation* of local heritage, *adaptation* rather than uncritical *acceptance* of the urban mores, and *continuity* more than *novelty*, should more satisfactorily explain the burgeoning of provincial natural knowledge during the eighteenth century.

6

Rustic seasons

Many people have observed, that when ants wander carelessly from the seat of the republic, in the spring of the year, a drought almost invariably ensues; but when they daub and plaister the sides of their inhabitation, and confine themselves nearer home, a very dripping, wet summer is known to follow. Swallows flying low, occasioned by the weight of the atmosphere pressing down their prey, denote speedy rain. In a drizzly morning, when the whole village is in doubt, whether it will be a thorough wet day, or clear up before noon, the sheep will often tell them.

Nathaniel Kent, *Hints to Gentlemen of Landed Property.*[1]

The meteoric tradition launched the parochial environment into the national sphere of learning. It also provided experiential evidence for Aristotle's theoretical treatment of meteors because it observed rarely occurring halos, fireballs, and earthquakes, not continuous states of weather. Parochial chronicles of meteorological rarities reflected the widely held classical axiom that meteorology explained *meteora*, not the weather; events, not processes. This axiom was integral to the theory which by the 1750s (and often beyond) enjoyed a consensus and defined meteors as the result of opposing forces and combinations of mineral exhalations. During the second half of the eighteenth-century, however, theoretical and cultural changes made a considerable impact on this view. By the end of the century these changes would begin to challenge the foundations of both the meteoric tradition and theoretical meteorology.

The present chapter will begin to explore the trends through which meteorology became a science of *weather*, not individual meteors. Grounded in the idea of providential self-regulation, this new science concentrated on recurring, predictable, and long-lasting changes in the atmosphere. However, rather than discovering the causes of these changes, its champions searched for those rules, correlations, and patterns underlying the *evolution* rather than the *state* of weather. Through discovering the weather in flux naturalists diminished the significance of individual, short-lived, meteoric events and undermined both the traditional theory of exhalations and its empirical expression in meteoric reportage. As will become clearer in what follows, the emphasis changed from the experimental philosophy of air to the rationalization of prognostic signs, from theoretical knowledge to prediction, from causal explanation to forecasting rules, and, more generally, from the science of meteors to the science of weather and seasons. The following addresses the domains within which this redefinition occurred and the role of natural theology in reinforcing both the significance of traditional prognostics of weather and the emphasis on the regularity of the seasons. In the process of legitimizing these concerns, naturalists made claims about the aim, method, and social relevance of meteorology.

An impasse in theory

Early in the eighteenth century, naturalists expressed recurring doubts as to the possibility of any deterministic knowledge of the sublunary sphere. Aware of the Newtonian syntheses in astronomy, some hesitated to say whether a meteorological Newton was even possible. The period's theoretical meteorology was thus both a repetitive exposition of the exhalation theory and the proverbial lament over its hopelessness. The Achille's heel of the exhalation theory was meteors: uncertainty about names, shapes, their duration, extent, speed, and the causes of meteors made them "insubstantial, shifting, and literally groundless," with "a different ontological status from natural objects" such as rocks, stones, and trees.[2] Deterministic explanations or reliable predictions looked chimerical given meteors' deceptive and sudden appearances, the confusion surrounding their description, and the failure to ascertain patterns of their return. The reasons for this were partly theological; early in the century, some contended that "no Man can Find out the Work that God maketh from the beginning to the End; nor trace the Order of remote Causes, which lie involved in endless combinations, to their first Origin, in the immense Regions of Air and Water."[3] The winds wandered through the "Wilderness of Instability," and the wet and dry weather changed "alternately at every turn of the Wind, without any prospect of settlement."[4] More practically, an author of instrument manuals warned that meteorological causes were "very difficult for Human Wit to Fathom,"[5] a belief shared by the Rev. Edward Saul for whom no certain judgment could be made of a "thing so unstable as the succeeding State of the Weather; because the sudden opposition of winds, confusions in the Streams of Air and Vapours driving above are usually attended with Changes equally sudden and irregular."[6]

Dr Johnson rendered his judgment on the issue alluding to the madness of an Egyptian astronomer claiming the power of being able to regulate the seasons. The astronomer eventually gives up the power, realizing that weather-government is the hardest gift one could receive from the gods because such power exceeds the knowledge required for its use: "I have found it impossible," says the astronomer, "to make a disposition [of the Earth's axis] by which the world may be advantaged; what one region gains, another loses by any imaginable alteration."[7]

Neither did measurements offer hope. In 1709, William Derham thought fourteen years of observations not long enough to explain the origins of a single frost of 1708: "not as yet having Observations enough to clear and demonstrate my Hypothesis, I must beg leave to defere [sic] what I might have said."[8] Observations thwarted attempts to correlate earthquakes with the attending weather: in 1742, Jurin read an account to the Royal Society sent by an Italian ecclesiastic about a series of earthquakes of which "some happen in cloudy, some in serene, some in still, and others in quite stormy Weather."[9] Late in the century, such doubts stretched the meaning of causation itself:

> Why do we not find the same regularity in meteors that we do in other phenomena of nature? Most of the atmospheric phenomena are so various and uncertain, that no person can pretend to reduce them to any kind of rule. Some of [their] causes may be supposed to be fermentations and other commotions within the bowells of the earth itself; but as all fermentation is a regular process, . . . why are not regular effects observed in proportion? It does not indeed appear, that the immense variety which occurs in meterological appearances can by any means be accounted for but by the interference of some causes in their own nature *irregular*; that is, capable of such endless variety, that no assignable space of time is sufficient to exhaust it.[10]

These problems were not restricted to this period and were insoluble by methodological measures alone. In fact, the limits of meteorological knowledge could be seen as being determined by the nature of meteors themselves. Plot used antiquarian methods to make meteors collectable items of natural history. Furthermore, the inability of many observers to come to terms with what they saw translated into a language of ambiguities, metaphors and periphrastic detail of frustrating vagueness. Barlow explained that, when heated, the earth released vapors to "the highest pitch" which afterwards fell with great speed in the form of rain, hail, and storms. But the earthly effluvia of a more refined composition were left behind to "sport themselves in *Meteors* of all sorts in that upper region, which by the uneasy crowding together of their Heterogenous Matter into diverse situations, condense, and fall in as different *Postures*; and taking Fire by their agitation shoot into *Stars* that vanish each into a Line of Smoak; which leaves a dusky Track of Opacuous Stuff in the Air of three to four Minutes of Continuance behind it, Undispersed."[11]

Meteors seemed to require such an idiom. Barlow's contemporaries spoke of "impressions made upon the Elements, exhibiting them in different Forms,"[12] or defined the meteor as "a mixed, moveable, crude, inconstant, imperfect body, or Semblance of a Body, formed out of the Matter of the common elements, altered a little, but not transformed."[13] Problems of terminology plagued classifications: Aristotle wrote about flatulent apparitions, Renaissance commentators about "affections," "impressions," or "passions" while a nineteenth-century author called them airy "exhibitions."[14] John Morton wrote that the ancients divided meteors into "Fiery Darts and Spears, of Flying Lances, Fire Dragons, Burning Beams, Pyramidal Pillars and the like, as if they really were of determinate form, yet in truth, the figures of them are very uncertain and irregular."[15] To avoid linguistic confines, Mr George Costard opted for illustrations (FIGURE 19).[16]

The recognition of these difficulties was in part a recognition that reports suffered from the incompetence of their authors and their lack of linguistic and observational discipline. Halley fought against the "fanciful reports" about the meteor of 1718, as did John Pringle in 1759, who wrote that the deception of senses, the short duration of the unusual meteors and the astonishment "among people unaccustomed to think of these subjects, will sufficiently account, not only for the variety, but the contradictions, in the several observations."[17] To solve these theoretical, methodological, and observational difficulties naturalists began to question classical heritage, thinking of "meteors" as not given to any form of known empirical scrutiny, certainly not to the kind which the planets enjoyed in astronomy. Some of the critics asked if new avenues of knowledge would be more successful. Central to these new developments was the understanding that the sublunary world – to which Aristotle assigned decay and probabilism – possessed the order, regularity, and self-regulation which had made other domains of nature amenable to scientific scrutiny. This understanding appeared within late-seventeenth century natural theology.

[523]

the 27th, 1744; I shall now inform you of another, seen by myself on *Sunday, July* the 14th, 1745.

As I was coming from my Living, just before I reach'd a Place called *Stanlake broad,* and a little before 8 o' Clock in the Evening, I was on a sudden surprised to see a long Stream of Fire, of a Colour resembling molten Glass, and of a Figure like that in the Margin, which shot down from *A* to *B*, in Length, I guess, about twenty Degrees, and seemed immediately to run up again from *B* to *A*; where it turned to a sort of Smoke, or rather to a fine lambent Flame like that of an *Aurora borealis*; which continued for some time in a sort of oblong Shape,

but afterwards by degrees, changed into

this, 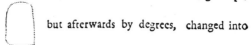 and at last into this

other Form, under which, parallel to the Horizon, it grew fainter and fainter, till it intirely vanish'd about nine o' Clock.

There was a fine gentle Breeze all this time; but I could not observe that it affected the *Phænomenon* so far as to make it change its Place, which

X x x 2   was

FIGURE 19 An explosion of a meteor. George Costard, "Concerning a Fiery Meteor seen in the Air on July 14, 1745," *Philosophical Transactions* 43 (1745): 522–4.

The weather in order

By emphasizing the average and uniform in meteorology, natural theology was a foundation for a *science* of weather, in contrast to the probabilism of mineral meteorology and its "gunpowder" analogies. Contrary to the beliefs of the irreducibility of airy "affections," and contrary to the notion of meteors as *intrinsically* local, the discourse of natural theology opened a possibility for treating meteors within a framework of order, making them liable to contemporary methodologies of science. The belief in nature as a *system* could in this sense "regulate scientific thinking both in the choice of problems and in the construction of acceptable solutions."[18] Nature as a system was at the heart of natural theology which, drawing upon the Mosaic narrative and providential wisdom, provided the basis for the works of Edmund Halley, John Arbuthnot, Willem s'Gravesande, Richard Kirwan, Jean DeLuc and even Darwin.[19] In the history of meteorology, this principle shaped the way in which practitioners defined the aim and content of the study when they presupposed that the weather was governed by a rational law, a self-sustained, patterned, and finite system of interconnected processes. In contrast to the meteoric tradition, weather science, as shaped by theology, normalized the extraordinary by stressing the predestined and benevolent. This normalization eliminated the heart of the troublesome orthodoxy associated with agnostic arguments and the ontological problems of *meteora*.

Only within this framework could the Georgian natural philosophers put forward an alternative to the understanding of meteors as "semblances of bodies." William Derham thought it a particular challenge to subject just *this* class of phenomena to the scrutiny of natural theology. Given the kinds of problems naturalists encountered while explaining meteors, Derham's decision was especially demanding: if meteoric activity could point to a benevolent "designer," so could anything else. In sermons preached in St. Mary le Bow Church in London, he endeavored to show that meteoric activity can be understood as governed by general providence; meteors were no less subject to the uniformity of nature than were tides or eclipses. His concern to stabilize meteoric activity was shared by Whiston, Halley, and Pointer – discussed in the context of the reaction to the ominous meaning of the Northern Lights – but the emphasis here is on the methodological, rather than the socio-cultural effects of his views.[20]

Derham argued that the "uses" of the atmosphere, for life, health, comfort, and pleasure could be shown to prove the coordinative virtue of God's *general* providence, which sustained the overall wholesomeness of the world, animate and inanimate. He therefore understood the manifestation of *special* providence – which had played the central role in polemical meteorology in the late-Stuart era – to be following the dictates of general providence. In storms, earthquakes, or lightning, God was not punishing the sins of humanity, but was doing it good by releasing noxious exhalations and merely delivering warnings to the impious.[21] Whatever the Reverend Divine could have said about God's anger manifested in convulsions of nature, Dr George Cheyne's volcanoes were a "bountiful Contrivance in Nature," serving to lessen the "tumultuous Steams [of fermenting vapors] which otherwise might make much Greater Havock than they do."[22] It required proper reflection and a larger point of reference to know that while "all weathers are at sometimes seasonable," they are good in themselves, and only *accidentally* evil: "We ought not to measure Things of a general Nature, by particular Rules."[23] This idea led Samuel Johnson's astronomer to admonish the poet Imlac – to whom he had bequeathed an imagined power to rule the weather – not to take pride in innovation and disordering of the seasons which are as equally fair to all nations as they could possibly be. The world could not have a better weather.[24] Nothing could have presented a greater challenge to the moralism of the homiletic meteorology of special providence. This more comprehensive benevolence undermined the socio-political opportunism of meteoric marvels by assimilating it within the framework of God's true intention.

Thus the gulf broadened between the naturalists' interest in individual and unique occurrences like floods, storms, and aurorae and the regular, cyclical, and wholesome processes embodied in the predictably unfolding seasons. This gulf also represented the parting of the ways between meteorological "enthusiasm" and a new science of the weather. We have traced the origins of the former approach – that of the "meteoric tradition" – to chorography, popular homiletics, and Bacon's classification of knowledge. The origins of the latter approach, however, lay in the conjuction of the view of weather as an orderly system of laws, and the tradition of weather prognostication. The contributors to the emerging theory of weather recognized that the agenda of a theologically sanctioned science of meteors had to do with identifying long-term changes, recurrent patterns, and, if possible, the methods of forecasting.

With classical origins, these concerns became the subject of intensified interest among meteorologists, physicians, and agricultural writers and practitioners. For many in these groups, knowledge preserved in the ephemera weather-rules – agricultural and gardening manuals, ephemerides, almanacs – conveyed a theologically acceptable idea of a reciprocity between the prognostic signs and the changes of weather that these signs stood for. In these sources, the experience and the vocabulary purportedly used by shepherds, sailors, farmers, and gardeners became a catalogued "folk" wisdom whose content persisted unchanged over the centuries. In contrast to meteorological theory, which had a comparable record of continuity, the tradition of prognostics never developed a *theoretical* account of how the rules worked.[25] Their reliability was in their use, not in a *reasoned* defence of their validity or in a causal connection between the signs and predictions. The *telos* of the prognostics lay in correlating the observed sign with the predicted outcome, which meant that such signs did not entail causality and neither a compiler nor reader would have conflated the certainty of the "a-causal" practical wisdom with the uncertainty of the causal philosophical conjectures.

Shepherds as experts

The earliest known Greek text dealing with the intricacies of weather prognostics is the *Phaenomena* by Aratus of Soli, written early in the third century BC. It is an astronomical poem with *Diosemiai*, a section on prognostic signs. *Diosemiai* and the anonymous *De Signis Tempestates* (attributed to Theophrastus) represent two chief sources of the most influential ancient work in the genre, the first book of Virgil's *Georgics* (37 BC). The *Georgics* and its later paraphrases do not pretend to an exhaustive treatment of all sublunary phenomena; they rather provide comprehensive guides on weather signs and agrometeorological knowledge.[26] The *Georgics* discussed the seasonal changes of weather as far as they affect the ordinary people rather than the philosopher; it lists the signs of stormy and fair *weather*, dry and wet, windy and calm, warm and cold, clear and foggy. These are foretold by plants, the behavior of ants, swallows, frogs, and ravens, or by the characteristics of the moon, sky, clouds, and stars. These signs were correlations, but correlations with an important consequence for their eighteenth-century revival: they testified to regularities, regularities to order, and order to reason. For the optimists, meteorological knowledge was merely about reaching the last step in the equation.

Together with William Derham's *Physico-Theology*, John Pointer's *Rational Account of the Weather* (1738) and Claridge's *The Shepherd of Banbury's Rules* (1748) validated meteorological thinking that disregarded the abnormal and unknowable, initiating in this way a tradition in which a knowledge of everyday weather could claim precedence over theory. Meteorology was hitherto often associated with the knowledge of rules (or laws) to foretell the weather.[27] In a conjectural history of science Richard Kirwan, an Irish meteorologist, equated meteorology with the formulation of the relationship between consecutive seasons and "species of weather" in a single locale. With Claridge, Pointer, Kirwan, and John Mills – mentioning the most influential – a meteorology of seasons emerged as the epistemically and socially accepted form of meteorological knowledge.[28]

The originality of this approach was its overruling concern with agriculture. The concern was two-tiered: the seasons were known to determine the periods of agricultural practice and, even more pertinent, the farmers and shepherds were supposed to possess a knowledge of seasons superior to that of the city-dwelling philosophers. Very much in the manner described in the previous chapter, an unmediated access to information overshadowed all other criteria which could be otherwise considered paramount to the acquisition of knowledge. But why agriculture and why at this point? If the knowledge of sowing and ploughing were of the most ancient provenance, why did this agricultural bias appear so late in meteorological reasoning? First, the eighteenth-century was not the first period of time that the agricultural meteorology of seasons attracted considerable interest; Virgil, Columella and early modern writers on husbandry had often praised the value of weather rules. However, such practical concerns – for they were perceived as such – could find no sympathy among philosophers who asked about meteors, not seasons. Second, the visibility of agricultural treatises during and after the mid-century was in considerable part due to the literary and political phenomenon which Anthony Low has called the Georgic Revolution – the new emphasis on the role of agricultural labor in the moral and material economy of English society – and which he considers to be the framework of late eighteenth-century agricultural change. [29]

In the literary culture of Hanoverian England, Virgil's *Georgics* provided the model for a number of didactic and descriptive poems, celebrating in various ways the cult of farming, gardening, and, importantly, the civic value of the improvement of agriculture and its significance for the national commerce and public welfare.[30] During the eighteenth century the value of Virgilian agricultural theory was acknowledged and recommended to English farmers in works such as John Mortimer's *The Whole Art of Husbandry* (1707), Richard Bradley's *Survey of the Ancient Husbandry and Gardening* (1725), and *The Science of Good Husbandry* (1727) and the Rev. Walter Harte's *Essays on Husbandry* (1764). Columella's *Rei Rusticae* was published in 1745, while in 1740, in the first edition of the new translation of the *Georgics*, John Martyn maintained that even "though the soil and climate are different from those of England; yet it has been found by experience, that most of his [Virgil's] rules may be put in practice, even here, to advantage."[31] Adam Dickson was of the opinion that "notwithstanding the difference in climate, the maxims of the Ancient Roman farmers" were like those of modern English farmers. Views like Martyn's and Dickson's together with the comparative studies of ancients they contained, served to put classical doctrines and rules "into the hands of English readers who were without Latinity."[32]

The two early works which espoused this comparison in the meteorological context were those of John Pointer and John Claridge. Both aimed at a larger audience, but without popularizing intentions. Both authors championed the traditional, even oral, knowledge of weather-rules and so sought to bestow more respectability on the discourse marginalized in the theory- and city-driven English enlightenment. Somewhat disingenuously, they employed the rhetoric of science for that purpose. Pointer claimed that his attempt was to *ground* traditional weather prognostication in the modern natural philosophy of meteors,[33] making his "rational account" an elaboration of the "philosophical reasons" of prognostic rules collected not only from observations, but from "the most Celebrated Philosophers, and the most Judicious Naturalists of this and former Ages."[34] The "rational account of *weather*" was in his view a *science* of rules by which to judge the weather, not a new theory of meteors. In this new framework, Pointer defended his eclectic borrowings from Derham, Halley, Seneca, Pliny, and Aristotle as grounds for the *practical* knowledge of weather changes required to legitimize the project in respect of the "Sick Man [who is always] inquisitive what Weather it will be," the "Country-man [who would like] to be assur'd of Good Weather to sow or reap his Corn and hay," and "the Travellers [who] depend upon the Weather."[35] Such practical concerns were as far a cry from theoretical issues as they were from the exotic narratives of the provincial clerics. But these concerns offered a possibility that meteorology perhaps had no claims to the theoretical grounding of the kind shared by rational mechanics or astronomy. This possibility thus seemed to open a new area of research: a phenomenology of weather-rules found among the ancient autodidacts and modern shepherds.

The option found a more specific expression in John Claridge's *Shepherd of Banbury Rules*. It appeared in 1670,[36] had three mid eighteenth-century editions and was considered a "classic" of weather wisdom in the nineteenth century. Colonel James Capper of Cardiff, credited with the discovery of rotating storms, observed in 1809 that "few meteorologists are unacquainted with the valuable practical hints on this subject [weather prognostics] by the honest Shepherd of Banbury."[37] Claridge's little book was a compendium of aphoristic weather-rules based on the appearance and color of the celestial phenomena and the behavior of plants and animals. The preface, a veritable manifesto of folk epistemology, elaborated on the public character of weather, outlining in detail its relevance for outdoor occupations. Claridge urged the reader to recognize the connection between these occupations and practical knowledge: "it is amazing how great a Progress [the shepherd] makes in [making observations] and to how a great Certainty at last he arrives by mere dint of comparing Signs and Events."[38] Certainty is the key issue here not least because of its obviously controversial association with contentious meteorological knowledge. In justifying it, Claridge ingeniously appeals to the authority of General Providence which gives the weather a nomological stability and provides a benevolent "Ballance [sic] of the Weather" in contrast to "a supernatural interposition of almighty power in extraodinary natural events."[39] It follows that a theological "normalization" of weather requires neither meteoric tradition nor Aristotelian jargon, but the knowledge of the slow and recurrent seasons. This is where it is possible to reinstate the shepherd's epistemic credentials because he or she is *in* the seasons and because it is the "*Manner* in which we perform, and not the *Character* that makes the Player, and in this sense What Man is not a Player? Here is an Instance of one who has for many Years studied his Part, and now communicates his Discoveries freely. In a Physician, in a Philosopher, in a Mathematician, this would be highly commendable, and why not in a Shepherd?"[40] (FIGURE 20).

As long as one's abilities were put to maximum use, the individual's social station had little effect on his epistemic credibility. Claridge's democratic theory of knowledge is here grafted on to the Georgian social hierarchy, validating the shepherd's expertise on the premise of his or her vocation and diligence. His business is "to observe what has a Reference to the Flock under his Care, who spends all his days and many of his Nights in the open Air, [and] is in a Manner obliged to take particular Notice of the Alterations of the Weather."[41] The knowledge so gleaned highlights the shortcomings of the "Pedantry in Desarts [sic] as well as Colleges" of those "half wise People" who too easily discount the prognostication of "a [lay]Man of a larger compass of Knowledge."[42] But to the shepherd, "the cawing of the Ravens, the chattering of Swallows, and a Cats washing her Face are not superstitious Signs, but natural tokens of a Change of Weather, and as such . . . they have been thought worthy of Notice by *Aristotle*, *Virgil*, *Pliny*, and all the wisest and gravest Writers of Antiquity."[43]

The claim transforms "popular" prognostics from "superstition" to true knowledge. Claridge does that by pulling together the authority of the ancients and that of weather-beaten shepherds, reinstating the value of a science proved in time-hallowed praxis, against that of a science based on hypothetical reasoning, experimental techniques, or instrumentation. Indeed, the forecasting ability of the barometer was a point of contention at least from around 1700, when some astrologers put forward objections to the barometer's presumed infallibility. Even if the barometer *could* predict immediate changes, an anonymous writer argued, it could be of no help to someone concerned with long-term atmospheric conditions. What gain could there be from a knowledge of what weather will be tomorrow? "It is not the Prescience of Frosty or Foul; of a Shining, or Showry Day but a forward or backward Spring, Summer etc., that proves the advantage to the Husbandman or Natural Philosopher."[44] Pointer

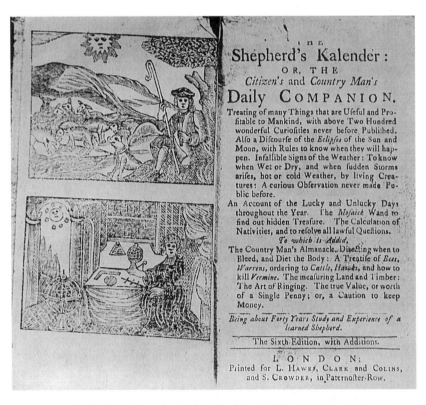

FIGURE 20   The illustration accompanying the title page of an early eighteenth-century calendar contrasts the stereotypes of a shepherd and a city astronomer. *The Shepherd's Kalender* [1712]: *or the Citizen's and Country Man's Daily Companion.*

also saw instruments as means by which "men of Ordinary Capacities" may forecast the weather,[45] but pointed out their secondary role in the prognostics of *long-term* changes. Others questioned the accuracy of instruments and, given the existing weather-rules, even saw them as redundant. At the height of the barometer-craze, the chorographer Charles Leigh recorded the skill of the inhabitants of rural Lancashire to "make as early and certain a Prognostick of the Changes of Weather, as the Modern Virtuosi can do by their Mercurial *Tubes.*"[46]

Claridge was especially vocal in this respect. He complained that the use of the thermometer and hygrometer were limited by the range and precision of their scales. The air had to become "hot to such or such a Degree before it is discerned" by the thermometer. The hygrometer operated only in the air whose changes in humidity were sufficiently pronounced to be indicated by the instrument. The barometer might indicate alterations in the weight of the atmospheric air, "but then those Alterations are the Cause" of barometric readings, which for that reason cannot be presumed to have anything to do with forecasting future changes. In short, the instruments were *a-fortiori* indicators of changes and, however useful they might be in other respects, they contributed very little to "prognosticating of a change of Weather at a Distance, and it is from the Experience of this, that they are so little esteemed, so lightly regarded by the common people."[47]

In minimalizing its relevance, the georgic authors disassociated the barometer from ordinary people and portrayed it as a piece of furniture locked up in London's philosophical parlors.[48] In contrast, weather-rules based on natural signs were advertised as having an *application* beyond philosophic *amusement*, as useful to all ranks of people, including convalescents, travellers, sportsmen in pursuit of their game, "as well as to the industrious Husbandman who constantly follows his Labour."[49] A split was thus created between the countryside-based producers of meteorological knowledge and the metropolitan pedants of curiosity. The split is explicit in the work of several authors. Thomas Short, a Sheffield physician with an interest in study of mineral waters, asked

> But which of all these modern Discoveries can supply us with Hints what the ensuing season will be? Whether the next Winter will be mild or severe, short or long? Whether the Spring will be late or early, Summer hot or cold, rainy or droughty, the Product of the Earth next Season plentiful or scarce? Yet all these, and much more, have been known and told without them. Wherefore it would be more advisable, prudent, and profitable for the honest Countryman to study and be better acquainted with his Book of Nature, to which he has daily access; and if he closely observes it, he may oftener depend on it than his Book of Art; whilst the Citizen, who wants his Opportunities, may attend to his Barometer and Thermometer, which often deceives him. But the former, if he understands his Book of Nature rightly, may often learn Intimation of extraordinary Sets of Weather and Seasons, sometimes before they come, of which the latter, or Citizen's Tackle, cannot inform him.[50]

Others said that it could not be expected that natural philosophers – who principally resided in cities and adopted habits of life which confined them indoors – were able to watch the sudden changes which take place in the atmosphere. Mariners, farmers, gardeners, and shepherds were those to be trusted in solving meteorological problems.[51] Short even believed that the savants "deserve[d] to be ridiculed" for setting aside observations of rural countrymen "as not flowing from any natural connection that he knows of; and because he knows them not, they cannot be." Instead of mercury, the "honest countrymen" may well be justified in trusting the predictive properties of organic fluids of animals and vegetables, because those were, as Pointer put it, "a Contexture of Hygrometers, Barometers, and Thermometers."[52] The farmer had surer ways than "philosophy" to presage changes in weather: the sap of trees, leeches, sea-weed, and the numerous improvizations based on the hygroscopic qualities of woolen cord.[53] Through the image of the rustic environment as the source of the knowledge, these authors voiced their discontent with the generally accepted methods of science production. They were arguing against the "mundane means" involved in the making of science when they rejected the class and political manners essential to the "modish" urbanity and the philosophy nurtured in the laboratories, elite societies, lecture halls, study rooms, or boudoirs. In the view of the apologists of rustic ways, there was something socially, morally and even literally suffocating about these venues of knowledge.[54]

If the Virgilian meteorologists steered away from philosophical reasoning and its material culture, they also mistrusted astrology, arguing that natural signifiers were not of astral origin. Claridge wrote that his use of celestial "lumin-aries" in the discourse of practical weather knowledge had an entirely rational basis, devoid of numinal and political intentions. In the second edition of his *Account* (1739), Pointer exposed the astrologers' failure to establish even "prob-able Conjectures" in relation to the weather, because "if they cou'd, we had long before this time had certain and infallible Schemes of the weather deliver'd down to us, from the experience of some of the best Astronomers." Apart from granting the influence of the sun and moon on the weather by reason of their proximity, Pointer saw no other causes of weather changes but those in "our Sublunary World," and no other alternative than to "deduce Prognostications from the Animals and Vegetables of this our Terrene Globe." Benjamin Stillingfleet – discussing the mistaken link between stars and farming – wrote that the Egyptians, Greeks, and Romans were misled into it by the fair Mediterranean weather. The notion that celestial phenomena influenced the weather "would never have begun in such uncertain climates as are found in England."[55] Only natural signs, having natural causes, can be presumed to have natural effects. It was up to the com-mon people to decipher them and put them to practical use.

Henceforth, knowledge of the weather would be presented as established on a new authority – that of outdoor lay people, uniquely situated to discover the regularities hidden in the changes of weather. But the shepherds and farmers were able to do so not simply because they lacked education, and were thereby free from the snares of hypothetical reasoning. Nor did they become meteorological masters because they were exposed to the elements and were thus *occupationally* prone to acquire such knowledge. Rather, it was in the nature of the divinely sustained order to reveal the patterns of weather to humankind. In Short's view, the justification for georgic meteorology was the holistic connection which it maintained between the organic and inorganic domains, in which Providence would not leave inhabitants of the world without the understanding of "Presages, Marks, and Signs of Weather, especially when a total Incapacity to prepare for some extraordinary Changes might be of the worst Consequence."[56] Weather knowledge was a "biological" adaptation to changes in the environment with prognostic rules serving as mental tools of survival: humankind thrives because the rules work. [57] On the basis of the connection which God (in Virgil's *Georgics*, Jove) had established between the weather and the animate world (birds, animals, plants) the latter is *naturally* predestined to sense impending changes of the former.[58] The signs inferred from animals are above mere probability because they disclose the objective interdependence established by God. In Virgil's view, not only had God invented inclement seasons to spur the arts, but had also made sure that these arts were reliable and corresponded to human needs. The Virgilian farmer thus predicts weather because Jove himself has *revealed* to him the true correspondence between signs and signified events, and not because the farmer has established a link on an independent humanly manufactured basis – by, say, a statistical manipulation of observations or by means of an "empirical" method of discovery.

The importance of this argument is clear in retrospect. In nineteenth-century natural theology, it was held that the processes of the earth and atmosphere had been designed for human benefit and were open to understanding. Charles Bell, author of one of *The Bridgewater Treatise*s wrote that by a series of terraquaeous revolutions, God had prepared the earth for the reception of "man" and that, as a consequence, "the strictest relation is established between his intellectual capacities and the material world."[59] Within this scheme, foretelling weather meant extrapolating man's place in nature. Everyone could be a meteorologist – convalescents especially[60] – because everyone was placed in a state of dependence upon the elements. Observing their changes was considered a necessary part of the everyday life of mariners, shepherds, and farmers. Indeed, the obligation of constant attention to the alteration in the atmosphere "has endowed *the most illiterate of the species* with a certain degree of prescience of some of its most capricious alterations."[61] Such claims presented meteorology as a matter of instinct, routine, and necessity while entirely dispensing with theory. The only other authority which they sanctioned was that of the canonical ancients.

A return to the ancients

In arguing for the epistemological priority of millennia of outdoor experience, the exponents of the Virgilian study of weather linked modern farmers with ancient lore. The return to the ancients was based on the growing disrepute of astrology, meteorological theory and instrumentation.[62] Without astrology, without quantitative methods, and without alternatives, the ancients spoke with great force: "it would be wantoness to throw down a work of ages without submitting a better substitute," argued William Marshall.[63]

In agriculture, the value of ancient knowledge was repeatedly praised until very late in the eighteenth century. The comparative study of classical maxims which John Martyn and other "classicists" recommended as the basis of agricultural science differed from the methodical precepts of the natural history of weather in that it sanctioned practically oriented "classicist" meteorology. In this change of emphasis, the "new" agricultural meteorologists were allowed to forsake the principle of causation and focus on the "statistical" correlation between prognostic techniques and observed phenomena. Apart from attentive observation, the business of a meteorologist was now to *decipher* predictive signs and to establish a connection between them and the weather. Claridge pointed out that in this "Art it is always allow'd as a point of great Consequence, when several Masters therein agree as to the meaning of a Character, and it is from thence very justly presum'd that this character is rightly decypher'd."[64]

Several authors emphasized the necessity of using classical prognostic texts. They argued that if the same rule appeared in several independent sources, its prognostic value was greater than that of a rule appearing in only one or two sources. For example, whereas only Aratus included "chickens crowing more than usual" as a sign of rain, "ants bringing out eggs from nest" was noted not only by him, but also by Theophrastus, Virgil, Pliny, and Leonard Digges.[65] The second sign, consequently, had better "empirical" support. Embracing this logic of cumulative probability, agro-meteorological writers from the second part of the eighteenth century acknowledged the advantage of consulting traditional prognostics and approved the use of the weather-wisdom of classical works – for example, that of Aratus, Pliny, Theophrastus, Columella, and Virgil.[66]

The return to antiquity, however, was not only a result of the skeptical attitudes toward the "indoor" philosophy of meteors, nor merely a fill-in for astrometeorology. In fact, mid eighteenth-century authors interested in the state and utility of agrometeorological knowledge would have found the ancients' knowledge of weather as astute as their political acumen, because both bodies of knowledge were meant to bring about the well-being of civil subjects: a just distribution of labor derived from knowledge of seasons benefited the citizens as much as a just government. The analogy can be pursued in the idea, sanctioned in Virgil's *Georgics*, that if Augustan Rome should be admired for its prosperity and good government, it was because the moral basis of both was the Appenine land and the tilling of it by the Roman farmer. The presumed intention of the *Georgics* was to mirror the destiny of Rome in the virtuous farmer, whose work made him the epitome of the productive element of his society. In the poem, the Roman farmer comes to represent good health at the heart of Roman life, the source and justification for its imperial energies. Richard Feingold calls this line of reasoning "bucolic" because it expressed a "judgment of the moral quality of civil society by explicit and implicit reference to rural experience."[67]

As bucolic writers established the farmer's centrality in the acquisition of wealth in a society, they also favoured the farmer's common-sense over the "indoor" process of reasoning of the city *philosophe*. A pointed expression of such a conviction occurs in the pages of Claridge's manual, where, in the chapter on the signs of frosty winters, the author sees the rustic's autopsia as nothing short of an epiphany. The author explained that the knowledge of nature is the light acquired by the study in which "Experience is a Kind of Revelation, that is to say, it is a sort of Knowledge that comes to us from without, and is infallible in itself."[68] Where Virgil envisioned agriculture as an art forged from practice, Claridge's shepherd explained *all* knowledge as sensual in origin: certainty followed from trials, rules from observations, knowledge from facts. Even though it might have sounded too obvious for Claridge's educated contemporaries, the origin of such a contention differed from both Bacon's and Locke's systems. Unlike in those, Shepherd's "empiricism" represented a move *towards* the canonical, not a break from it.

Two premises underwrote this claim. The most significant was physico-theological: nature was taken to exhibit a regenerating "system" of events unfolding under the laws of a creator. The consequence of this constancy was that, with qualifications, eighteenth-century authors, like Martyn and Dickson, assumed that the Roman tokens of fair weather, earthquakes, storms, and the coming seasons, were also the tokens applicable to the weather in the British Isles. No lapse of time could invalidate them; to the contrary, with the passage of time they would receive further confirmation. The minor premise stressed the public character of the weather. If Greek and Roman political wisdom emerged in public places – on the Agora, the Forum, and the Senate, – weather-wisdom was shaped in even more open realms such as the fields, pastures, and seas. Public accessibility of the subject and participation in its study entailed a communal knowledge whose con*sensual* quality approached the truth more closely than the speculations of individuals. That the weather was repeatedly labeled as universally "accessible" may seem obvious to note, but it was partly on this premise that the ancient and lay naturalists deserved the attention of their modern, "scientific" peers.

The career and writings of John Mills (d.1784) vividly exemplifies these concerns. Reclaiming the contemporaneity of the ancients, Mills not only highlighted pitfalls of modern natural philosophy, but professed a method that went beyond scientific truths to reach national interests. During the 1760s, Mills was a member of the French Society of Agriculture, the Economical Society of Berne and the Pallatine Academy. From 1763 to 1765, he published five volumes of *A New and Complete System of Practical Husbandry*, designed to aid the farmer and contribute to "the ornament and improvement of the country gentleman's estate."[69] He also translated Duhamel du Monceau's treatise on husbandry (1762), wrote on bee-management (1766), and in 1767 completed the unfinished "Memoirs of the Court of Augustus," begun by Thomas Blackwell, the Greek classicist from Aberdeen. In *An Essay on the Weather* Mills wrote that despite two centuries of philosophical work, no account of the weather had been proposed that could in certainty match those of the earliest authors, "though it may be presumed that the operations of nature are set in a much clearer light to us, than they could be to them. . . . Perhaps the philosophers have not had opportunities from their own observation of laying down any certain rules of the changes of the weather, and either despised and neglected the remarks of illiterate country people.[70]

Mills shared the beliefs of Claridge and Short, but went on to develop a social rationale for a more comprehensive Virgilian science. He asserted that country people can judge the weather because, like the ancients, they were "long practiced in [the] business." They would do even better if they observed and recorded the atmospheric conditions in critical periods of the vegetative cycle, as they would thus reap a "crop of useful knowledge." Mills reiterated the metaphor of meteorological knowledge as a good crop in the maxim that "good crops depend on this observation, as the welfare of a country depends on good crops." This was nothing less than an attempt to redeem the public, economic, and even scientific relevance of farmers – "the much neglected class of mankind" – and to present rural *autopsia* and hands-on experience as the source of patriotism, improvement and commerce. The observations collected by a work of many individuals, he argued, would afford infallible rules to guide the farmer: "I cannot too strongly recommend it to the public spirited inhabitants of the British dominions in particular, as means by which the power and opulence of this happy state cannot fail to be considerably increased, and the felicity of individuals to be consequently confirmed."[71]

Central is Mills's intention to use the bucolic rationale to ally *emperia* with the *Imperia*: a suggestion that observations should serve the farmer's needs mirrors the more general resolve to apply the empirical method to a branch of economy, rather than, as the physician James Jurin did, to a philosophical history of the air. Within this context, Mills's urging for standardization resembles Pointer's "rationally" derived prognostics, an early attempt to validate the ancients by modern standards. In so conceiving an art of *profitable* observing, Mills praised the ancients, because they, in determining the onset of agricultural seasons, had arrived at better solutions than the moderns. He has an interesting example to illustrate this claim. The ancient peoples, he explains, observing that the weather of each season started at a known time of the year, attributed its quality to the influence of the stars which happened to be visible at that time. In succeeding ages, "monks and designing priests" ascribed this influence to the saints, but the moderns, rejecting both stellar and saintly powers, rejected with them the observations of the ancients, without thinking that the facts might have been established first, and the stars and saints only called into account for them. "The ancients acted more rationally than monks, in not fixing these changes to particular days, but only to a stated time of the year, as appears from Pliny and other writers on this subject."[72]

The ancients, agriculture, and application — these concerns increasingly organized meteorological practice in the latter part of the eighteenth century. In agricultural texts and didactic manuals, theoretical science lacked the urgency which was associated with the "empiric" knowledge of the rules of local prediction. The sheer amount of late eighteenth-century publications on weather-rules indicates that an alternative voice had been increasingly heard among various audiences.[73] In 1791 – the year that marks the end of the enthusiasm which *Philosophical Transactions* showed for strange weather – the eminent George Adams asserted that "there is no part of meteorology which interests mankind so much, as the prediction of the weather."[74] For George Gregory, procuring prognostics of the weather "is the great desideratum in the scale of useful knowledge."[75] James Capper claimed that meteorologists had "to look how far *Georgics* correspond with modern discoveries"[76] and that sudden changes in the weather "escape the notice of theoretical meteorologists, whilst at the same time they still afford very useful information to the practical observer."[77]

With all the emphasis on observation, many of the authors did not seem, nor feel a need to be, familiar with the requirements and meaning of a natural history of meteors – the project continued for the larger part of the eighteenth century in the pages of the *Philosophical Transactions* – nor did they seem aware of previous schemes for weather recording. When his *Minutes of Agriculture* appeared in 1783, William Marshall acknowledged, as if out of the blue, that it was not his intention "to enter into an analytic Investigation of the Weather," but that he could not "refrain from communicating such observations and Reflections as have occurred to me, and which, I flatter myself, will be considered as at least an Overture towards founding a foreknowledge of the weather on the durable basis of science."[78] The fact that Marshall thought meteorological "foreknowledge" lacked even its "overture," speaks about how much he and his contemporaries found it hard to imagine anything in the history of weather investigation deserving the title of meteorological theory.[79] Such a sentiment recurred in almost every meteorological work in the period but – in contrast to the skepticism pervading early eighteenth-century meteorology – it stemmed from a new methodological agenda. The latter involved the spatial network of quantitative histories of weather in which theoretical enquiry into the nature of meteors assumed only a secondary role.

Having looked at the origins of rural meteorology, it must be asked whether it had an impact on the development of the meteoric tradition. The two approaches were dissimilar but they also thrived side-by-side in the interests of men like Thomas Barker, William Borlase and Gilbert White. In this sense, the agrometeorology of the seasons did not directly challenge the genre of uncommon meteors, but it challenged its unspoken assumptions about the value – either cultural or epistemological – of uncommon meteorological situations. This challenge showed the way for more comprehensive criticisms against the meteoric tradition which are addressed in the next chapter. Drawing upon theological notions of order, regularity and self-regulation and the epistemological necessity of long-term exposure to the elements, rural meteorology deflected meteorologists' attention from the preternatural to natural, from esoteric to ordinary, and from rare to common. When a democratic knowledge of signs claimed precedence over the elitism of theory, experiment, and barometry, it was the moment when down-to-earth empiricism questioned the social and scientific rationale of Augustan curiosity and its elaborate culture of the extraordinary, wonderful, and unique. Here was the true challenge of the eighteenth-century meteoric tradition. "We do not bring meteorology forward as a pursuit adapted for the occupation of tedious leisure," wrote John Ruskin in 1839, "or the amusement of a careless hour. Neither do we advance it on the ground of its interest or beauty, [nor is it] to be learned among the gaseous exhalations of the deathfull laboratory. It is science of the pure air [that] involves questions of the highest practical importance, and the solution of which will be productive of most substantial benefit to those classes who can least comprehend the speculation from which these advantages are derived."[80]

7

Laboratory atmospheres

> I would give to Aristotle the electrical shock: I would carry Alexander to see the
> experiments upon the Warren at Woolwich . . . I would shew to Julius Ceasar,
> the invader of Britain, an English man of war; to Archimedes a fire engine, and
> a reflecting telescope.
>
> William Jones, *Physiological Disquisitions*[1]

The late eighteenth-century naturalists announced a new meteorology. Results
of chemical, electrical, and pneumatic research during the second half of the eighteenth century defined new thematic and practical priorities, while the discovery
of the extraterrestrial origins of meteorites narrowed the scope of traditional
subject matter. Quantification displaced the narratives of meteoric tradition,
averages were more relevant than extremes, and recurring phenomena more telling
than singularities. The sublunary region, the ontological crucible of meteors, was
remade into a fluid of predictable behavior. These changes were inaugurated by
a group of chemists, physicians, natural philosophers, and university professors.
Their expertise, methods, and reinterpretations challenged the Woodwardian
meteorology of terrestrial vapors and problematized Halley's and Whiston's
explanation of upper-atmospheric fiery meteors. In the relevant literature of the
period (particularly in encyclopedias and monographs), meteorology ceased to
be a science of meteors by becoming a constellation of physico-chemical enquiry
into the nature of atmospheric air and its planetary circulation. The unifying
ontology of the exhalation theory gave way to the heterogenous matrices of
electro-chemical research, on the one hand, and theories of the extraterrestrial
origins of fiery meteors on the other. As a result, the semantic fields of "meteors"
and "meteorology" underwent alterations that put an end to the classical understanding of the discipline's subject matter.

This chapter will examine the origins and the impact of the theoretical and
social underpinnings of these changes. It will be necessary to move away from
the dominion of clerical naturalists, their networks, and their provincial setting
and draw upon the writings of natural philosophers, contributors to encyclopedias and textbook writers. As we have seen, provincial meteorological enthusiasts
took some genteel pride in abstaining from speculation and remained faithful to
what they regarded as reportage based on non-controversial matters of fact. If
in some cases they ventured into theory, the fate of such exercises was left in the
hands of their philosophical mentors at the universities or at the quarters of the
Royal Society. Conventions which obliged the secretary of that body to give due

consideration to all contributions – however insignificant and implausible they might appear – were not inviolable and clerical theorizing often had to be turned down for publication. For these, and reasons of actual geographical residence, late eighteenth-century meteorology gradually became an urban-based pneumatic, chemical and electrical investigation of the aerial atmosphere, despite contemporary neo-Virgilian weather prediction and its idealization of rustic episteme.[2]

How did this turn come about? It has been observed that, as far as the Royal Society's editorial policy with regard to the *Philosophical Transactions* was concerned, the second half of the eighteenth century witnessed a slow disappearance of the event-narratives which had largely informed activities of the early Hanoverian members and had shaped the content of the journal during the presidencies of Sloane and Folkes. In 1751, the controversial botanist John Hill satirized the excesses of curiosity in his *Review of the Works of the Royal Society*. In 1752, the Council of the Royal Society issued a statement about its editorial policy which was to tighten the standards in the selection of the journal's contributions. The result was an increase in the number of articles on the mathematical sciences,[3] and a recognition of the necessity of hypothetical reasoning which had been all but banned since Sloane's announcement that he would have "so little regard for" hypotheses, because the "future accidents, and observations, will make them go off."[4] Direct reporting of individual experiences – event-narratives – which had defined the Royal Society's policy since its foundation, and the Society's promotion of Bacon's natural history program, had been gradually dropped as an institutional role after Joseph Banks's accession to the presidency in 1778. Although a prominent natural historian, Banks strongly endorsed the practical application of mathematical natural philosophy and promoted the research of such individuals as Henry Cavendish and William Herschel.[5]

David P. Miller and Simon Schaffer have recently shown that this shift in priorities from literary and experimental "curiosities" to a search for quantifiable laws reflected three parallel developments. First, the interest in natural philosophy as a public spectacle involved the Society in virtual police action over experimental philosophy of a sensational character. As a result, meetings in the second half of the eighteenth century entailed reading's of the (notoriously dry) papers that were used to "demarcate the 'serious' purveyors of natural philosophy from the pretenders."[6] Second, the Society of Antiquaries (founded 1754) provided an alternative forum for the activities of a number of dilettante antiquarians. The disappearance of antiquarian papers from the *Transactions* affected the way in which the "natural philosophic" constituency of the Royal Society looked upon meteoric reportage which, as shown in chapter 4, used antiquarian idiom as the main vehicle in its descriptive techniques. Finally, the new principle in the management of the Royal Society became in the last two decades of the century "a matter of managing the representation of [the Society's] constituencies."[7] There was a considerable increase from 1766 to 1820 in the representation of merchants, and army and navy officers among the members of the Society.[8]

Parallel to these developments, British natural philosophers had in the second half of the century achieved considerable institutional and theoretical consolidation in fields such as electrical, pneumatic, and chemical research. All of this was congruent with the emerging self-image of the Royal Society philosophical constituency. As a result a major redefinition of traditional meteorology occurred, which may be summarized as follows. In classical meteorology, the subject matter was defined with reference to the place in which meteors occurred, for example, in the region between the surface of the earth and the lower limits of the celestial region. Anything suspended (*meteoros*) in this region was subject to a science of suspended things (*meteorologia*). In the late eighteenth-century, however, meteorology was defined with reference to air and the atmosphere. These new "atmospheric" phenomena were meteorological because they appeared in a physico-chemical body of matter, not because they were suspended in the sublunary region. This perception questioned the legitimacy of both the quasi-Aristotelian meteorology and the qualitative meteoric tradition.

For example, in the 1784 edition of Chambers *Cyclopædia*, the article on "atmosphere" featured a detailed account, three double-column folio pages long, on research into the determination of the height of the atmosphere. The topics included calculation of the vertical distribution of air pressure, barometry, and studies of humidity and condensation. "Meteorology," on the other hand, received a single-sentence treatment in which it was described as "the doctrine of meteors; explaining their origin, formation, kinds, phenomena &c." According to the author, its most important theorists were Descartes, Aristotle, Gassendi and Woodward: it appeared as if meteorology ended with the controversial London fossil collector.

As will be shown in what follows, the contrast reveals that developments in late eighteenth-century meteorology cannot be explained in terms only of a "transformation" within the discipline which took place during a time of vigorous chemico-pneumatic and other investigation. The idea of an internal transformation of the discipline cannot bear scrutiny because the results of research into the physico-chemical properties of air had been known since John Mayow, Walter Charleton, Robert Boyle, and John Wallis who, like their successors, considered such research *outside* meteorology proper. It was this separation of strictly "physico-chemical" enquiry into the nature of air from the "meteorological" enquiry into the nature of meteors that justified Boyle's distinction between the "lax" and the "strict" meaning of "air."

The contrast also illustrates that meteorology as a science of *meteors* had been cultivated throughout the eighteenth century *without* directly being challenged by the Boylean study of air, Hales's researches on "fixed" airs, or by the recent physico-chemical research. For these reasons, meteorology and the older physico-chemical research into air could *coexist* apart from each other,[9] just like the meteoric tradition coexisted with the instrumental diaries of weather. In this sense, no "fatal" (fatal for "meteorology" as a science of meteors) theoretical transformation could occur *within* traditional meteorology as a result of the intrusion of a presumably more fertile approach to the study of the atmosphere. Rather, the situation was one in which during the last quarter of the eighteenth century, naturalists increasingly associated meteorology with chemical expertise, education, training, and status in the scientific forums of those who continued to approach weather from a global and analytical perspective.

The brevity of the Chambers's meteorology entry and its reference to Aristotle and Woodward illustrate a *cessation* of the public presence and the scientific value of classical theory, not a mere theoretical transformation of a discipline. Classical meteorology was increasingly described as a matter of the past, indeed, a science of irredeemable antiquity, and, ultimately, something of a dogma. It was seen as a system with its place in the history of ideas, beginning to share the fate of its alphabetical neighbor, "meteoromancy" – "a species of divination by meteors." This change would have an important repercussion on the fate of meteoric reportage.

Electrification

Following the research of Boerhaave, Musschenbroek, and Gravesande, and the posthumous publication in 1744 of Newton's "ether" letter to Boyle and Oldenburg, theories of subtle (imponderable) fluids proliferated in European scientific forums. The Dutch naturalists had discussed the expansibility of fluids as the property of elemental "fire," manifested in sensible heat, light, and electrical fluid. Earlier in the century, Francis Hauksbee had conceived electricity as an effluvium emanating from a spinning glass globe. In 1729, Stephen Gray discovered the transmission of electrical effects by contact. In France, Charles Dufay distinguished the "resinous" and "vitreous" electrical fluids, whereas in Philadelphia Benjamin Franklin argued for the existence of a single electrical "atmosphere."[10] In 1747, Franklin demonstrated a "leakage" of the fluid from an electrified sphere to a sharp metal point, and suggested that the pointed metal rods would "draw off" electrical fire from the thunderclouds. Franklin proposed that the lightning could be conceived as a discharge of electrical fire between non-electrical clouds raised from the sea and electrical clouds raised from the land. The idea reached the English public through Peter Collinson, a London merchant, who promptly submitted Franklin's letters for publication to the *Philosophical Transactions*.[11] Within two decades, the electrical explanation of lightning became commonly accepted.[12]

Meanwhile, after large electrical charges were made possible by the discovery of the Leyden jar in 1746, and after Albrecht von Haller's well-known paper on electrical experiments was published in the *Gentleman's Magazine*, electrical experiments increased in popularity and theoretical import.[13] In mid eighteenth-century London and in provincial towns, instrument makers sold electrical machines and manuals in great numbers, and public lecturers established "a standard repertoire of electrical phenomena" exhibited during well-attended demonstrations. Once the word spread about the electrical nature of meteors, lectures included the electrical emulation of meteors such as "Lightning, the Ignis Fatuus, the Shooting Stars, and *Aurora Borealis*."[14]

Almost universally, electrical fluid became a central explanatory concept in meteorology. The jargon of exhalations seemed to be utterly abandoned while the traditional repertoire of meteors received a swift "electrification." "The little snap," wrote Stukeley, "which we hear, in our electrical experiments, when produced by a thousand miles compass of clouds, and that re-echoed from cloud to cloud, through the extent of the firmament, makes that thunder, which afrightens us."[15] Is not the aurora borealis, asked John Canton, "the flashing of electrical fire from positive, toward negative clouds at a great distance."[16] John Perkins, Franklin's correspondent from Boston, suggested that shooting stars might be "passes of electrical fire from place to place in the atmosphere."[17] In 1755, the Irishman Henry Eeles claimed that the rarefaction caused by sub-terranean heat was insufficient to raise exhalations – the issue which bothered Halley – and proposed that the electrical fire was the cause of rendering them lighter than the air. When the vaporous particles underwent coagulation, Eeles reasoned, the electricity was lost, the particle weight increased, and the result was rainfall. Consequently, the electrically governed oscillations of vapors were the prime causes of the wind.[18]

Others theorized that electrical matter suffused throughout the atmosphere. It could be detected at times of extraordinary meteoric activity. The physician Thomas Henry wrote that after an earthquake in Manchester in 1777, "[m]any people complained of nervous pains and hysteric affections, and of sensations similar to those of persons who have been strongly electrified."[19] In the next decades even large-scale weather changes were linked to variations in the quantity of electrical fluid, leading some to propose that the inclemency of the British climate could be improved by constructing machines that would "electrize the whole of [the] atmosphere of Great Britain one mile in height."[20]

Some naturalists combined electrical and mineral theories. William de Brahm set out in 1774 to complete a "physico-systematic" picture of the atmosphere and its "passions from different meteors" using electrical fluid as the primary cause and the vertical motion of air as the secondary one. Electrical fluid, in de Brahm's opinion, "volatiles" the gross matter and carries it into the atmosphere where the mixture becomes "known by the appellation of vapours, [i.e.] those brought out of the bowels of the earth," which are mineral and ponderous, while those raised from the sea, are ammoniacal and light. De Brahm paid lip service to the Aristotelian distinction between the dry and moist exhalations but used the idea of friction between the *electrical* streams and the globules of air to argue that the latter created "bales" and "globes" that at length "burst and discharge[d] their enclosed ether in the form of *winds*."[21]

Early enthusiasm for electrical meteorology was such that Horace Walpole called it "the fashionable cause, and everything is resolved into electrical appearances, as formerly everything was accounted for by Descartes' vortices, or Sir Isaac's gravitation."[22] From the 1750s, many naturalists agreed that at least two meteors, aurora borealis and lightning, result from the accumulation and discharge of the electrical fluid and that the "nitro-sulfurous" material could not be a sole cause of the fiery and other meteors. In an "electrical" critique of Halley's account of the fireball of 1718 (which he thought was "a train of combustible vapours, accumulated in those lofty regions, [and] suddenly set on fire"), Charles Blagden, MD, Physician to the Army, argued that Halley could have no notion of the nature of inflammable vapors nor of the manner in which they could be raised to a said height. Neither could he account for the meteor's straight trajectory, nor explain why the vapors remained dense without diffusing into the high atmosphere.[23] After reviewing descriptions received from his country correspondents or those published in local newspapers, Blagden was led to believe that the only agent which "seems capable of producing such phenomena, is ELECTRICITY," and that the hypothesis ought to be accepted on the basis of the electrical experiments reported in Priestley's *History of Electricity* and the "concurrent testimony of many observers."[24] If the atmosphere were permeated by the electrical fluid, as Blagden hoped further observations would demonstrate, then practically all fiery meteors could be accounted for as manifestations of the uneven distribution of the fluid at different heights.[25]

Electrical researchers began in this way to challenge the theory of mineral vapors. The notion of electrical fluids appeared to provide the answer to the cause of lightning, earthquakes, northern lights and falling stars. It answered objections put forward against the improbable height which the meteoric material needed to reach before it changed into meteors. Electrical research thus resulted in a diversification of theoretical meteorology; while not all, then certainly a great number of meteors lent themselves to electrical treatment, losing their mineral identity in the process. Electrical meteorology also opened a way for a simulation and quantification of atmospheric phenomena. Electrical fluid was produced inside the laboratory and, with the help of imagination, a spark could be seen as miniature lightning. This appealing analogy reduced the scale of meteors from public events to events on a table top and, if electrical theory would prove to be true, meteorology could turn into a laboratory practice, leaving its outdoor community of "real-weather" observers suspended in an epistemic vacuum. It will be shown how this exclusion actually took place in the methodological rhetoric at the end of the eighteenth century.

At another level, however, there was an obvious parallel between, for instance, Halley's sulfurous and Blagden's electrical model of fireballs: both posited a fluid of unequal distribution susceptible to erratic ignition to explain the production of meteors. Electrical, like mineral, theory maintained an element of the traditional framework, one in which meteors appear *within* the sublunary region and are generated through modifications of the exhaled material. On other points of similarity, electricians replaced exhalations with the similarly defined fluids: the interaction of two exhalations (wet and dry) was replaced by the interaction of "negatively" and "positively" charged electrical ethers. And importantly, electrical meteorologists, like their predecessors, investigated individual meteoric phenomena, not the weather and the seasons. The subject matter and the explanatory tools of electrical meteorology in this sense continued the quasi-Aristotelian basis of meteorological theory.

Other sets of issues, however, becoming more prominent in the latter part of the century, affected the status of meteors from the point of view of definition. It has been argued that neither Aristotle's nor Newton's notions of the cosmos, nor electrical theory, allowed for "interplanetary" meteoric activity: meteors could not possibly be engendered in empty spaces beyond the atmosphere, or, more precisely, beyond the region of elemental corruption. The stones found in the wake of thunderstorms had, since antiquity, been explained as generated in air during periods when a "very great exhalation composed of many terrestrial particles is hardened and confined in a small volume by the cold of the clouds that envelop it."[26]

In the early eighteenth century, the convergence of two trends made it increasingly difficult to maintain the atmospheric origins of stones from the sky. First was the growing disbelief among mineralogists and chemists that stones were created in clouds by a fermentation of terrestrial vapors. The French chemist Nicolas Lemery claimed that even though in principle stones might be produced in the atmosphere, it was more probable that they resulted from the ignition of the minerals *in situ* occasioned by the action of lightning. Second, researchers in paleontology and archeology discovered that so-called thunderbolts could be explained as "either fossils, ancient stone implements, or crystal masses of common mineral."[27]

Despite these criticisms, however, the number of fireballs observed in Europe in the 1750s and 1760s led the London physician John Pringle, later President of the Royal Society, to challenge both archeological and "*in situ*" hypotheses.[28] Pringle collated impressive evidence to reconstruct the path of the fiery meteor of 26 November 1758, concluding that, among other factors, the great explosion which followed its fall meant that the "substance of the meteor was of a firmer texture than what could arise from mere exhalations" (as maintained by Halley and the majority) and that "there seems to be more ground for believing this body was solid."[29] (FIGURE 21). In a long rebuttal of Halley's ideas, Pringle doubted both that inflammable vapors could rise to great heights in the atmosphere and that the sulfurous vapors could, in such an ascent, preserve their volatile character, especially once they had arrived in the colder regions. Vapor's convection to those heights was highly improbable in the light of experiments which demonstrated that inflammable matter possessed a "proportion of an acid salt," too ponderous to allow convection.[30] Pringle wrote: "Are we not led to this notion of the innate levity and *vis centrifuga* of igneous matter, from finding, that heat has a greater tendency upwards than downwards?" When Halley and other "exhalation" theorists used the notion of innate levity, Pringle argued, they resorted to occult notion and, worse yet, had no suggestion as to the cause of the purported ignition of dry exhalations, the axiom of the theory of fiery meteors: "what is to set this train on fire; since the ferment, according to the chemists, implies a mixture of heterogeneous parts, contrary to the doctor's [Halley's] *hypothesis?*"[31]

Pringle invoked these formidable objections because he possessed the means to answer to them. He thus began by using the electrical explanation of lightning to expose the error of those who had argued that fiery meteors were a kind of lightning (something Whiston had claimed for the *aurora* and Halley for fireballs), but was cautious to interpret the premises of exhalation theory as a product of its time:

> [B]efore the matter of lightning was discovered to be of electrical kind, it was natural to suppose it to be formed of the sulphureous vapours arising from the earth; and if the earth was found proper for producing such exhalations, of course it was judged capable of furnishing materials for all lucid phenomena in the aetherial regions.[32]

If the previous century thought that even comets carried terrestrial exhalation, Pringle noted, it should not come as a surprise to find "the best naturalists" of the early eighteenth century applying the same ideas to fiery meteors like shooting stars and fireballs. In mid-century, however, Pringle thought these ideas were untenable, for the evidence went a long way to show that balls of fire had to be solid objects coming from "regions beyond the reach of our vapours," having "a motion of their own," [i.e. independent of the earth's motions], and moving with great velocity towards the earth.[33] They were bodies "of the nobler origin"

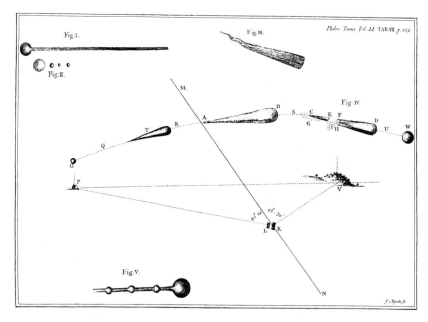

FIGURE 21   John Pringle's reconstruction of the path of the 1758 fiery meteor seen throughout England. John Pringle, "Several Accounts of the Fiery Meteors which appeared on Sunday Nov 26, 1758 between 8 and 9 at Night.", *Philosophical Transactions* 51 (1759): 265–81.

than they received in the exhalation theory, "possibly revolving about some center, formed and regulated by the Creator for wise and beneficent purposes; even with regard to our atmosphere," (for example, supplying the air with a wholesome cosmic matter, or removing the noxious emanations from the atmosphere).[34] Translated into Aristotelian idiom, the fiery meteors of the upper region of air were not of sublunary origin.[35]

Pringle's ideas required half a century to gain hard currency. One of the reasons for this gestation period was a widespread suspicion about the reliability of the evidence based on fallen stones.[36] In the period between 1794 and 1803, however, a change of opinion occurred, the result of which was an acceptance of the cosmic origins of meteoric stones. The story of this acceptance has been told in detail elsewhere and here we will only note the most influential factors contributing to these developments. As the most important, Burke singles out Ernst F. F. Chladni's essay on the cosmic origins of fireballs. Chladni's controversial paper was soon supported by the publication in 1802 of Edward Howard's results demonstrating the similar chemical composition of a sample of the "sky" stones. The third factor was Jean-Baptiste Biot's report to the Institut National de France in 1803 about the stones found after the "French" fireball of the year before. Finally, these investigations were all spurred by the unprecedented interest in meteoric falls developed during the last quarter of the century.[37]

The acceptance of the extraterrestrial origin of fireballs and shooting stars directly impinged on meteorological subject matter. If the latter included phenomena *originating* in the sublunary realm, then a theory of fiery meteors which explained them as mere "cosmic debris," removed them from meteorological subject matter. Fireballs, new researchers asserted, were cosmic stones pulled into the atmosphere, not ignited terrestrial vapors. The traditional notion of meteorology which dealt with the "affections" raised in the sublunary realm – regardless of the height – lost its full meaning with late-century proof that some upper-atmospheric meteors had supralunary origins. Who was to investigate those origins and explore the extraterrestrial realms? Classical meteorology lost its comprehensive content and scope and, when the remaining lower meteors fell within electrical and chemical investigations, it could be perceived as fragmented, obsolete and undeserving of the status of a discipline with a well-defined domain. On the other hand, a new analytical approach continued to move meteorology indoors and associate it with measurement, tables, graphs, and averages. If these techniques and this expertise now became dominant, it was inevitable that those who failed to follow the chorus of change – because they might lack incentive, knowledge, or the equipment – found themselves declared inadequate and estranged from an aspiring community of chemists, electricians, and calculators.

Chemical hegemony

During the same period, another series of studies came to dominate British natural philosophy, after Joseph Black, Joseph Priestley, Henry Cavendish, and

James Watt conducted experimental research in what is known as pneumatic chemistry. Following Stephen Hales's early work on extracting gases from organic substances, and that of Daniel Fahrenheit and Jean-Jacques d'Ortons de Mairan on the thermal phenomena attending freezing, the attention of chemists had been directed to "vaporous state of matter."[38] In 1754 William Cullen began his studies on the nature of vaporization and discovered its existence in the vacuum. In 1757 Joseph Black, Cullen's student, described his experiments on the amount of heat "fixed" in substances and released from them during freezing or condensation, i.e. the "latent heat." The study of phase-transitions – between solid, liquid, and gaseous phases of matter – was advanced by Black's students William Irvine and James Watt, whose investigation of the efficiency of steam engines led to the further elaboration of the quality of latent heat and the first calculation of the value of heat "fixed" in water vapor.[39]

In France, Robert Jacques Turgot conjectured that vaporization processes applied to all substances subject to a sufficiently intense heat ("matter of fire"). Other chemists suspected that the vaporous – "aeriform" – state might be a combination of "fire" and ordinary matter, which, if true, would point to a number of unobserved airs in the atmosphere, produced in combinations of a fluid of fire and exhaled materials. The common atmospheric air was thus imagined as a heterogeneous mixture of other airs present in the atmosphere.[40] The actual discovery of one such air was made by Joseph Black in 1754 when during his research on the medicinal properties of *magnesia alba* (magnesium carbonate), he found that the substance would release a quantity of air whose weight exactly compensated the loss of the *magnesia alba's* initial weight. The released air differed from the atmospheric air; Black called it the "fixed air." Following these early studies, Cavendish isolated "inflammable air" in 1766 and by 1772, Priestley isolated airs he termed nitrous, alkaline, marine acid, and phlogisticated.

A year after Priestley announced his results, the Swiss Alpine meteorologist and instrument maker Jean Andre Deluc, after meetings with Watt and Priestley, argued that the hygrometrical problems he had encountered in the Swiss Alps – such as the puzzling decrease in air humidity before and during the mountain storms – could be accounted for by assuming that water vapor "hides itself in the atmosphere in the form of an aeriform fluid."[41] He rejected the solution theory of evaporation which treated water vapor as a spontaneous solution of the water into a circumambient medium. Instead, Deluc supposed that "fire" had a greater affinity for water than air and when the temperature decreased with the height, so did the humidity. Deluc's idea that water vapor might be a union of water and the fluid of "fire" (heat) opened a possibility for the inclusion of the chemistry of subtle fluids into meteorological theory, a theory which had an influential debut in his *Idées sur la météorologie*, published in London in 1786. This work was in great measure the result of Deluc's exposure to Scottish pneumatics and the ideas he adopted from the work of the Swiss natural philosopher Jean Senebier, influences that led him to prophesy that chemistry might be "one of the greatest aids for the acquisition of true knowledge in terrestrial physics."[42]

By the late-eighteenth century, the meteorological application of subtle fluids became increasingly de rigeur. Another Swiss meteorologist, Horace Benedict de Saussure, wrote in 1783 that evaporation – a process which classical meteorology treated as a simple rise of a wet exhalation – was the effect of a union of fire and water."[43] Saussure thus claimed that the rise of air in general was due to the presence in it of a "warm body" (i.e. the fluid of heat, or caloric), rather than to an "innate lightness" of the warm air.[44] Early in the nineteenth century, an author of a study on the atmosphere wrote in the opening paragraph that "independently of exhalations from the earth's surface, atmospheric air is likewise intimately combined with certain proportions of heat, light, and electricity."[45]

Within a decade following Deluc's publication, the belief that meteorological phenomena should be explained with reference to subtle fluids began to dominate the study of hydro-meteors in Britain.[46] In his influential *Lectures on Natural and Experimental Philosophy*, George Adams, the mathematical instrument-maker to George III, insisted that Deluc's *Idées* "should be considered by all who mean to understand [meteorology],"[47] and that "only by letting meteorology and chemistry go hand in hand . . . we can hope to be secure from error in either pursuit, or to be enabled to make advances toward real knowledge."[48] By 1800, chemistry did not just *explain* meteorological phenomena, but, in the words of Antoine François Fourcroy, "meteorological phenomena are manifestly caused by immense chemical operations."[49] It is important to note that none of these writers referred to the "older" chemistry of sulf-nitrous vapors. As noted above, electrical meteorologists showed that the most successful explanation of the older theory – that of ignition of the exhaled material – could also be solved by the electrical theory. Consequently, the new meteorology was a chemistry of fluids rather than that of the exhalations of sulf-nitrous particles.

The important consequence of the new "chemical agenda" was the loss of disciplinary "integrity" which meteorology as a classical discipline suffered in the process of changing into a species of chemistry. The process is discernible in the 1809 abridgment of the *Philosophical Transactions*, in which the editors placed meteorology under the general heading of "Chemical Philosophy," where it shared space with Chemistry and Geology. Pneumatics, Hydrostatics, and Thermometry (among other subdisciplines) belonged to "Mechanical Philosophy." Chemists like George Shaw, William Nicholson and Thomas Thomson, believed that meteorology as an *autonomous* enquiry into the origins of "meteors" – the phenomena characterized by the "disorder" intrinsic to the region in which they were generated – could and should be replaced by a set of chemical studies into the constitution of atmosphere. These investigations, it was argued, should allow only those variables which were identifiable chemically. From now on, meteorological autonomy was to be recognized merely in the *scale* of the processes under investigation, not in their unique physical composition or the character of their behavior.

These radical forays of chemistry into meteorological research were consolidated by a series of remarkable extended metaphors. For instance, as early as 1771, the French meteorologist Le Roy anticipated these disciplinary ramifications by calling the atmosphere "a vast chemical laboratory, in which there are a thousand different chemical combinations,"[50] and Deluc went as far as to envision the atmosphere as "a chemical laboratory, as important as the bowels of the earth for the physical phenomena of our globe." These analogies invalidated the distinction between outdoor and laboratory experiences, as well as the distinction – maintained by Virgilian meteorologists – between the respective merits of these two sources of experiences in the study of weather.[51] Richard Kirwan, the Irish chemist, and friend of Priestley and Cavendish, suggested that meteorologists differed from laboratory chemists only in their inability to alter and control the spontaneous course of nature. In methods and aims, however, meteorology and chemistry were identical.[52]

In the early 1790s, these issues were taken up by George Adams. In the methodological preamble to a chapter on meteorology in his *Lectures on Experimental Philosophy*, he expanded upon Kirwan's idea of the meteorologist's inability to control the course of nature but, surprisingly, thought that this inability could be an advantage. He proposed that the discrepancies between theory and laboratory results might proceed from unidentifiable factors, such as the "foreign influences" of the vessels used, the accuracy of the instruments, or unknown operations of the substances under observations. Because natural philosophers usually assumed the presence of these factors, they seldom questioned their theories because they ascribed disagreements between facts and theory to one of these undetectables influences. But in "the laboratory of the atmosphere," phenomena could not be produced by anything that is a "foreign" influence; here, "every thing has a reference to the vessel itself, i.e. to the surface of our globe, whose distinct parts, as minerals, vegetables, and animals, offer masses perpetually changing; here, therefore, the disagreement of theory with facts must give us great and important lessons." Realizing, however, that the argument could endanger the crucial analogy between the atmospheric and laboratory events, Adams warned that these observations were by no means intended to lessen the "esteem for our laboratories," because meteorology advanced only through a comparison between outdoor observations and indoor experiments.[53]

As Adams, Kirwan, and others made clear, meteorological enquiry was a complex affair, its subject matter unmanageable with available knowledge. Yet they also affirmed that this state of affairs was temporary, not axiomatic. In principle, meteorological knowledge remained a realistic goal attainable through cross-disciplinary examination. This conviction was a significant departure from classical meteorological agnosticism: "brought to perfection, [meteorology] would, like astronomy, enable us at least to foresee those changes we could not prevent."[54] The laws of weather were chemically defined, and, notwithstanding drawbacks, chemists were to become the Newtons of meteorology.

What is in a name?

If the series of studies surveyed above suggest an irrevocable change, the problem remains as to how and why the term "meteorology" nevertheless survived the end of its original meaning. If it had lost its status as a theory of sublunary meteors and, as such, ceased to exist by the early 1800s, we need to understand how its name, if nothing more, has endured in the scientific world until our own times. Alchemy and astrology did not survive similar challenges. We may begin to answer this question by looking at the 1771 and 1784 editions of the *New Royal* and Chambers's encyclopedias in which the authors borrowed the entries "meteor" and "meteorology" from the first edition of Chambers (1728): meteorology was "the doctrine of meteors," and meteors were "changeable, movable, imperfect" bodies divided into three categories (airy, watery, fiery).[55] Twenty-five years later, however, William Nicholson's *British Encyclopedia* identified "meteor" as "a moveable *igneous* body"[56] and meteorology as "the science of studying the phenomena of the *atmosphere*," including the study of the weight of atmosphere, temperature, quantity of evaporation and rain, and electric phenomena. Emphasizing the separation of igneous and other kinds of meteors, other scientists described meteors as transient and aleatory as shooting stars, fireballs, ignes fatui and aurorae boreales. But these authors claimed that clouds, rain, hail, earthquakes, and wind were not meteors "proper." Paradoxically, meteorology was described as a science of the weather, or "of various affections and phenomena of the atmosphere, as winds, clouds, rain, hail, snow, dew, etc." Its subject was therefore defined with reference to the weather and the atmosphere. The meaning of the term "meteor" was within the same context narrowed from an inclusive tri-partite domain to just its "igneous" part.[57]

In encyclopedias, meteors ceased to define the subject of meteorology. This was stressed by other sources; in 1814, a dictionary author wrote that "Meteor is used by some authors to denote *all* the various phenomena of the atmosphere, while others apply it exclusively to denote those luminous bodies which appear at a considerable height above the earth."[58] This is confirmed by the juxtaposition made in the 1839 edition of *The Penny Encyclopedia*: in which meteorology – in its "extended sense" – embraced all physical causes of the atmosphere connected with the phenomena of heat and cold, dew, rain, hail, parhelia, halos, and so on. However, the author wrote, Aristotle used the term in a still more extensive way to include what "are now called meteors, every affection (*pathos*) common to the air and water, with the characters of the different parts of the earth, and their affection, as winds and earthquakes, and everything incident to such kind of motion."[59] Like the Virgilians, new meteorologists were after a knowledge about the "weather," but unlike the former, this notion of weather was now defined with reference to "chemical combinations and decompositions" taking place in the atmosphere.[60] As this new outlook was becoming widespread, meteors and meteorology were increasingly linked to "atmospheric" phenomena specifically – airy and watery meteors, in classical idiom.[61] Those who pursued the "electro-chemical" science of the atmosphere thus disregarded investigations concerning fireballs, shooting stars, earthquakes, as well as the optical and magnetic phenomena, that had hitherto defined meteorological theory. That a phenomenon had a spatial dimension – as the classical meteor had in the sublunary region – was unimportant for those who required that it be quantified and analyzed in chemical terms. From now on, only these new procedures could warrant a scientific validity of meteorological and other practices.[62]

The key word in these new emphases was the *atmosphere*. Indeed, by the early nineteenth century, the alliance between "meteorology" and the atmosphere (or the atmospheric *air*) had become familiar. In the 1770s, John Walker, Cullen's student of chemistry and the Professor of Natural History at the University of Edinburgh, lectured that "meteorology is to be understood as the natural history of Atmosphere," only marginally concerned with meteors in the classical sense.[63] For Walker, the history of the atmosphere evolved as a science only after the discovery of the barometer and the early experiments on air. These and electrical theories, "reduced into a Scientific form what was formerly a mass of vague matter with erroneous deductions."[64] In 1801, Murdo Downie, a Master in the Royal Navy, published *Observations Upon the Nature and Properties of the Atmosphere*,[65] Henry Robertson, M. D., wrote a two-volume *Natural History of the Atmosphere* in 1808, and Thomas Ignatius Maria Forster, M. D., finished his *Researches About Atmospheric Phenomena* in 1813.[66]

These authors emphasized the medical and agricultural benefits of meteorology, devoting major sections of their works on the constitution of the atmosphere, and its barometric, thermal, and hygrometric characteristics. They made meteorology a science of atmospheric, not "meteoric" phenomena. By the 1820s it was increasingly difficult to think of meteorology in connection with Aristotelian *meteora*. Because it was now associated with a study of the atmosphere, meteorology continued only after a complete overhaul of its original meaning. This had a profound effect on how quickly naturalists adopted the change and with what degree of understanding they approached a field now defined without reference to either the exhalation theory or the sublunary region. Critical to this shift of emphasis was a removal of phenomena from a *spatially* defined region to a *physico-chemically* defined body of air. For the meteorologist to follow this new idea it was therefore necessary to determine whether the object of study belonged to a spatial or physical domain and whether it lent itself to analysis framed in terms of the atmosphere. This meant that the modern approach required a prior recognition of the atmospheric basis of meteors. From a professional perspective, new meteorologists had to become acquainted with the new methods and theoretical jargon of proper "atmospheric" investigation, whose requirements will be examined in the next section.

Weather remodeled

We have seen that in the eighteenth-century meteoric tradition, provincial naturalists derived their scientific authority from parochial affiliations and access to local facts. This socio-cultural constellation sustained and was sustained by the cultures of genteel curiosity which channeled exploratory interests into what was remarkable and worthy of recording. Extraordinary phenonema were such partly because they were rare and, within the eighteenth-century ethos of natural enquiry, rarity possessed an epistemological cachet within the cultures of esthetic curiosity, chorography, and clerical natural history. The vehicles which bestowed national visibility on these loco-centric and topophilic studies were the scientific press and a network of scholars in correspondence. Thus until the 1790s, the secretaryship of the Royal Society kept its gates open – although progressively narrower – to articles written by provincial naturalists. Simultaneously, the networks of correspondence enabled a circulation of facts, whether in the form of learned commodities, pieces of conviviality, or items for comparative analysis.[67]

On the other hand, the professional expertise of the early nineteenth century looked toward method, abstraction, and training. Metropolitan specialists – doctors, chemists, professors, or instrument-makers – represented a community extending beyond the limits of parish, region, and even the capital itself. These individuals thought of their practices as divorced from their social, geographical, linguistic, and other "ephemeral" identities. This new community drew its collegiality from the participation in a discourse bounded by a common goal, common means, and common forums of the information exchange. The sense of a common goal which united these people was nurtured by the notions of synchronization, standardization and quantification. These practices provided a syntax for a new identity which represented the opposite of Georgian museological meteorology. However, in a manner similar to chorographic curiosity, these practices defined an identity for a community which had no clout – or felt no need for it – within the context of the waning relevance of a land-centered society.

In the present section we will explore the means whereby new meteorological "experts" argued against the unsystematic and inconclusive histories of extraordinary weather. We have seen that the material and cultural preconditions of the meteoric tradition evolved around the notion of "stationary residence" and the cultural status of learned "residents" on the cusp between the parish and the nation. The position of metropolitan chemists, physicians, and an increasing number of public servants engaged in meteorological observation differed considerably, because in the capital "local" information had no genuine meaning. Moreover, the social identity of metropolitan residents in a city parish had little to do with their laboratory equipment and expertise. In their social milieu, metropolitan chemists did not have a stake in an activity, which they could not perform well on account of their location, and from which they could neither derive nor enhance professional status. Their attempt to extend the laboratory to encompass the entire atmosphere – years before Pasteur – may appear like a conspiracy born out of professional arrogance or the disadvantages of being in a city. In reality, however, this attempt could also be justified by pointing out that chemical meteorology possessed theoretical potential comparable to that of electrical theory and pneumatics. Chemical, pneumatic, and electrical expertise also promised to engage with practical issues, with the advantages of accuracy, organization and a *causally* based forecast. Such promise harnessed a public-spiritedness that individual curiosity lacked as soon as it moved outside the communal culture of a parish or shire.

The mechanics of imperial data collection during Banks's presidency required the help of civil servants and professionals who, by the nature of their occupations, moved between places or had a "stationary residence" in an alien environment in which their judgment of what counted as "remarkable" mattered far less than back home. Some of these individuals were officers of the navy, hydrographers, or employees of the East India Company. For such individuals, it made little sense to send news about rare heats in India, where the levels of humidity and heat almost daily exceeded the most "extraordinary" summers in England and where the definition of "prodigious rain" needed to come to terms with the onslaughts of monsoon showers.[68] Such climatological (dis)placement required the fashioning of transnational and objective credibility supplied by instruments, standardized registers and regular observations. In 1805, James Horsburgh sent his daily observations of the barometer in the tropics to London, and Matthew Flinders reported his barometric measurements on the coast of New South Wales. James Renell presented his measurements on the effects of westerly winds on raising the level of the "British" Channel, while John Davy reported the series of thermometric measurements on his voyage to Ceylon (Sri Lanka). Horsburgh was a hydrographer, Flinders a captain in the navy, and Renell an officer in the navy as a young man, and, in his later years, became Surveyor-General of the East India Company. John Davy, brother of Humphry Davy, was an army surgeon and inspector of hospitals in Ceylon. These men allied an ethic of measurement to professional practice, commerce, and colonial administration in a manner similar to that in which eighteenth-century military engineers worked on improving the accuracy of the barometer for the hypsometric topography.[69]

True, journals were still kept by the gentry – for example a James Fox, Esq. from Plymouth, and the Right Honorary Lord Gray from Kinfauns Castle in Yorkshire – but permanent residence was not crucial for the new ethos. Journals kept in single locations for long periods of time could be useless if they were not comparable with others. Local weather was useless if collected for its own sake. Thomas Thomson dismissed such practice when, in 1812, he wrote that a large number of papers on meteorology in the *Philosophical Transactions* consisted of "bare diaries of the weather in particular spots, or of insulated meteorological observations, which, though they may all have their utility and be of service toward constructing a system of meteorology, cannot with propriety be noticed in a work of this kind."[70] In an ironic turn of events, meteorological observations that Borlase, White, Cole, and other clerical diarists treated as valuable scientific contributions were described by Thomson as "bare diaries of the weather in particular spots." Where eighteenth-century meteorologists considered "particular spots" the *sine qua non* of meteorological reportage, Thomson looked upon them as a limitation.

Concerns about instruments, standardization, and the discipline of measurement were shared by metropolitan chemists and physicians such as Luke Howard, Thomas I. Forster, George Shaw, George Adams, Thomas Thomson, Humphry Davy, Charles Babbage, John Herschel, and John Daniell, all of whom at one time or another expressed discontent with the descriptive meteoric tradition.

These people were not chorographers of county airs, nor were they interested in the traditional natural historic approach to meteorology. They advertised a "removal" of meteorological practice from places of life to places on the map.[71] Accounts of individual meteorological appearances, which before that time represented the majority of meteorological papers published in the *Philosophical Transactions*, rapidly declined in number (BOX 7). *Philosophical Magazine* featured

---

BOX 7   Early nineteenth-century contributions to meteorology in the *Philosophical Transactions* of the Royal Society of London

Thomas Barker, Esq. 1800. "Abstract of a Register of the Barometer, Thermometer, and Rain at Lyndon, in Rutland, for the year 1798", 46.

J. Horsburgh, Esq. 1805. "Abstract of Observations on a diurnal Variation of Barometer between Tropics", 163.

Matthew Flinders, Esq. 1806. "Observations upon the Marine Barometer, made during the Examination of the Coast of New Holland and New South Wales, in the Year 1801, 1802, and 1803", 239.

W. H. Pepys. 1807. "A New Eudiometer", 247.

James Rennell, Esq, FRS. 1809. "On the Effect of Westerly Winds in raising the Level of the British Chanell", 400.

Francis John Hyde Wollaston. 1817. "Description of a Thermometrical Barometer for measuring Altitudes", 183.

John Davy, MD. 1817. "Observations on the Temperature of the Ocean and Atmosphere, and the Density of sea water, made during a Voyage to Ceylon", 275.

Humphry Davy, Bart, FRS etc. 1819. "Some Observations on the Formation of Mists in particular situations", 123.

James Anderson, Capt. 1819. "Some Observations on the Peculiarity of the Tides between Farleigh and the North Foreland", 217.

F. J. H. Wollaston. 1820. "On the Measurement of Snowdon, by Thermometrical Barometer", 295.

Luke Howard, Esq, FRS. 1822. "On the late Extraordinary Depression of the Barometer", 113.

Captain E. Sabine. 1823. "On the Temperature at considerable Depths of the Caribbean Sea", 206.

Captain E. Sabine. 1824. "A Comparison of Barometric Measurement, with the Trigonometrical Determination of a Height at Spitzbergen", 290.

Mr Thomas Jones. 1826. "Description of an Improved Hygrometer", 53.

W. Heberden, MD, FRS. 1826. "An Account of the Heat of July, 1825, together with some remarks upon sensible Cold", 69.

John Dalton, Esq, FRS etc. 1826. "On the Constitution of the Atmosphere", 174.

William Ritchie, AM. 1827. "On a new form of the Differential Thermometer with some of its applications", 129.

James Prinsep, Esq. 1828. "On the Measurement of High Temperatures", 79.

James Prinsep, Esq. 1828. "Abstract of a Meteorological Journal kept at Benares during Years 1824, 1825, 1826", 251.

W. H. Wollaston. 1829. "On a Differential Barometer", 133.

"extraordinary" meteorology in the first dozen of its volumes, but from 1804 "some of our philosophical friends expressed a desire to see in our work a regular monthly register of the barometer and thermometer," the meteoric genre was relegated to the "Miscellaneous Intelligence" and eventually dropped altogether. Meteorological journals were frequently printed in other periodicals. *Annals of Philosophy*[72] announced that meteorological tables "will be considered an important department in the Journal."[73] There was a conspicuous neglect of unusual meteorological occurrences. What were the implications of these trends for meteorology? What was the meaning of this replacement of qualitative description by instrumental measurement? What was the nature of the socio-cultural baggage that was inherent in this new focus on the common and repeatable, rather than the rare and the unrepeatable?

At the first approximation, the success of instrumental meteorology can be viewed as the result of the hackneyed rhetoric of the science of the enlightenment's ideal of communication and rational progress of knowledge. It can also be viewed as supported by the "conjectural" histories of scientific disciplines which followed Adam Smith's "History of Astronomy." In 1787, for instance, Kirwan introduced a tradition which would persist until our own time, in which the birth of meteorology began with the invention of meteorological instruments in the early sixteenth century. The "arrested development" of the meteorological enquiry of Kirwan's own day was, symmetrically, due to the "discordant multiplicity, as well as imperfection of the instruments, the narrow limits to which individual observations were confined [and] the interruption, either by death, sickness, or the common cares of life, of a series."[74] Kirwan distinguished between the "Empyric" method – vague and uncertain[75] – and the "Scientific," still in its infancy, but "grounded on *a long series of observations* accurately taken of all the changes of the atmosphere, from whence some general law may at length be deduced."[76]

In contrast to the Virgilian emphasis on the visceral understanding of seasonal change and the necessity of ancient knowledge, Kirwan maintained that "however desirous the ancients might have been of cultivating this science; the want of instruments necessarily denied them all access to it." This was a strong claim. No earlier author, and not many among Kirwan's contemporaries, would have rejected the value of ancient maxims or qualitative meteorology. These forms of knowledge were considered a prerequisite for a "philosophical" history of the atmosphere. But with the late eighteenth-century interest in the atmosphere as a *plenum* of physico-chemical events, the practice of extraordinary meteorology met with a sustained skepticism. If the atmosphere was a grand laboratory where no irrelevant substances and no "foreign influences" could be found, as Adams held, this space was then an *isotropic* whole of the intersecting processes – watery, airy, electrical, and perhaps even lunar. The naturalist would therefore give no credence to place, time, nor any other aspect within the system of atmospheric processes, pursuing a disciplined and even mechanical measurement of everything which occurred in the atmosphere and which could be recorded in a quantitative way.

"There is no science," Kirwan wrote, "in the whole circle of those attainable by man, which requires such *conspiracy*, if I may so call it, of all nations, to bring it to perfection as Meteorology."[77] The quest for total quantification shows Kirwan's "conspiracy" as a goal defined by the discovery of atmospheric laws. He did not believe – as did the virtuosi and gentleman naturalists – that theoretical knowledge ought to come only after completing an *entire* history of nature. Instead, he and his contemporaries held that "only" a quantitative history would suffice. Contrary to Bacon's belief that nature spoke clearly to the enquiring mind when it "sported" itself in the extraordinary, Kirwan thought that in these instances nature was most misleading. The globalization of meteorology was an antidote to early Baconianism.[78]

As much as it reflected theory, the discipline of quantification shaped practice. It constituted a culture of distancing, exclusion, and rejection. Its power lay in making alternatives intellectually inferior and embarrassing for their expositors. Just as Whiston explained the northern lights by means of a critique of vulgar politicking, so late eighteenth-century quantifiers of science defended their position by a critique of the uneducated amateurism of meteoric reportage. The atmosphere was to be reinvented through numbers but this could only occur if the observers were made to think in numbers. The pitfalls of narrative meteorology ought to be avoided by a readjustment in attitudes and a practice, which was, in turn, mandated by attempts to split meteorological totalities into individual processes of chemistry, heat theory, and electricity. Quantification was as much a methodology as it was a social technology: "the quest of precision has been sustained in science for reasons having more to do with moral economy than theoretical rigor. Precision has been valued as a sign of diligence, skill, and impersonality."[79]

The business of philosophers, argued George Adams, was to discover the connection between the parts of the world and to "explain why they appear to us as separated." But the philosophers' work was not continually to "accumulate unconnected facts" and impose "disjointed links" on to nature chains. He found Thomas Young's opinion especially pertinent:

> But alas, you find the philosopher continually losing sight of the true construction of nature, and endeavoring to build systems upon matter *independently* considered. You find physicians treating of the nature and causes of diseases, of blemishes, of preternatural appearances in the body; but wholly indifferent, and altogether inattentive to the proceedings of the healthy economy: you will find a hundred dissertations on fevers, for one upon life.[80]

Young's philosophers and physicians were not just any philosophers and physicians: they were those who were admonished to change or perish. The collectivization of the meteorological mind directly impinged on the entry into the discipline. In the wake of these criticisms, the Royal Society established a committee for the handling of matters to do with instrumental techniques, training, and the publication of diaries. During the 1810s, commentators observed that the committee failed in its mission to handle such matters efficiently which led the doyen of science in Manchester to quip that if the Royal Society deemed the weather unworthy of its investigations, it would have been better "at once to dismiss the register from the transactions."[81] For John F. Daniell, the neglect was especially harmful because the "meteorological observations" were the only part of the journal officially acknowledged by the Council of the Royal Society, and, if it fell short of international standards of accuracy, the recognition of such a fault by foreign naturalists would amount to an "almost national disgrace." Daniell, a chemist at the Royal Institution, inventor of a hygrometer, and author of the influential *Meteorological Essays and Observations* (1823), was especially critical of the Royal Society's neglect of instruments.[82]

But Daniell also stressed that the source of these problems was less due to a lack of understanding than to the idiosyncrasies of observational practice. He thus blamed the observers, not philosophers. A most embarrassing factor, Daniell argued, was the lack of discipline: hours of observations seemed to depend on nothing but "the observer's night cap," or, in the words of the editors of *Annales de Chimie*, "la commodité de l'observateur."[83] Furthermore, unreliable instruments and the arbitrary routine of observation were coupled with "the detached labours of individuals" who were "utterly incompetent to effect that advancement of the science." This was a deplorable situation for a science as complicated as meteorology which after centuries of investigation continued to depend on "the observations of the vulgar." The complicated nature of the processes in the "immense laboratory of nature" had been simply "too much for the mind to comprehend."[84] Linking the threat to national reputation to the "night cap" of incompetent individuals could only serve as a powerful deterrent against methodological slackness and personal irresponsibility.

The issue of "detached labours" was of preeminent value in this rhetoric. No doubt it was applicable to curious clerics who, if they considered Daniell as an expert, either gave up or bought better barometers. But as an occasional article in the scientific press still testified, the choice was neither this simple and nor were Daniell's words listened to in earnest. Daniell's and Young's complaints were being recycled throughout the first half of the nineteenth century.[85] For the German meteorologist Heinrich Dove, the new meteorological method should seek "the general in particular phenomena" rather than collecting particulars for their own sake. But "by a singular misunderstanding just the opposite is demanded in meteorology. When in an unusually hot spell everything threatens to wither, when a very severe winter in our geographical latitude very nearly drives us mad, when floods and earthquakes destroy fertile regions, everyone says what an interesting year for meteorology!"[86]

Dove pointed out that in experimental research mere enhancement of phenomena, such as the generation of great electrical charges or the use of strong magnets, was of no use in understanding the true nature of the processes. Because the same was true of meteorology, it was unwise to "seek in the unusual and the strange an explanation of the phenomena," or to scrutinize "the totality of phenomena" at the expense of "individual events." Dove illustrated this dichotomy with the example of the thunderstorm as a totality of a set of inter-acting "individual events" (electrical, barometric, hygrometric) which could be broken down into their physico-chemical parts. Understanding these elements was a prerequisite for a knowledge of isolated "totalities," but to reach such understanding one needed to discard ordinary intuitions about the "elementari-ness" of meteorological events. Whereas ordinary minds perceived elementary phenomena in specific storms, fireballs, or halos, Dove's new meteorology found them in the measurement of electrical charges, condensation, evaporation, and congelation. All storms were the same, their numbers differed.[87]

This bid for quantification was not simply another appeal for data accumu-lation. No doubt Kirwan, Adams, Daniell and Dove subscribed to the notion of international "conspiracy" in the study of weather, but they also thought that mere tabulation, reduction, and calculation of averages, would hardly warrant a science of weather. In Daniell's opinion, for instance, one needed to proceed "*synthetically* [by] making more general inferences from the established simple truths" and in the long run one would reach the explanations. This wasn't a trite inductive philosophy, but a heuristic of approximation to reality through a series of models of increasing complexity. Instead of attacking the atmosphere as it was – a "confusedly jumbled chaos of effluvia" – a new science of weather should approach its subject by examining the idealized "habitudes" or models: "the atmosphere of perfectly dry, elastic fluid," "of pure aqueous vapours," and "of the compound mixtures of the two."[88]

These syntheses of weather were embroiled in the quantification agenda just as the latter was predicated upon the analogy between laboratory and the atmo-sphere. But there was also a theological moment. The new demands were tied to the assumptions – now uncontroversial – that nature was a whole of knowable order in which isolated facts would have neither methodological nor doctrinal rationale. Kirwan's and Adams's recognition that a "general meteorological provid-ence" sustained atmospheric cycles represented a context in which they considered it possible to discredit the secular vestiges of "specially" providential meteors and, more significantly, exclude the practitioners of the meteoric tradition on the grounds of knowledge and piety. In the work of the London scientific writer John Read, prodigious weather was a doubtful issue that did not conform to the theology of benevolent atmosphere, while the self-regulatory circulation of the electrical fluid was nature in her "best attire." In thunder storms, however, she was "in a state of apparent disorder."[89]

From a practical perspective, the mutually sustaining triad of epistemolo-gical, professional and theological exigencies was an assault on a subjective, inchoate, and "situated" meteorology of "pre-disciplinary" individuals living within public weather. Jan Golinski has suggested that this and similar translations may be analyzed in terms of the difference between eighteenth-century "experience" and nineteenth-century "expertise." The notion of experience connotes "accu-mulating factual information and the individual autonomy necessary for this." In the case of Georgian meteorological practices, scientific autonomy drew on the same resources which sustained social and political autonomy. These were "local" in the sense that they stemmed from the individual's social, geographical, political, associational and scholarly placement in nature and society rather than from a level of compliance with pre-fashioned rules of action. The notion of expertise, on the other hand, requires "exclusive knowledge and discourse, intensive practical training, elaborate and expensive apparatus." The nineteenth-century meteorological expertise jeopardized the autonomy of the individual and entailed a personal renunciation on his or her part of the credentials granted by non-expert agencies.[90]

This series of new emphases did not reflect a disembodied rationality of an ideal meteorological expertise. The new knowledge of weather was promoted by a community with the same expectations about what meteorology should become. Importantly, this community also shared a distrust in traditional meteorological experience because such experience could not satisfy new priorities, nor make sense of what new knowledge had become available in the last decades of the eighteenth century. These new priorities were publicized by those whose training, practice, and prestige revolved around the ideals fashioned in a specific geographic and pedagogic environment: the city and the laboratory. By changing the scale of meteorological phenomena from life-shapers to parameters, this new environment worked to replace the place-centered and curiosity-driven authority of meteoric reportage by an indoor computation of atmospheric "tides," and storm paths. If the latter concerns define modern meteorology, it is because modern societies value the priorities imposed by late eighteenth-century naturalists. It does not mean, however, that such priorities make the same impression on all, nor that they should.

Conclusion

After the decline of traditional *meteorologia* in the early nineteenth century, meteorologists turned toward investigating the atmosphere as a laboratory of chemical, thermal, barometric, physical, magnetic, and electrical phenomena. The method they favored was the comparison of a series of instrumental observations leading conceivably to a discovery of the laws of atmospheric behavior. The response was as enthusiastic, as the results ambiguous. In 1832, James D. Forbes proposed nothing less then "a total revision" of these meteorological principles, because, as his first *Report on Meteorology* to the British Association for the Advancement of Science maintained, "meteorological instruments have been for the most part treated like toys, and much time and labour have been lost in making and recording observations utterly useless for any scientific purpose." In effect, "the infant science of Meteorology" resembled "patches of cultivation upon spots chosen without discrimination and treated on no common principle."[1] Both Forbes and John Herschel deplored the diffuse boundaries of empirical meteorologizing. In his second *Report on Meteorology* (1840), Forbes commended those who opposed "the popular call for the bare *quantum* of information gleaned from desultory experience," while Herschel asked for "*thought*, steadily directed to single objects, with a determination to eschew the besetting evil of our age – the temptation to squander and dilute upon a thousand different lines of inquiry."[2] With his assistant William Radcliffe Birt, Hershel embarked upon research on "atmospheric waves" designed to contribute to the investigation on the dynamics of storms, but their results were inconclusive. For the time being, meteorology was to remain an emphatically empirical science.[3]

Historians have claimed that the surge of geophysical empiricism which worried Forbes and Herschel was typical of the Humboldtian sciences. Susan Faye Cannon showed that the Humboldtian approach to nature was dominant in the first half of the nineteenth century and described it as "the accurate, measured study of widespread but interconnected real phenomena" such as "geographical distribution, terrestrial magnetism, meteorology, hydrology." Through their commitment to work on location, the Humboldtians, Cannon claimed, opposed "the study of nature in the laboratory."[4] In Theodore Merz's terminology, they pursued "morphological" sciences by investigating phenomena as they appeared in nature, rather than as they were constructed in the laboratories. An ethos of outdoor collection and measurement was at the heart of disciplines like geology, mineralogy, hydrology, and the like. In light of what has been said in preceding chapters, we must ask whether a distinction made between the "laboratory" and the "field" can be usefully applied to nineteenth-century meteorology. We may ask whether nineteenth-century meteorology, with its emphasis on measurement and location, was indeed an anti-laboratory Humboldtian discipline, or a "morphological" field study, as Martin Rudwick claimed for Victorian geology.[5]

As this study has attempted to show, any characterization of eighteenth-century meteorology must take into account several traditions of investigation. For the most part, I have sought to explain the phenomenon of qualitative meteorological reportage as a practice inextricably linked with the notion of locale. It has been demonstrated that the majority of the eighteenth-century studies of the weather originated from the social and cultural significance of regional investigations of nature. In that respect, "pre-Humboldtian" meteorology epitomized a commitment to the "real" location and "real" processes of nature.

Second, it has been noted that since the seventeenth century, naturalists pursued "laboratory meteorology." Boyle, Lister, Derham, Lemery, Wasse, and others sometimes resorted to experiments to simulate outdoor phenomena. John Whiteside at the Ashmolean Museum demonstrated by experiment that sulfurous vapors collected at the top of a receiver and thus provided support for the theory of the sulfurous origins of fiery meteors in the upper regions of air. However irrelevant to meteorological theory it might have been, the recreation of "laboratory weather" was a familiar eighteenth-century experimental technique. It would be difficult, without considerable distortion of the evidence, to perceive meteorology as an anti-laboratory discipline. As discussed in the last chapter, it was the laboratory that, in the last decades of the eighteenth century, figured as the guiding principle of quantitative meteorology. In the methodologies of Deluc, Adams, Kirwan, Daniells, and others, the laboratory was the prime icon of meteorological research, the site in which all the phenomena interconnected in a meaningful way. The image of laboratory space became the central vehicle in arguing for the crucial relevance of instrumentation. Instrumental measurements were justified *because* the atmosphere was conceivable as a laboratory, not because these observations, as Cannon put it, symbolized opposition to the study of nature in the laboratory. To paraphrase Adams, meteorology represented the laboratory research whose object of study was the entire atmosphere.

Finally, we may consider the effect which meteorology conceived in this fashion had on the older tradition of empirical meteorology. In the minds of provincial naturalists, antiquarians, schoolmasters, gentry and clergy, weather happened to, and in, a place. The weather of a place – its meteorological "identity" – was one of many faces of locale. Interest in patterns of local weather, insofar as they derived from this culture of chorographic individuation were subservient to a broader psychological bias in regard to habitat. Within a culture of topophilic partiality, differences in weather, like differences in landscape and history, could be important to the extent they could reflect differences between the "individualities" of geographic regions.

Later in the eighteenth century, however, this chorographic impulse ceased to sustain the meteorological enterprise. With the research into *local* weather being presented by new meteorological experts as a prerequisite for a knowledge of *globally* evolving systems, the locality became part of a larger entity, not a domain of its own. Scrutiny of local weather, whether in the form of a forecast or a series of observations, mattered only to the extent to which the atmosphere could manifest itself in a place. A rain-gauge in Cornwall was not intended to describe Cornwall, but to aid a construction of a map of European iso-lines.

The community of provincial observers, which in the past had included men like Thoresby, Borlase, and White, had been asked to deliver a set of data which reflected new priorities and more comprehensive undertakings. Such undertakings promised knowledge which the new meteorologists thought had eluded their predecessors. But their predecessors did not see the need for such a knowledge. They did not attempt to produce it and therefore they cannot be seen as those who failed. Eighteenth-century meteorology was an engagement with the human rather than the physical environment. It embodied a culture of providentialism and an ethos of locality. It served party politics and landed society. Most importantly, it was a sentiment rather than a science. Borlase did what he liked, not what he thought would be ideal. Thoresby did what he thought would be best within the context of his love of Leeds. Arderon observed weather because it gave him pleasure. Cockin wrote about an unusual appearance in the mist because it was "curious." Their contemporaries did what their lives and resources allowed them to do.

Appendix

Meteoric phenomena reported by individuals between 1694 and 1795

	AB	eff. l/t	EQ	meteors	optical	storm	waterspout
1694							Mayne
1695							
1696		Garden					
1697		Mawgridge			Halley (iris)	Halley	
1698		Wallis					
1699		Thoresby			Gray (parhelia)		
1700		Thoresby			Gray (parhelia+halo)		
1701							Gordon
1702					Halley (parhelia)		Stuart
1703						Fuller	de la Pryme
1704						Derham	de la Pryme
1705							
1706					Derham (glade of light)		
1707		Molyneux			Derham (pyramidal app.)		
1708		Bridgman					
1709		Nelson					
1710							
1711				Thoresby	Thoresby (lunar iris)		
1712						Thoresby	
1713						Chamberlayne	

This appendix page is a chronological chart (printed sideways). The year rows run 1714–1729, with names placed in several columns according to year.

1714		Halley			Richardson
1715	Halley				
1716	Barrell				Thoresby
1717	Folkes				
1718		Barham Halley			
1719	Halley Cruwys Hearne		Halley (parhelion) Whiston (parhelia) Pemberton (iris)		
1720	Percival	Cotes	Dobbs (parhelion) Langwith (color arches)		
1721	Wasse Cruwys				
1722	Burrman				
1723				Wasse	
1724					
1725	Dobbs Langwith Huxham Hallet Hadley Calandrini				
1726					
1727	Lynn Rastrick Langwith	Derham		Whiston (parhelia)	
1728					
1729	Derham				

	AB	eff. l/t	EQ	meteors	optical	storm	waterspout
1730		Davies					
1731							
1732							
1733							
1734							
1735							
1736		Logan	Bayley Wasse			Forth	
1737					Neve (parhelia) Folkes (mock suns)		
1738	Bevis			Cooke			
1739	J. Short	Clark	Temple	Vievar		Dorby	
1740	Mortimer Martyn			Crocker T. Short			
1741	Neve		Johnson	Beauchamp Fuller Gostling Mason		Fuller	
1742		Petre	Pedini	Cooke Gordon Gostling Milner Lovell	Miles (parhelia)		
1743							
1744		Baker					
1745		Costard					

1746	Martyn (2), Miles, Huxham, Baker
1747	
1748	Miles, Waddell; Palmer; Huxham, Brander, Child; Smeaton; Montaine
1749	
1750	Perry, Borlase
1751	Stukeley, Mortimer
1752	Chalmers; Barker, Smith, Baker; Hirst; Pringle, Forster, Colebrooke, Dutton
1753	
1754	
1755	Webb (iris); Edwards (noct. iris)
1756	Collinson (gleam); Daval (extra rainbow), Arderon (halo, parheli)
1757	
1758	Henry, Borlase; Dyer, Miller; Cooper, Whitfeld
1759	
1760	Ray

	AB	eff. l/t	EQ	meteors	optical	storm	waterspout
1761				Swinton	Barker (halo)		
1762					Swinton (anthelion)		
1763		Heberden		Dunn			
1764		Watson		Swinton			
1765		Delaval		Swinton		Griffith	
1766		Lawrence					
1767							
1768							
1769	Swinton	Paxton	Visme				
1770				Swinton			
1771		Henly				Williams	
1772		Kirkshaw					
1773		Kings		Brydone			
1774	Winn	Hamilton				Nicholson	
1775							
1776		Lambert/green					
1777							
1778			Henry				
1779							
1780							

1781	Blagden/Nairne	Lloyd		Cavallo (luminous)
1782				
1783				Brydone
				Stanhope
1784			Cavallo	
1785			Aubert	
1786			Cooper	
1787		More	Edgeworth	Baxter (halos)
			Blagden	
			Piggott	
1788	Withering			
1789				Hey (arches), *also:*
				Wollaston
				Hutchinson
				Franklin
				Piggott
				Cavendish
1790				
1791				
1792		Turner		
1793				Stuges (rainbows)
1794				
1795				

Notes

Notes to introduction

1 John Williams, *The Climate of Great Britain* (London: C. and R. Baldwin, 1806), 252.

2 Jonathan Swift, "A Description of a City Shower" in Robert Greenburg and William Bowan Piper, eds, *The Writings of Jonathan Swift* (New York, London: W. W. Norton, 1973), 518–9; *The Anti-Jacobin Review and Magazine* (December 1806), 339.

3 William Borlase to Henry Baker, 18 July 1763, quoted in P. A. S. Pool, *William Borlase* (Truro: The Royal Institution of Cornwall, 1987), 253. Richard Kirwan, *A Comparative View of Meteorological Observations Made in Ireland since the Year 1789, with some Hints towards Forming Prognostics of the Weather* (Dublin: George Bonham, 1794), 3–4.

4 Sara Warnecke, "A Taste for Newfangledness: The Destructive Potential of Novelty in Early Modern England," *Sixteenth-Century Journal* 26 (1995): 881–96, 886.

5 It should be noted that the present work does not pretend to be a comprehensive account of eighteenth-century meteorological science. I have relatively little to say about organized programs of instrumental observations, partly because the theme has already attracted considerable attention from historians, and partly because the present argument tends to play down the importance of these programs for the protagonists of this study. Nor does this study explore other activities that were affected by the weather, such as navigation, commerce, and medicine.

6 Keith Thomas, *Man and the Natural World* (Harmondsworth, London: Penguin, 1984), 242–69; Simon Schama, *Landscape and Memory* (London: Fontana Press, 1996), 517–79.

7 Robert Plot, *Natural History of Staffordshire* (Oxford: Theatre, 1686), 23–4.

8 Lorraine Daston, "Marvelous Facts and Miraculous Evidence in Early Modern Europe," *Critical Inquiry* 18 (1991): 93–124; Michael McKeon, *The Origins of the English Novel 1600–1740* (Baltimore: The Johns Hopkins University Press, 1987); Simon Schaffer, "Natural Philosophy and Public Spectacle in the Eighteenth Century," *History of Science* 21 (1983): 1–43; Andrea Rusnock, "Correspondence Networks and the Royal Society, 1700–1750" *British Journal for the History of Science* 32 (1999): 155–69; David P. Miller, "'Into the Valley of Darkness': Reflections on the Royal Society in the Eighteenth Century," *History of Science* 27 (1989): 155–66;

9 Adi Ophir and Steven Shapin, "The Place of Knowledge: A Methodological Survey," *Science in Context* 4: 1 (1991): 3–21, 9.

10 See however David E. Allen, *The Naturalist in Britain: A Social History* (Princeton, N. J.: Princeton University Press, 1994); James A. Secord, *Controversy in Victorian Geology: The Cambrian-Silurian Dispute* (Princeton, N. J.: Princeton University Press, 1986); Martin Rudwick, *The Great Devonian Controversy: The Shaping of Natural Knowledge Among Gentlemanly Specialists* (Chicago: University of Chicago Press, 1985); Henrika Kuklick and Robert E. Kohler, eds, *Science in the Field* (Ithaca, N. Y.: Cornell University Press, 1996); W. Conner Sorensen, *Brethren on the Net: American Entomology 1840–1880* (Tuscaloosa and London: The University of Alabama Press, 1995).

11 *Publications of the Surtees Society* (1836), 3: 129.

12 Peter Borsay, *The English Urban Renaissance: Culture and Society in the Provincial Town, 1660–1770* (Oxford: Clarendon Press, 1989).

13 W. J. Keith, *The Rural Tradition. A Study of the Non-Fiction Prose Writers of the English Countryside* (Toronto and Buffalo: University of Toronto Press, 1974).

14 White Kennet, *Parochial Antiquities attempted in the History of Ambrosden and Other Adjacent Parts* (1695), quoted in W. G. Hoskins, *Local History in England* (London: Longman, 1959), 27.

15 Hoskins, *Local History*, 18; James Rosenheim, *The Emergence of a Ruling Order* (London and New York: Longman, 1998), 206. See also John Walsh and Stephen Taylor, "Introduction: The Church and Anglicanism in the 'long' Eighteenth Century," in John Walsh et al., eds, *The Church of England c.1689–1833: From Toleration to Tractarianism* (Cambridge: Cambridge University Press, 1993), 1–67, 28; J. C. D. Clark, *English Society 1688–1832. Ideology, Social Structure and Political Practice during the Ancien Regime* (Cambridge: Cambridge University Press, 1985), 161–73; Jonathan Barry, "Cultural Patronage and the Anglican Crisis: Bristol c. 1689–1775," in John Walsh, et al., eds, *The Church of England*, 191–208; W. M. Jacob, *Lay People and Religion in the Early Eighteenth Century* (Cambridge: Cambridge University Press, 1996).

16 Richard Kirwan, *An Estimate of the Temperature of Different Latitudes* (London: J. Davis, 1787), iii; Aristotle, *Meteorologica*, translated by H. D. P. Lee (Loeb Classical Library, Cambridge, Mass.: Harvard University Press, 1952, reprinted 1962); W. E. Knowles Middleton, "Chemistry and Meteorology, 1700–1825," *Annals of Science* 20 (1965): 125–41; Theodore Sherman Feldman, "The History of Meteorology, 1750–1800: A Study in the Quantification of Experimental Physics," *Diss. Abstr. Int.* 45 (1984): 922-A, 1; Sir William Napier Shaw and Elaine Austin, *Manual of Meteorology, vol. I: Meteorology in History* (Cambridge: Cambridge University Press, 1932), 115; Gustav Hellman, *Meteorologische Beobachtungen vom XIV bis XVII Jahrhundert* (Berlin, 1901, Nendeln, Lichenstein: Kraus Reprint, 1969). H. Howard Frisinger, "Aristotle's Legacy in Meteorology," *Bulletin of the American Meteorological Society* 54 (March 1973): 198; James R. Fleming, *Meteorology in America, 1800–1870* (Baltimore and London: The Johns Hopkins University Press, 1990), 1. John G. Burke, *Cosmic Debris: Meteorites in History* (Berkeley: University of California Press, 1986), 6–10 passim. See also Arthur Hughes, "Science in English Encyclopaedias, 1704–1875, III, Meteorology," *Annals of Science* 9 (1953), 233–64.

17 John Cairns Mitchell, *Chester: Its Situation and Climate with a Record of Remarkable Weather in 1893* (Chester: Courant Press, 1894), 8.

Notes to chapter 1

1 Seneca, *Naturales Quaestiones*, translated by Thomas H. Corcoran (Loeb Classical Library; Cambridge, Mass.: Harvard University Press, 1971), 1.15.

2 George Adams, *Lectures on Natural and Experimental Philosophy* (London: J. Dillon and Co., 1799), 471.

3 William Nicholson, *An Introduction to Natural Philosophy* (London, J. Johnson, 1796), vol. 2, 63; "Extraordinary Fog," *Annals of Philosophy; or Magazine of Chemistry, Mineralogy, Mechanics, Natural History, Agriculture and Arts* 3 (1814): 155; Patrick Murphy, *Meteorology Considered in its Connection with Astronomy, Climate, and the Geographical Distribution of Animals and Plants, Equally with the Seasons and Changes of Weather* (London: J. B. Baillière, 1836), 3.

4 For the theoretical basis of Elizabethan meteorology, see S. K. Heninger, *A Handbook of Renaissance Meteorology* (Durham, N. C.: Duke University Press, 1962), v.

5 A history of the ancient meteorological theories is Otto Gilbert, *Die Meteorologischen Theorien des Griechischen Altertums* (Leipzig: Teubner, 1907). *Meteoria* in Latin means "the state of having one's head in the clouds," or "absent mindedness," P. G. W. Clare, ed., *Oxford Latin Dictionary* (Oxford: Clarendon Press, 1982), 1105. For other meanings see *A Greek-English Lexicon*, compiled by Henry George Liddell and Robert Scott (Oxford: Clarendon Press, 1968), 1120.

6 If not otherwise indicated, I shall use the term "meteor" in its widest connotation – i.e. as the subject matter of the science which treats all the "elevated" phenomena.

7 Aristotle, *Meteorologica*, 338a 26. References to this work will be made in the main text. Apart from Lee's preface to the Loeb edition, I have relied on the following accounts of the work: Roger French, *Ancient Natural History: Histories of Nature* (London and New York: Routledge, 1994), 27–37; W. D. Ross, *Aristotle* (London: Methuen, 1945), 108–11. It is agreed that *De Caelo, Physics, De Generatione and Corruptione* and the first three books of the *Meteorologica* comprise a coherent theoretical opus. The fourth book (on "chemistry" of the elements) is loosely linked with the first three, and its addition was attributed to Eudemus. See Carnes

Lord, "On the Early History of the Aristotelian Corpus," *American Journal of Philology* 107 (1986): 137–61. The recent authoritative treatment of the fourth book is Eric Lewis's introduction to Alexander of Aphrodisias, *On Aristotle's Meteorology 4*, translated by Eric Lewis (Ithaca: Cornell University Press, 1996), 1–61. Lewis argues that *Meteorology 4* is a logical transition from meteorological to biological phenomena.

8 Aristotle, *Metaphysics* 1025b 13.

9 See however Aristotle, *Mete.* 379b 25, 381a 1. On the "imperfect mixtures," see Ross, *Aristotle*, 108–9.

10 Aristotle, *Mete.* 344a 5 ff. Lloyd uses this proposition as an outstanding instance of Aristotle's willingness to loosen the requirements of demonstrative knowledge, and "settle for less than total elucidation of the problem." G. E. R. Lloyd, *Aristotelian Explorations* (Cambridge: Cambridge University Press, 1996), 25. In *Nichomachean Ethics* 1094b 24 (and in *Metaphysics* E), Aristotle assumes that the degree of exactness required by a science ought to reflect the nature of the subject. The same exactness cannot be asked from a mathematician and a rhetor.

11 Aristotle, *Mete.* 340b 23, passim.

12 Aristotle, *Mete.* 346b 11, 344a 9.

13 It is worth mentioning here that for Aristotle (some) comets belong to the sublunary sphere. This is often a reliable criterion for distinguishing Aristotle's doctrines from those of Seneca, for instance, for whom comets move exclusively above the sublunary sphere.

14 For the reasons behind this inclusion, see French, *Ancient Natural History*, 33.

15 Lucretius, *De Natura Rerum*, translated by W. H. D. Rouse (Loeb Classical Library, Cambridge, Mass.: Harvard University Press, 1982), VI. 50, VI. 61.

16 F. H. Sandbach, *Aristotle and the Stoics* (Cambridge: Cambridge Philosophical Society, 1985).

17 L. Edelstein and I. G. Kidd, eds, *Posidonius* (Cambridge: Cambridge University Press, 1972), 503–15.

18 I. G. Kidd, "Theophrastus' *Meteorology*, Aristotle and Posidonius," in William Fortenbaugh and Dimitri Gutas, eds, *Theophrastus: His Psychological, Doxographical and Scientific Writings* (New Brunswick and London: Transactions Publishers, 1992), 294–306. See also Hans Daiber, "The *Meteorology* of Theophrastus in Syriac and Arabic Translation," in ibid., 166–293. Democritus is credited with introducing the method of multiple causation. It is considered to be a consequence of regarding natural phenomena as signs. See Elizabeth Asmis, *Epicurus' Scientific Method* (Ithaca and London: Cornell University Press, 1984), 175–227.

19 Theophrastus, *Meteorology*, I. 2–23, quoted in Daiber, "The *Meteorology* of Theophrastus," 262.

20 See Diogenes Laertius, *Lives of Eminent Philosophers*, translated by R. D. Hicks (Loeb Classical Library; Cambridge, Mass.: Harvard University Press, 1979). In chapter X, Laertius reports on Epicurus's meteorology as favorable to non-exclusive multiple causality.

21 Theophrastus, *Meteorology*, 14. 16, quoted in Daiber, "The *Meteorology* of Theophrastus," 270 (my italics). Cleombrot, one of the characters in Plutarch's *Dialogues*, makes a distinction between divine, demonic, and mortal natures, and shows the ways in which Nature has affirmed the difference: sun and stars represent divine nature, moon has demonic properties, and fiery meteors, shooting stars, and comets all symbolize mortal, corruptible nature.

22 Theophrastus, *Meteorology*, 14. 14–18, quoted in Daiber, "The Meteorology of *Theophrastus*," 270. This premise was elaborated in a dramatic form in Lucretius's *De Natura rerum*. Lucretius is following Aristotle's explanatory scheme, but his poetry suggests an apocalyptic atmospheric excess. Using striking imagery to enhance the sense of strife, he writes about winds fighting against each other, thunder loudly roaring, clouds scraping their bodies as they drag on, and the cloud-imprisoned winds menacing "like wild beasts in the cages" (*De Natura rerum*, VI. 196). Discussing thunderbolts, he reiterates the separation between the divine order and natural disorder and asks "why again [the Gods] aim at deserts and waste their labor? Or are they practicing their arms and strengthening their muscles?" (ibid., VI. 396).

23 Seneca, *Naturales Quaestiones*, 1.15; 2.1.

24 Seneca explains that the region of air represents an essential part of the universe in that it separates and connects the celestial and the earthly. Moreover, it "transfuses to earthly objects the influence of the stars" (*Naturales Quaestiones*, 2.4). On the Etruscan origins of Seneca's divination theory see ibid., 2.32.

25 Pliny the Elder, *Natural History*, translated by H. Rackham (Loeb Classical Library; Cambridge, Mass.: Harvard University Press, 1986), I.45, I.49.

26 A. Locher, "The Structure of Pliny the Elder's Natural History," in Roger French and Frank Greenaway, eds, *Science in the Early Roman Empire: Pliny the Elder, His Sources and Influence* (Totowa, N. J.: Barnes and Noble Books, 1986), 22–3.

27 The distinction between meteors as signs rather than causes of natural disasters is Etruscan in origin (reported by Seneca, *Naturales Quaestiones*, 2.32). Pliny says: "[These misfortunes] did not happen because marvelous occurrences took place but that these took place because the misfortunes were going to occur." (*Natural History*, I, 26). In this context, Pliny also writes of two kinds of thunders: one is produced from "on high" and is prophetic because it originates from the fixed natural causes and "from their own stars." But the other kind, produced by the earthy exhalation, is accidental – "they cause mere senseless and ineffectual thunder-claps as their coming obeys no principle of nature" (*ibid.*, II, 43).

28 L. P. Wilkinson, "The Intention of Virgil's *Georgics*," *Greece and Rome* 55 (1950): 19–28.

29 L. A. S. Jermyn, "Virgil's Agricultural Lore," *Greece and Rome* 53 (1949): 49–69; idem., "Weather-Signs in Virgil," *Greece and Rome* 58 (1951): 28–59.

30 Gustav Hellmann, *Beiträge zur Geschichte der Meteorologie* (Berlin: Behrend & Co, 1917), II, 16–45; 68–90. In the essay on the development of meteorological textbooks, Hellmann cites more than 150 (mostly Latin) commentaries and translations of Aristotle's works dealing with meteors published 1474–1807. In addition, he lists 153 textbook titles printed between 1500–1800; 61 were published in the seventeenth century and 35 in the eighteenth century. For the English context see Heninger, *A Handbook of Renaissance Meteorology*, 16–20. During the sixteenth and seventeenth centuries, the university curricula remained largely classical in their disciplinary divisions and pedagogical methods. See also William T. Costello, S. J. *The Scholastic Curriculum at Early Seventeenth-Century Cambridge* (Cambridge, Mass.: Harvard University Press, 1958), 83–102.

31 On the social, religious and political context and uses of almanacs, see Bernard Capp, *English Almanacs, 1500–1800: Astrology and the Popular Press* (Ithaca, N. Y.: Cornell University Press, 1979). There are a number of authoritative studies on the impact of print on early modern society, the relationship between the oral and printed word, and the religious uses of print following the Reformation. An impressive secondary literature is marshaled in Alex Walsham, "Aspects of Providentialism in Early Modern England" (Ph.D. Dissertation, Cambridge University, 1995), chapter I. The standard sources are Elizabeth Eisenstein, *The Printed Press as an Agent of Change: Communications and Cultural Transformations in Early Modern Europe* (Cambridge: Cambridge University Press, 1979); David Cressy, *Literacy and the Social Order: Reading and Writing in Tudor and Stuart England* (Cambridge: Cambridge University Press, 1980); Tessa Watt, *Cheap Print and Popular Piety* (Cambridge: Cambridge University Press, 1989); Margaret Spufford, *Small Books and Pleasant Histories: Popular Fiction and its Readership in Seventeenth-Century England* (Oxford: Oxford University Press, 1981). Other relevant sources will be cited in chapters 2 and 3.

32 Sister Patricia Reif, "The Textbook Tradition in Natural Philosophy, 1600–1650," *Journal of the History of Ideas* 30 (1969): 17–32.

33 Charles B. Schmitt, "Towards a Reassessment of Renaissance Aristotelianism," *History of Science* 11 (1973): 159–93, 154.

34 On the motivations for this unfair selectiveness in history of early modern science see Schmitt, "Toward a Reassessment," 164–5, where he describes the conflicting cultural and social significacion of the texts written in Latin and the vernacular.

35 Hugh Kearney, *Scholars and Gentlemen: Universities and Society in Pre-Industrial Britain, 1500–1700* (London: Faber and Faber, 1970), 77–90.

36 *Ibid.*, 103–5.

37 On the role of the Laudian Code in maintaining the authority of ancients at Oxford University, see Phyllis Allen, "Scientific Studies in the English Universities of the Seventeenth Century," *Journal of the History of Ideas* 10 (1949): 219–53, 219. For the developments within the Aristotelian core curriculum and its effects on the study of natural philosophy, see Allen G. Debus, *Science and Education in the Seventeenth Century* (London: Macdonald, 1970), 5–15.

The prominence of universities in new scientific developments is discussed by Robert G. Frank, Jr., "Science, Medicine and the Universities of Early Modern England: Background and Sources, Part I," *History of Science* 11 (1973): 194–216, 199, and John Gascoigne, "A Reappraisal of the Role of the Universities in the Scientific Revolution," in David C. Lindberg and Robert Westman, eds, *Reappraisals of the Scientific Revolution* (Cambridge: Cambridge University Press, 1990), 207–61.

38 On the Continental works incidentally dealing with meteors in the Aristotelian manner, see Lynn Thorndike, *History of Magic and Experimental Science*, 8 vols (London, New York: Macmillan, 1923–58); vol. VII, 48, 573, 604, 655, and vol. VIII, 131, 286–7, 313, 376, 607. On the availability and circulation of the literature on weather-signs, see Carroll Camden, Jr., "Elizabethan Almanacs and Prognostications," *The Library* 12 (1931): 100–8, and Don Cameron Allen, *The Star-Crossed Renaissance* (Durham, N. C.: Duke University Press, 1941), 190–246.

39 Hellmann, *Beitrage*, 39–41; Vitale Zuccolo, *Dialogo delle Cose Meteorologiche* (Venetia: Paolo Megietti, 1590).

40 William Gilbert, *De Mundo Nostro Sublunari Philosophia Nova* (Amstelodami: Ludovicum Elzevirium, 1651). The last three chapters of the *De Mundo* have the subtitle "Nova meteorologia contra Aristotelem."

41 Sister Suzanne Kelly, O. S. B., *The De Mundo of William Gilbert* (Amsterdam: Menno Hertzberger & Co., 1965), 45 ff.

42 Liberti Fromondi, *Meteorologicum Libri Sex* (Londini: Typis E. Tyler, 1656). Its subsequent editions were printed in 1639, 1646, 1656, and 1670. See Hellmann, *Beiträge*, 79.

43 "Raised, or suspended," meteors are "imperfect mixed bodies, made of raised vapors and exhalations." The material cause of meteors is "vapor, hot and humid and exhalations, cold and dry." Fromondus, *Meteorologicum Libri Sex*, 1, 22.

44 Similarly, Scipion Dupleix speaks of the division of air in three regions, of meteors of three kinds (fiery, watery and airy) and explains them in the order and terms borrowed from the *Meteorologica*. Dupleix also used Lucretius, Seneca, and Pliny. See Roger Ariew, ed., M. Scipion Dupleix, *La Physique* (1603) (Paris: Libraire Arthème, Fayard, 1990). Practically identical in content and organization is Daniel Widdowes' English translation of Scribonius' *Philosophia Naturales* (1583) which distinguishes between "liveless" and "living" "mixed natures," where "Livelesse [are] meteors which are a hot smoake lifted up by the attractive force of starres, some 15 German miles into the ayre and no higher." Daniel Widdowes, *Naturall Philosophy, or A Description of the World* (London: J. Dowson, 1621). A second edition was printed in 1631. Fromondus was disseminated in England in part through Johannes Amos Comenius, *Naturall Philosophie Reformed by Divine Light, or a Synopsis of Physics* (London: Robert and William Leybourn, 1651) and his *Orbis Sensualium Pictus. A World of Things Obvious to the Senses Drawn in Pictures* (1659) (reprint, Menston, England: The Scolar Press Limited, 1970). Comenius changed Fromondus' vocabulary slightly when he differentiated between "appearing meteors" (halo, rainbow and other optical phenomena), and "concrete substances" (stars, meteors and minerals). "A concrete thing [was] a vapour coagulated endued with some form," and the newly introduced term "concrete" or "concreture" expressed "this degree of creatures which confers nothing but coagulation and figure." Comenius: *Naturall Philosophie*, 114–15. In Comenius's rendition, meteors (proper) were synonymous with the "aerial concretes" and were defined as "things daily concreted" and characterized by "small continuance" (ibid., 128). The Aristotelian theory is clearly discernible in the treatise. In the entry on clouds, Comenius taught on the generation of the thunder and the watery meteors. Thunder was "made of a brimstone, like vapour, which breaking of a cloud with lightning, thundereth and striketh with lightening" (ibid., 19). Wind, storm and whirl-wind were defined in the chapter on air, and earthquakes explained as caused by "the wind underground" (*ventus subterraneus*) (ibid., 15). The often-reprinted meteorological tract of the Dutch experimenter and natural philosopher Cornelius Drebbel represents one of the most influential neo-scholastic renditions of classical ideas. See *Een cort Tractat va de Natuere der Elementen. Ghedaen dooe Cornelis Drebbel* (Harlem, 1604, 1608) and Cornelius Drebbel, *A Dialogue Philosophicall. Wherein Natures Secret is Opened and the Cause of All Motion in Nature Shewed out of Matter and Form* (London: C. Knight,

1612). There were ten imprints of the Dutch edition between 1621 and 1732. See Helmann, *Beitrage*, II, 77–86.

45 William Fulke, *A Goodly Gallerye with a Most Pleasaunt Prospect, into the Garden of Natural Contemplation, to Behold the Natural Causes of all Kynde of Meteors* (1563), edited and annotated by Theodore Hornberger (Philadelphia: The American Philosophical Society, 1979). Hornberger lists nine editions published from 1563 to 1670. The second work was *Meteors: or a Plain Description of all Kinds of Meteors, By W. F., Doctor of Divinitie* (London: William Leake, 1654).

46 Aristotle, Seneca, and Pliny are most frequently mentioned, and a note or two is given to Plutarch, Virgil, Aratus, Theophrastus and Posidonius. The contemporary source was Jerome Cardan's *De Subtilitate rerum* (1550). For Cardan's use of Aristotle's *Meteorologica*, see J. C. Margolin, "Cardan, interprète d'Aristote," in *Platon et Aristote a la Renaissance* (Paris: Vrin, 1976), 307–34. Information on the circulation of Aristotle's meteorological works in the medieval period is given in Bernard G. Dod, "Aristoteles Latinus," in the *Cambridge Companion of Later Medieval Philosophy* (Cambridge: Cambridge University Press, 1982), 45–80. Dod mentions 76 manuscripts of Gerard of Cremona's commentary of the first three books of the *Meteorologica*. Hornberger has the following information on the availability of other ancient meteorological sources prior to 1563: Aristotle's Latin *Meteorologica* (with or without commentaries) was printed at least twenty-eight times, Seneca's *Opera Omnia* eleven times, and Pliny's *Historia Naturalis* saw its thirty-seven Latin editions before Philemon Holland's English translation appeared in 1601.

47 Fulke, *Goodly Gallerye*, 25.

48 *OED* (*Oxford English Dictionary*). In the first English translation of Pliny's *Natural History* (1601), Philemon Holland described the second book as "the Discourse of the World, of coelestiall impressions and meteors." In Heninger, *A Handbook of Renaissance Meteorology*, 12. Petrus Gassendi gave an exhaustive exegesis of the meteorological terminology in his *Syntagmatis philosophici*, book 2, chapter 2. See Petrus Gassendi, *Opera Omnia* (Reprint, Stuttgart-Bad Cannstatt, 1964), vol. 2, 69. Bernhard Varenius solicited in his *Geographia generalis* (Amsterdam, 1664) an investigation of the five kinds of the "celestial affections," one of which consisted of "atmospheric events." See Margarita Bowen, *Empiricism and Geographical Thought* (Cambridge: Cambridge University Press, 1981), 280. And Thomas Stanley, in his rendition of the Aristotle's *Meteorologica* (I, v) introduces the term "Phasmes, such as are called gulfes, chasmes, bloody, colours, and the like." See Thomas Stanley, *The History of Philosophy*, 3 vols (London: Humphrey Moseley, 1660), 64.

49 Fulke, *Goodly Gallerye*, 26–9. Books 2–5 explain meteors individually. The treatment of fiery meteors is exhaustive. Fulke first distinguished between the flames and apparitions and explained them in terms of inflammable exhalations. He thus treated burning stables, dancing goats, round pillars, burning spears, fiery drakes, and wide gapings (in the sky). In the remaining books, he dealt with the airy, watery and earthy impressions as well as with comets, the Milky Way, rivers, and springs.

50 Henry Guerlac, "The Poet's Nitre," *Isis* 45 (1954): 243–55; Allen G. Debus, "The Paracelsian Aerial Niter," *Isis* 55 (1964): 43–61.

51 Guerlac, "The Poets Nitre," 254.

52 John Mayow, *Medico-Physical Works. Being a Translation of Tractatus Quinque Medico-Physici* (Reprint, Edinburgh: Alembic Club, 1907), 149, 151.

53 Thomas Browne, *Pseudodoxia epidemica* (1646), quoted in Guerlac, "The Poet's Nitre," 252. For the presence of the nitre theory in physiological researches, see R. G. Frank, *Harvey and the Oxford Physiologists: A Study of Scientific Ideas and Social Interaction* (Berkeley and Los Angeles: University of California Press, 1980), chapter 5.

54 Browne, *Pseudodoxia epidemica*, 54. The fireworks seemed to have spurred popular interest in fiery meteors, such as the spectacular "Flying Dragon," which inspired pyrotechnicians to model their fireworks on the knowledge and images of those phenomena. See also John Bate, *The Mysteries of Nature and Art* (London, 1635) and Thomas Hill, *Mysteries of Nature* (London, 1646).

55 Robert Boyle, *An Examen of Antiperistasis*, in Robert Boyle, *The Works*, edited by Thomas Birch (Reprint, Hildesheim: Georg Olms Verlagsbuchhandlung, 1966), II, 659–82, 663.

56 John Woodward, *An Essay Toward a Natural History of the Earth and Terrestrial Bodies* (London: Richard Wilkin, 1695). Woodward's involvement in the Mosaic controversy is described by Joseph M. Levine, *Dr. Woodward's Shield: History, Science, and Satire in Augustan England* (Ithaca and London: Cornell University Press, 1977), 37–47. In 1696, Woodward issued his *Brief Instructions for Making Observations* (London: Richard Wilkin, 1696), where this theory was made the basis for observing "whether there follow not great winds, rains, thunders and lightning after the Earthquake is over; whether lastly, Earthquakes happen in any, unless mountainous, cavernous, and stoney [sic], Countries, and in such as yield Sulphur and Nitre," Ibid., 8. Martin Lister's theory of earthquakes posits the same mechanism, but differs in chemistry of the material: "The break of the pyrites, as I have before shown, is sulphur, and it naturally takes fire of itself. The material cause of thunder and lightning, and of earthquakes, is one and the same, viz., the inflammable breath of the pyrites: the difference is that one is fired in the air, the other underground." Martin Lister, "On the Nature of Earthquakes," *Philosophical Transactions* 13 (1683): 512.

57 This is Aristotle's principle from *Meteorologica* I, ii. See W. E. Knowles Middleton, *A History of the Theories of Rain and other Forms of Precipitation* (London: Oldbourne Book Co. Ltd., 1965), 8.

58 John Pointer, *A Rational Account of the Weather Showing the Signs of its Several Changes and Alterations* (London: Aaron Ward, 1723), 103, 121; William Derham, "Observations on the Lumen Boreale, or Streaming," *Phil. Trans.* 35 (1727–28): 245. At different times, it was believed that the vapors released during earthquakes would rise into the upper layers of the atmosphere where they would ferment and produce Aurorae Boreales. This is explicitly asserted by Pasqual R. Pedini in "An Account of the Earthquakes felt in Leghorn," *Phil. Trans.* 42 (1742–43): 90–2.

59 Stephen Hales remarked that "[i]n the history of earthquakes it is observed, that they generally begin with calm weather, with a black cloud. And when the air is clear, just before an earthquake, yet there are then often signs of plenty of inflammable sulphureous matter in the air; such as *Ignes Fatui* or Jack-a-Lanterns, and the meteors called falling stars." Hales, "A Collection of Various Papers presented to the Royal Society concerning several earthquakes, felt in England, and other Countries, in 1750 and other Years," *Phil. Trans.* 46 (1749–50): 601. The preceding atmospheric condition could be disputed but never abandoned. Observations made by the chemist Thomas Henry, the early secretary of the Manchester Literary and Philosophical Society, contradicted Stukeley's review of the weather preceding the London earthquakes of 1750. Stukeley had argued that the phenomena preceding the earthquakes were a long drought, frequent aurorae boreales, fireballs, lightning, and corruscations. Henry's review for the 1777 earthquake produced a cold and wet season, rare thunderstorms and aurorae, and only a single waterspout and fireball. The connection between the occurrences of the Aurora and the weather was familiar throughout the eighteenth century. Samuel Cruwys from Tiverton, Devon complemented his ocular observations with excerpts from the weather journal. See his "Description of an Aurora Borealis," *Phil. Trans.* 32 (1723): 187, and 31 (1721): 1101. The barometer was surmised to be able to register the agitation of the thin auroral vapors. Timothy Neve in his account "[Concerning] the Aurora Australis of March 18, 1738," *Phil. Trans.* 41 (1739–41): 843, wrote: "The three preceding days the barometer fell, and we had great quantities of snow, hail and rain, most of that time; and if I remember right, the lights in the air, of late years, appear after such storms."

60 Thomas Robinson, *New Observations on the Natural History of the World of Matter and this World of Life. Being a Philosophical Discourse, Grounded Upon Mosaick System of Creation, and the Flood. To which is Added a Treatise on Meteorology* (London: John Newton, 1696), 181–9. Robinson here explained that fermented vapors contained inflammable material rising from the earth, supplying the atmosphere with meteoric substance. He summarized this cycle by comparing the earth with "the Great Animal [which] purges herself of gross Humours by Mushrooms and other Pinguid Evaporations" which in the form of "Mineral Spirits and Pinguid Respirations raise, condense in clouds or fall as fertilizing showers." Ibid., 100.

61 Robinson, *New Observations on the Natural History of this World*, 196–8. "If any should think this Account of Thunder to be rather Figment and Romance, than true Natural Philosophy,"

warned Robinson, "I advise him whenever he sees Thunder Packs rising White and Translucent in a South-East Point, when he feels the Air hot and Sulphurous, with some contrary Blasts of Wind coming whistling from the West, that he make haste on to the Top of Crossfelt, or some other high mountain, that gives prospect to both East and West, and he may be informed both as to truth and manner of this Aerial Battle," ibid., 198.

62 Edward Barlow, *Meteorological Essays, Concerning the Origin of Springs, Generation of Rain, and Production of Wind*. (London: J. Hooke, 1715), 76–8. Barlow also published *An Exact Survey of the Tide, with a Preliminary Treatise Concerning Origin of Springs, Generation of Rain, and Production of Wind* (London: Benjamin Tooke and J. Tooke, 1717). The gunpowder analogy can also be found in John Wallis, "On the Production and Effects of Hail, Thunder, and Lightning," *Phil. Trans.* 19 (1695–97): 653–65. The association of lightning, thunder, and uncommon fiery meteors was a trans-European phenomenon, both popular and learned. In 1697, a Jesuit, Franz Reinzer, published the *Meteorologia Philosophico-Politica*, an encyclopedic inventory of the meteo-military phenomena. For an omnipresence of gunpowder see, G. R. Hale, "Gunpowder and the Renaissance: An Essay in the History of Ideas," in Hale, *Renaissance War Studies* (London: The Hambledon Press, 1983), 389–420.

63 Isaac Newton, *Opticks* (Reprint of third edition; New York: Dover, 1979) 379–80. Substituting dry exhalations for the chemical terminology, Newton's summary reads like Aristotle's: "Any of the dry exhalations that gets trapped when the air is in the process of cooling is forcibly ejected as the clouds condense and in its course strikes surrounding clouds, and the noise caused by the impact is what we call thunder . . . [T]he ejected wind burns with a fine and gentle fire, and it is then what we call lightning." "[T]he same natural substance causes wind on the earth's surface, earthquakes beneath it, and thunder in the clouds, for all these have the same substance, the dry exhalation." Aristotle, *Mete.*, 369b 5.

64 See Arthur Hughes, "Science in English Encyclopedias, 1704–1875," *Annals of Science* 9 (1953): 233–64. Hughes shows how the hypothesis of an inflammable fluid (analogous with the Woodwardian subtle exhalation) in supra-atmospheric space was entertained by John Leslie as late as 1842. See also John Clare, *The Motion of Fluids, Natural and Artificial; in Particular that of the Air and Water* (Third edition; London: A Ward, 1747); John Rowning, *A Compendious System of Natural Philosophy* (London: John Rivington, 1779). Rowning's work had 32 editions before 1779. George Cheyne, *Philosophical Principles of Religion Natural and Revealed* (Fourth edition; London: George Strahan, 1734), 289 and William Derham, *Physico-Theology* (London: W. Innys, 1716).

65 John Harris, *Lexicon Technicum, or An Universal Dictionary of Arts and Sciences* (1704, New York and London: Johnson Reprint Corporation, 1966). The presence of nitrous and saline particles helped to "understand great Storms by alledging [sic] a greater quantity of those Particles collected in the Atmosphere at those times when these Storms invade us." Charles Leigh, *The Natural History of Lancashire, Cheshire, and the Peak in Derbyshire* (Oxford: The Author, 1700), 7. The antiquarian John Pointer wrote a survey of several strands of the doctrine in his *A Rational Account of the Weather*, 184–94; On Flamsteed, see Frances Willmoth, "John Flamsteed's Letter concerning the Natural causes of Earthquakes," *Annals of Science* 44 (1987): 23–70. William Derham, "Observations on the Late Storm," *Phil. Trans.* 23 (1702–3): 1530. In an unpublished sketch on the nature of lightning, Derham conceived the amount of inflammable air which was being ignited and in the expansion. During the expansion, he imagined the fiery ball to break a shell of ambient air, "much like that of a granado or a bomb. The flame growing bigger, it forces its passage thru its enclosing shell; and the lightning to stream out at one or more holes made throw the shell." William Derham, "Theory of Storms," *Classified Papers of the Royal Society* IV (1) (1705) 53. See also Martin Lister, "On the Nature of Earthquakes," *Phil. Trans.* 13 (1683): 512–19; Edmund Halley, "Account of the Evaporation of Water," *Phil. Trans.* 18 (1694): 183–90; On the Sulfurous Origins of *Aurora Borealis*, see William Whiston, MA, *An Account of a Surprizing Meteor, Seen in the Air, March 1715–16 at Night* (London: J. Senex, 1716), 57–8; The Rev. Roger Cotes, "A Description of the Great Meteor which was on the 6th of March 1715/6," *Phil. Trans.* 31 (720–21): 66–7; Richard Dobbs, "An Account of an Aurora Borealis Seen in Ireland in September 1725," *Phil. Trans.* 34 (1726–27): 366–7; 459; Edmund Halley, "An Account of the Extraordinary

Meteor Seen all over England, on the 19th March, 1718–19," *Phil. Trans.* 30 (1717–19): 978–92, 989.

66 *Encyclopedia [Britannica]: or a Dictionary of Arts, Sciences, and Misc. Literature* (Philadelphia, 1798), vol. 5, "Meteorology."

67 *Athenian Oracle,* 1718, I, 21, 25, 69, II, 297, III, 240. The same explanation may also be read in John Wallis's, "On the Production and Effects of Hail, Thunder and Lightning," *Phil. Trans.* 19 (1695–97): 653–65. According to these views, earthquakes were the winds escaping from subterranean caves. Above the earth's surface, these fatty underground emanations produced meteors of the lower region – Castor and Polux, Ignis Fatuus, and St Elmo's Fire; see also *Time's Telescope Universal and Perpetual and a Brief Discourse of all kinds of Meteors, or Appearances in Heavens. Natural Prognosticks of Weather* (London: J. Wilcox, J. Oswald, 1734). "Godfridus" equated wind with the dry exhalation, a quintessential Aristotelian concept. *The New Book of Knowledge. Shewing the Effects of Planetary and Astrological Constellations; A Brief Discourse of the Natural Causes of Meteors* (London: A. Wilde, 1758); "An Extraordinary Phenomenon," *Gentleman's Magazine* 20 (1750): 136; "An Extraordinary Meteor," ibid., 136; "On the Causes of Earthquakes," ibid., 89; "A New Hypothesis of Lightning," ibid. 22 (1763): 227; Benjamin Martin, *The Philosophical Grammar. Being a View of the Present State of Experimented Physiology or Natural Philosophy.* (Sixth edition; London: J. Noon, 1762). The first edition appeared in 1743, and the work was reprinted in eleven editions by 1778.

68 Robert Boyle, "An Experimental Discourse of some Unheeded Causes of the Insalubrity and Salubrity of the Air, being a Part of an Intended Natural History of Air," in *The Works,* v, 44.

69 Robert Plot, *The Natural History of Oxfordshire, Being Essay Toward the Natural History of England* (Oxford: Theatre, 1677), 4. Richard Blome, *The Gentleman's Recreations* (London: S. Rotcroft, 1686), 154.

70 Robert Boyle, "An Experimental History of Cold," *Phil. Trans.* 1 (1666): 8–14; idem., "A New Frigorific Experiment," *Phil. Trans.* 2 (1667): 255–62; Dr Nehemiah Grew, "Observations on the Nature of Snow," *Phil. Trans.* 8 (1673): 5193–96; Dr. Martin Lister, "Some Experiments about Freezing," *Phil. Trans.* 14 (1784): 836–39; William Derham, "An Account of the Great Frost in the Winter of 1708/9," *Phil. Trans.* 26 (1708–9): 454–57.

71 Halley, "An Estimate of the Quantity of Vapour Raised out of the Sea by the Warmth of the Sun," *Phil. Trans.* 16 (1686–92): 366; idem., "On the Circulation of the Watery Vapours at the Sea," *Phil. Trans.* 16 (1686–92): 468–81. See also idem., "Observations on Falling Dew, made at Middleburg in Zealand," *Phil. Trans.* 42 (1742–43): 112.

72 "An Unusual Storm of Hail at Lisle in Flanders," *Phil. Trans.* 16 (1686–92): 858. The explosion was to prove the existence of nitre in the hailstone and, by consequence, the nitric origin of the cold in upper regions of the atmosphere.

73 Joseph Wasse, "On the Effects of Lightning," *Phil. Trans.* 33 (1724–25): 369–73.

74 Halley, "An Account of the Extraordinary Meteor," *Phil. Trans.* 30 (1717–19): 989.

75 The experiment in question was titled "Sulphureous and Fresh Air make Efflorecence when mixed," Hales, *Statical Essays* (London: T. Woodward, 1733), 283. Hales's ideas on earthquakes are part of "A Collection of Various Papers Presented to the Royal Society Concerning Several Earthquakes, felt in England, and other Countries, in 1750 and other Years," *Phil. Trans.* 46 (1749–50): 601. He published his thoughts in a separate publication that also addressed the theological issues: *Some Considerations on the Causes of Earthquakes* (London: R. Manby, 1750). "The hand of God is not to be overlooked in these things, etc," ibid., 6. In the same year, the *Gentleman's Magazine* ran the article "Of the Causes of Earthquakes," indicating five possibilities: (1) subterranean waters can emit fumes and blasts if rarified by subterranean fires, (2) air compressed in the caverns can catch fire and explode, (3) fire may result from the collision of the subterraneous vapors, (4) the earth can be the cause and (5) "the inflammable breath of the Pyrites, and substantial sulphur, which takes fire of itself, may be the common material cause of thunder, lightning and earthquakes." *Gentleman's Magazine* 20 (1750): 89.

76 Woodward, *An Essay Toward a The Natural History of the Earth,* 133.

77 E. Chambers, *Cyclopædia: or an Universal Dictionary of Arts and Sciences* (London: James and John Knapton, 1728). Other olfactory reports appeared in Ralph Thoresby, "An Account of a Young Man Slain by Thunder," *Phil. Trans.* 21 (1699): 51; Garden, "Effects of a Very

Extraordinary Thunder," *Phil. Trans.* 19 (1695–97): 311; Wallis, "On the Effects of a Great Storm of Lightning," *Phil. Trans.* 20 (1698): 5; Molyneux, "Account of the Strange Effects of Thunder," *Phil. Trans.* 26 (1708–09): 36; Wasse, "On the Effects of Lightning," *Phil. Trans.* 33 (1724–24): 366; Thomas Lord Lovell, "Of a Meteor Seen near Holkam in Norfolk," *Phil. Trans.* 42 (1742–43): 184–85. Benjamin Cook from Isle of Wight wrote: "we find frequent Mention in the Description of Thunder Storms in hot Climates, that there falls often a flaming bituminous Matter to the Ground, which sometimes burns not to be soon extinguished, but more frequently spatters into an infinite Number of fiery Sparks . . . , always attended with a sulphureous Smell, commonly compared to that of Gunpowder, in "A Letter . . . Concerning a Ball of Sulphur Supposed to be Generated in the Air," *Phil. Trans.* 40 (1737–38): 428. Cook could have in mind Henry Barham's account from Jamaica of the holes in the ground produced by a ball of fire after which "a strong smell of Sulphur remain'd for a good while." Henry Barham, "A Relation of a Fiery Meteor," *Phil. Trans.* 30 (1717–19): 837. On the smell of another fireball, see Mr Chalmers, "Of an Extraordinary Fireball Bursting at Sea," *Phil. Trans.* 46 (1749–50): 366.

78 [Vincentius Bonajutus], "An Account of the Earthquakes in Sicilia," *Phil. Trans.* 18 (1694): 2–10. Four decades later, another Italian wrote that he observed, just before a strong earthquake, "a certain Dark cloud, which passed with a bad Smell." Pasqual R. Pedini, "An Account of the Earthquakes Felt in Leghorn," *Phil. Trans.* 43 (1742–43): 90–2.

79 During the earthquake of March 8, 1750, Hales noted that "there was a hollow, rushing noise in the house, which ended in a loud explosion up in the air, like that of a small canon. The soldiers who were on duty in St James's park, and others who were then up, saw a blackish cloud, with considerable lightning, just before the earthquake began; it was also very calm weather." "A Collection of the Various Papers Concerning Several Earthquakes," *Phil. Trans.* 46 (1749–50): 603.

80 Woodward, *The Natural History of the Earth* (London: Thomas Edlin, 1726), 133. The Hon. Henry Temple wrote about disorders following an Italian earthquake: "I and all the company where I was, as soon as the shock was over, were seized with the shaking, just as if we all had the palsy, our teeth chattering in our heads to such a degree that we could hardly speak; and I found that half the town felt the same effect from it." In Temple's "Concerning an Earthquake at Naples," *Phil. Trans.* 41 (1739–41): 340–3.

81 Whiston, *An Account of a Surprising Meteor* (1716), 54. Derham's correspondents experienced the same sensation during the lights of October 8, 1726: "As [a gentleman] was viewing this Appearance, he plainly perceived a Sulphurous Smell in the Air; and another Person did the same, on the Top of another House near him." William Derham, "Observations of the Lumen Boreale," *Phil. Trans.* 34 (1726–27): 251.

82 The latter was typically defined as "the different temperature of Air." In 1728, Chambers *Cyclopedia* also had two distinct entries for "meteorology" and "weather."

Notes to chapter 2

1 William Cockin, "Account of an Extraordinary Appearance in a Mist," *Phil. Trans.* 70 (1780): 157–62, 160.

2 The term "meteoric tradition" has been introduced by the literary historian Arden Reed. I am partly following his characterization of the tradition and its literary function. In agreement with Michel Serres, Reed claims that the "renegade significance" of meteors was their representation of the natural world in its most unstable, unpredictable, and indeterminate mode. The meteoric tradition was the literary and scientific expression of that understanding. See Michel Serres, *Naissance de la physique dans le texte de Lucrèce: Fleuves et Turbulences* (Paris: Editions de Minuit, 1975), 85–6; Arden Reed, *Romantic Weather: The Climates of Coleridge and Baudelaire* (Hanover and London: University Press of New England, 1983), 4.

3 Until the end of the eighteenth century, meteorological writings relied on the etiology of Aristotle's *Meteorologica*. See Arthur Hughes, "Science in English Encyclopaedias, 1704–1875, III, Meteorology," *Annals of Science* 9 (1953): 233.

4 Sir William Napier Shaw and Elaine Austin, *Manual of Meteorology. vol. I: Meteorology in History*, 113; Feldman, "The History of Meteorology," 218; Lorraine Daston, "Historical Epistemology," in Chandler et al., eds, *Questions of Evidence, Proof, Practice, and Persuasion across the Disciplines* (Chicago: University of Chicago Press, 1994), 285.

5 Feldman, "The History of Meteorology," 219.

6 Roy Porter, "Gentlemen and Geology: The Emergence of a Scientific Career, 1660–1920," *The Historical Journal* 21, 4 (1978): 809–36.

7 Nicholas Goodison, *English Barometers 1680–1860* (New York: Clarkson N. Potter, Inc., 1983), 234.

8 The remaining contributions on meteorology consisted of weather diaries and experimental and theoretical discourses. The custom of keeping weather diaries and its relationship with the rise of quantitative meteorology will be examined in chapter 5.

9 William Nicholson, *Introduction to Natural Philosophy* (London: J. Johnson, 1796), 63; [Alexander Tilloch], "Proceedings of Learned Societies," *The Philosophical Magazine* 24 (1806): 272.

10 L. F. Kaemtz, *A Complete Course of Meteorology*, translated by C. V. Walker (London: Hippolyte Baillière Publisher, 1845), 464.

11 Pascal, *Œuvres* (1963), quoted in Peter Dear, *Discipline and Experience: The Mathematical Way in the Scientific Revolution* (Chicago: The University of Chicago Press, 1995), 200–1.

12 Regarded as the "birth of empirical meteorology," these schemes are the most investigated aspect of early modern meteorology. See R. T. Gunther, *Early Science in Oxford*, vol. 4, (London: Dawson of Pall Mall, 1925), 34, 170; G. J. Symons, "The History of English Meteorological Societies, 1823 to 1880," *Quarterly Journal of the Meteorological Society* 7 (1881): 65–8; Gordon Manley, "The Weather and Diseases: Some Eighteenth Century Contributions to Observational Meteorology," *Notes and Records of the Royal Society of London* 9 (1952): 300–7; H. E. Landsberg, "Roots of Modern Climatology," *Journal of the Washington Academy of Sciences* 54 (1964): 130–41; A. K. Khrgian, *Meteorology: A Historical Survey* (Jerusalem: The Israel Program for Scientific Translations, 1970); D. J. Schove and David Reynolds, "Weather in Scotland, 1659–1660: The Diary of Andre Hay," *Annals of Science* 30 (1973): 165–77; Karl Schneider-Carius, *Weather Science, Weather Research*. Washington (New Delhi: The Indian National Scientific Documentation Centre, 1975); Howard Frisinger, *The History of Meteorology: to 1800* (New York: Science History Publications, 1977); J. Kington, *The Weather of the 1780s over Europe* (Cambridge: Cambridge University Press, 1978); Lisa Shields, "The Beginnings of Scientific Weather Observation in Ireland (1684–1708)," *Weather* 40 (1984): 304–11.

13 Lucinda McCray Beier, "Experience and Experiment: Robert Hooke, Illness and Medicine," in Michael Hunter and Simon Schaffer, eds, *Robert Hooke: New Studies* (London: The Boydel Press, 1989), 235–51; see also, Kenneth Dewhurst, *Dr. Thomas Sydenham (1624–1689)* (Berkeley and Los Angeles: University of California Press, 1966), 60–7. Samuel Jeake kept his weather journal for this purpose. Boyle and Plot thought likewise. See M. Hunter and A. Gregory, eds, *An Astrological Diary of the Seventeenth Century: Samuel Jeake of Rye, 1652–1699* (Oxford: Clarendon Press, 1988). Plot wrote to Lister concerning the latter's scheme: "thence too in time we may hope to be inform'd how far the positions of the planets in relation to one another, and to the fixt stars, are concerned with the alteration of the weather." Quoted in Hellmann, *Meteorologische Beobachtung*, vol. 3, 167.

14 James Jurin's, "Invitatio ad Observatione Meteorologicas Communi Consilio Instituendas," *Phil. Trans.* 32 (1723): 422–7; see also Rusnock, ed., *The Correspondence of James Jurin (1684–1750)* (Amsterdam: Rodopi, 1996) and Jan Golinski, "Barometers of Change: Meteorological Instruments as Machines of Enlightenment," in William Clark, Jan Golinski, and Simon Schaffer, eds, *The Sciences in Enlightened Europe* (Chicago: The University of Chicago Press, 1999): 69–93.

15 [Isaac Greenwood], "A New Method for Composing a Natural History of Meteors Communicated in a Letter to Dr Jurin," *Phil. Trans.* 35 (1727–28): 390–402.

16 Feldman, "The History of Meteorology," 1983, 205. In this work, Feldman has documented the history of the failure to establish a lasting network of weather observers. See also in this respect a somewhat optimistic analysis by James E. McClellan III, *Colonialism and Science:*

Saint Domingue in the Old Regime (Baltimore and London: The Johns Hopkins University Press, 1992), especially 163–80; for the French undertaking see J. P. Desaive et al., *Médecins, climat and épidémies à la fin du XVIIIᵉ siècle* (Paris: CNRS, 1972).

17 J. A. Kington, "The Societas Meteorologica Palatina: An Eighteenth-Century Meteorological Society," *Weather* 30 (1974): 416–26. Of special interest is this respect is the recent study of Henry E. Lowood, *Patriotism, Profit, and the Promotion of Science in the German Enlightenment: The Economic and Scientific Societies, 1760–1815* (New York: Garland Publishing, 1991), 205–79. Lowood points out that, in this period, weather became an ideal subject for the "regional science," *Landskunde*: "The accumulation of data on temperature, sky conditions, the influence of weather on crops, and distinctive meteorological events had an obvious place alongside the facts and figures of topography, economy, geology, botany and zoology pertaining to a particular region," ibid., 247.

18 François Neufchâtel, "Recherches historiques sur les années froides et humides," (1816), quoted in Feldman, "History of Meteorology," 273.

19 The most thorough discussion about English early modern genre of "strange news" is Alex Walsham, "Aspects of Providentialism in Early Modern England," chapter 2. My discussion is also based on Keith Thomas, *Religion and the Decline of Magic* (New York: Charles Scribner's Sons, 1971), chapter 4; Katharine Park and Lorraine J. Daston, "Unnatural Conceptions: The Study of Monsters in Sixteenth- and Seventeenth-century France and England," *Past and Present* 92 (1981): 20–54; Dennis Todd, *Imagining Monsters: Miscreation of the Self in Eighteenth-Century England* (Chicago: The University of Chicago Press, 1995), 153–6; Sandra Clark, *The Elizabethan Pamphleteers: Popular Moralistic Pamphlets: 1580–1640* (Rutherford, N. J.: Humanities Press, 1983); John Redwood, *Reason, Ridicule and Religion* (London: Thames and Hudson, 1976), chapter 6; T. D. Kendrick, *The Lisbon Earthquake* (Philadelphia, New York: J. B. Lippincott Co., 1955); Andrew Dickson White, *A History of the Warfare of Science with Theology in Christendom* (1896, reprint; Gloucester, Mass.: Peter Smith, 1978); Matthias A. Shaaber, *Some Forerunners of the Newspaper in England 1476–1622* (Philadelphia: University of Pennsylvania Press, 1929), chapter 6; Llewellyn M. Buell, "Elizabethan Portents: Superstition or Doctrine?" in idem., *Essays Critical and Historical* (Berkeley and Los Angeles: University of California Press, 1950), 27–41; J. Paul Hunter, *Before Novels: The Cultural Contexts of Eighteenth-Century English Fiction* (New York: W.W. Norton, 1992); McKeon, *Origins of the English Novel*, chapters 1 and 2.

20 Daston, "Marvelous Facts", 93–124, 100. Daston provides a useful analysis of the differences between the supernatural (*bona fide* miracles), preternatural (marvels) and prodigious natural productions. Some early commentators thought that an interest in such phenomena especially reflected the "newfangledness" of the English. Thomas Sprat, for instance, was embarrassed to note that "[t]his wild amusing of men's minds, with *Prodigies*, and conceits of *Providences*, has been one of the most considerable causes of those spiritual distractions, of which our Country has long bin the *Theater*." Sprat, *The History of Royal Society* (1665, Reprint; St. Louis: Washington University Press, 1966), 362.

21 Quoted in Buell, "Elizabethan Portents," 29.

22 *Mirabilis Annus (Eniagtos Terastios) or the year of Prodigies and Wonders* (London, 1661), ii.

23 Thomas Hill, *The Contemplation of Mysteries: Contayning the Rare Effects and Significations of Certayine Comets, and a Briefe Rehearsall of Sundrie Hystorical Examples* (London: Henry Denham, 1571); Thomas Willsford, *Natures Secrets, or, the Admirable and Wonderfull History of the Generation of Meteors* (London: Brooke, 1658).

24 On the link between political and natural turbulence, see Daston, "Marvelous Facts," 251. Specimens of the political pamphleteering included *Strange Fearful & True News, which Happened at Carlstadt, in the Kingdome of Croatia* (London: George Vincent and William Blackwall, 1606); L. Brinkmair, *The Warnings of Germany, by Wonderful Signes* (London: J. Horton, 1638); *A Report from Abbington towne in Berkshire, being a relation of what harme Thunder and Lightning did on Thursday last upon the body of Humphrey Richardson, a rich miserable farmer: with Exhortation for England to repent* (London: William Bowde, 1641); *A Strange and Wonderful Relation of the Miraculous Judgment of God in the late Thunder and Lightning* (London: Bernard Alsop, [1620–50]); *More Warning Yet. Being a True Relation of A Strange*

and most Dreadful Apparition Which was seen in the Air By several persons at Hull (London: J. Cottrel, 1654).

25 James Howell, *Dendrologia. Dodona's Grave, or the Vocall Forest* (London: T. B[adger], 1640).

26 *A Strange Wonder or The Cities Amazement. Being a Relation occasioned by a Wonderful and unusual Accident that happened in the River of Thames, Friday Feb. 4, 1641* (London: John Thomas, 1641).

27 Thomas, *Religion and the Decline of Magic*, 141–4. Stephen Baskerville discusses social uses of the millennium in puritan sermons in his *Not Peace but a Sword: The Political Theology of the English Revolution* (London: Routledge, 1993), 63–7; see also Christopher Hill, " 'Till the Conversion of the Jews'," in Richard Popkin, ed., *Millenarianism and Messianism in English Literature and Thought, 1650–1800* (Leiden: E. L. Brill, 1988), 12–37; on millenarian publications in the seventeenth-century, see Le Roy Edwin Froom, *Prophetic Faith of Our Fathers: The Historical Development of Prophetic Interpretation* (Washington: Review and Herald, 1946–54), vol. 2. In seventeenth-century "methaphysical sermons," *exempla* of extraordinary events served to elicit wonder and to prepare a congregation for the wonderful works of God. See Horton Davies, *Like Angels from a Cloud: The English Metaphysical Preachers, 1588–1645* (San Marino: Huntington Library, 1986), 58–66.

28 *The English Chapmans and Travellers Almanack for the Year of Christ 1697* (London: Thomas James, 1697). The uses of comets and mock-suns in late Stuart political culture are the subject of Sara Schechner Genuth, *Comets, Popular Culture and the Birth of Modern Cosmology* (Princeton, N. J.: Princeton University Press, 1997); Ann Geneva, *Astrology and the Seventeenth-Century Mind: William Lilly and the Language of Stars* (Manchester and London: Manchester University Press, 1995).

29 D. H. Atkinson, *Ralph Thoresby the Topographer; his Town and Times* (Leeds: Walker and Laycock, 1885), 58, 91.

30 *A True Relation of the Wonderful Apparition of a Cross in the Moon, visible at Wexford and other places, kingdom of Ireland upon April the 4th, 1688 for two Hours together, to the Admiration of many Eminent Persons, an others, that were Spectators thereof* (Dublin n.d.).

31 *A True and Wonderful Narrative of Two Particular Phenomena, which were seen in the sky in Germany* (Philadelphia: Anthony Armbruster, 1764). The relationship between the popular culture and millenarian ephemera, published during the eighteenth century, is discussed in J. F. C. Harrison, *The Second Coming: Popular Millenarianism, 1750–1850* (New Brunswick: Rutgers University Press, 1979), 39–54. For a recent survey of famous "aerial anomalies," see Jerome Clark, *The UFO Encyclopedia, vol. 2, The Emergence of a Phenomenon: UFOs from the Beginning through 1959* (Detroit: Omnigraphics Inc, 1993), 55–64.

32 Hunter, *Before Novels*, 208.

33 Edward Ward, *British Wonders: or, A Poetical Description of the Several Prodigies and Most Remarkable Accidents that have Happened in Britain since the Death of Queen Anne* (London: John Morphew, 1717), 52.

34 John Dunton, *The Life and Errors of John Dunton* (London: J. Nichols 1818), II, 435. For Crouch's bibliography see Hunter, *Before Novels*, 211–12.

35 These were subtitles of Crouch's works *History of Earthquakes*, *History of the Kingdoms of Scotland and Ireland*, and *Kingdom of Darkness*. Quoted in Robert Mayer, *History and the Early English Novel: Matters of Fact from Bacon to Defoe* (Cambridge: Cambridge University Press, 1997), 132. Mayer describes Crouch's enterprise as "the secret history of *Britannia*: a history known to many but rarely reported."

36 *God's Marvellous Wonders in England in 1694* (London: P. Brocksby, 1694).

37 Ibid., 16.

38 *London Journal*, September 24, 1725, quoted in Hunter, *Before Novels*, 217.

39 On hybrids, see Bruno Latour, *We Have Never Been Modern* (Cambridge, Mass.: Harvard University Press, 1993), 29–32.

40 Hans Frei, *Eclipse of Biblical Narrative: A Study in Eighteenth and Nineteenth Century Hermeneutics* (New Haven and London: Yale University Press, 1978), 77–8; Lorraine Daston, "Marvelous Facts," 263–70; McKeon, *Origins of the English Novel*, 83–4.

41 McKeon, *Origins of the English Novel*, 84. For the eighteenth-century versions of the argument from miracles, see Leslie Stephen, *History of English Thought in the Eighteenth Century*, 2 vols (New York: Putnam's Sons, 1927), vol. 1, 228–53.

42 John Redwood, *Reason, Ridicule and Religion*, chapter 6; Peter Dear, "Miracles, Experiments, and the Ordinary Course of Nature," *Isis* 81 (1990): 671.

43 Richard Baxter, *The Certainty of the Worlds of Spirits* (1691), quoted in McKeon, *Origins of the English Novel*, 84.

44 On the charges of sedition see Hill, " 'Till the Conversion of the Jews," 16.

45 Descartes, *Meteorology*, 330. The "military" subspecies of the apparition genre continued into the eighteenth century, with its acme in Franz Reinzer's elaborate *Meteorologia Philosophico-Politica* (Augsburg, 1697, 1712).

46 W. K., *News from Hereford; or, a Wonderful and Terrible Earthquake* (London, 1661) is laden with circumstantial detail and ends with "a list of names of the [nine] persons that witnesseth the truth of this . . . And divers others, too many to be here inserted."

47 Walsham, "Aspects of Providentialism," 93, n. 123.

48 William Averell, *A Wonderful and Strange News which Happened in the County of Suffolk and Essex the first of February, where it rained wheat the space of six or seven miles compass* (London, 1583).

49 *Looke Up and See Wonders. A miraculous Apparition in the Ayre, lately seen in Barke-shire at Bawlkin Greene neere Hatford, April 9, 1628* (London: Roger Michell, 1628).

50 William B. Ashworth Jr., "Natural History and the Emblematic World View," in David C. Lindberg and Robert S. Westman, eds, *Reappraisals of the Scientific Revolution*, 324.

51 Ibid., 4: 94–6.

52 Roger French uses the term "paradoxographers" to speak of the Hellenistic collectors of natural oddities and wonders. See his *Ancient Natural History*, 299–303.

53 Daston, "Marvelous Facts," 262.

54 Francis Bacon, *The New Organon*, in James Spedding et al., eds, *The Works of Francis Bacon*, 8 vols (London: Longmans and Co., 1870), 5: 274.

55 Ibid., 5: 274.

56 Ibid., 5: 29.

57 Ibid., 5: 275–7.

58 Ibid., 5: 285.

59 Ibid., 5: 438–9.

60 Ibid., 5: 139.

61 Similar questions are addressed in Aristotle's three-chapter discourse on wind in the second book of *Meteorologica*, as well as in Theophrastus's separate treatise on wind. Bacon's analysis uses the information from these and other classical works, such as Pliny's and Seneca's, and subscribes to the Aristotelian view that winds represent individual aerial rivers "controlled by the [vaporous] nurseries." Ibid., 5: 175.

62 K. Theodore Hoppen, "The Nature of the Early Royal Society," *British Journal for the History of Science* 9 (1976), 6.

63 Charles Webster described Childrey's book as a "miscellaneous collection of travelers' tales, designed to entertain the rich and over-awe the poor with its pseudo-erudition." Charles Webster, *The Great Instauration: Science, Medicine and Reform, 1626–1660* (London: Duckworth, 1975), 429. Joshua Childrey, *Britannia Baconica, or The Natural Rarities of England, Scotland, & Wales* (London: the Author, 1660).

64 For the primacy of Oxford in geographical studies see J. N. L. Baker, *The History of Geography* (New York: Barnes and Noble, 1963), 15.

65 Childrey apparently used the chronology of catastrophes for astrological purposes, which is made clear from his general interest in the astrology of tides and violent weather. Ibid., 109–10.

66 Michael Hunter, *John Aubrey, and the Realm of Learning* (London: Duckworth, 1975), 94.

67 Fairfax to Oldenburg, 25 April 1667, in A. R. Hall, and M. B. Hall, eds, *The Correspondence of Henry Oldenburg*, 6 vols (Madison, 1965–86), vol. 3, 401.

68 The chorographic sources include Camden, Lambarde, Dugdale, but Pliny, and other ancients also appear.

69 Childrey, *Britannia Baconica*, 78.

70 Childrey, *Britannia Baconica*, 79.

71 *History of Geography*, 23.

72 Childrey, *Britannia Baconica*, Preface, 2.

73 Ibid., 2. "We have Baiæ at Bath; the Alpes in North-Wales, Mount Baldus under the Picts Wall, the Spaw in Yorkshire," and so on.

74 Ibid., Preface, 10–11.

75 Ibid., Preface, 14. Childrey had such a strong predilection for the bloody rains that he suggested the investigation of their causes to the officers of the Royal Society.

76 R. F. Jones portrays Childrey as an eccentric philosophic fop who "presents an example of the struggle between the budding scientific spirit of the age and the inherited chaos of superstition and fear," Richard Foster Jones, *Ancients and Moderns* (St Louis: Washington University Press, 1961), 162.

77 Thomas Sprat, *History of the Royal Society* (1667), 167.

78 Joseph Glanvill, *The Vanity of Dogmatizing* (London, 1661), and *Plus Ultra or, The Progress and Advancement of Knowledge Since the Days of Aristotle* (London, 1668), quoted in McKeon, *Origins of the English Novel*, 69, 70.

79 McKeon, *Origins of the English Novel*, 71.

80 Sprat, *History of the Royal Society*, 90–1. Amadis de Gaula was the hero and the title of an Arthurian prose romance of chivalry (1508), possibly Portuguese in origin. The romance narrates Amadis's incredible invincibility and his great love for the daughter of the King of England, Oriana. Numerous sequels and imitations appeared throughout the sixteenth century. The first English adaptation appeared in 1567. *Encyclopedia Britannica, Micropaedia* 1, 287.

81 Glanvill, *The Vanity of Dogmatizing*, quoted in McKeon, *Origins of the English Novel*, 69. Along similar lines, the Restoration virtuoso John Aubrey stressed that the accounts of marvels should be collected and they could be true because "[t]he matter of this collection [of apparitions] is beyond human reach, we being miserably in the dark as to the economy of the invisible world." quoted in Mayer, *History and the Early English Novel* (Cambridge: Cambridge University Press, 1997), 121–2. Interest in the so-called fabulous stories had other sources too. In the science of *fossilia*, as Paolo Rossi noted, the students could seek traces of the history of nature in civil histories, because the actual catastrophes could have been disguised in the "fabulous" accounts. Robert Hooke desired that those "better versed in ancient Historians" should be equipped to discover past climatic changes. Paolo Rossi, *Dark Abyss of Time: The History of the Earth and the History of Nations from Hooke to Vico* (Chicago: The University of Chicago Press, 1984), 16–17.

82 The return was almost nil, however. The single weather record he received came from a "young melancholicke Gentleman, a Cadet, who you may see makes it a great part of his businesse to converse with ye aire," and who was especially apt for the task being astrologically predisposed for impartiality. Childrey feared the ambitious number of queries was the setback. *The Correspondence of Henry Oldenburg*, vol. 6, 604, 626, 108.

83 Glanvill, *Plus Ultra*. Quoted in McKeon, *Origins of the English Novel*, 68.

84 Sprat, *History of the Royal Society*, 195–6.

85 Robert Moray, "Extraordinary Tides in the West Isles of Scotland," *Phil. Trans.* 1–2 (1665–67): 53; Thomas Neale, "A Sad Effect of Thunder and Lightning," *Phil. Trans.* 1–2 (1665–67): 247; "Of Four Suns and Two Uncommon Rainbows observed in France," *Phil. Trans.* 1–2 (1665–67): 219.

86 "Strange Frost about Bristol," *Phil. Trans.* 8 (1673): 5138; Dr John Wallis, "Of an Unusual Meteor," *Phil. Trans.* 11–12 (1676–78): 863; Dr Nathaniel Fairfax, "Hail Stones of Unusual Size," *Phil. Trans.* 1–2 (1665–67): 481.

87 Zachary Mayne, "Concerning a Spout of Water that Happened at Topsham," *Phil. Trans.* 18 (1694): 28; "On a Whirlwind," *Phil. Trans.* 18 (1694): 192; George Garden, "A Very Extraordinary Thunder," *Phil. Trans.* 19 (1695–97): 311; Robert Mawgridge, "Surprising Effects of a Terrible Clap of Thunder," *Phil. Trans.* 19 (1695–97): 782; Robert St Clair, "On an Erruption of Fire," *Phil. Trans.* 21 (1699): 378; Stephen Gray, "An Unusual Parhelion and Halo," *Phil. Trans.* 22 (1700–01): 535.

Notes to chapter 3

1 Edward Ward, *British Wonders: or, A Poetical Description of the Several Prodigies and Most Remarkable Accidents that have Happened in Britain since the Death of Queen Anne* (London: John Morphew, 1717).

2 G. Gregory, *The Economy of Nature: Explained and Illustrated on the Principles of Modern Philosophy* (London: J. Johnson, 1796), 517–18.

3 "It was the people who gaped in blank astonishment at a prodigy and soon forgot it; it was the endeavor of the leaders of the nation to make them remember it and interpret it as a sign from God." Buell, "Elizabethan Portents," 41. For William Fulke, the ultimate purpose of meteors was also to "threaten [God's] vengeaunce, to punishe the world, to move to repentance." Fulke, *Goodly Gallerye*, 30.

4 Quoted in Buell, "Elizabethan Portents," 37.

5 A century later, "The Protestant Wind" carried some 500 hundred sailing ships of Prince William's Dutch Fleet into a favorable haven in Devon, while keeping the pursuing English Royal Navy at bay for a crucial three days, during which time the Dutch troops landed to rescue the Protestant cause. This time the wind was easterly. See J. L. Anderson, "Combined Operations and the Protestant Wind: Some Maritime Aspects of the Glorious Revolution of 1688," *The Great Circle* 5 (1983): 13–23. See also, Jonathan I. Israel and Geoffrey Parker, "Of Providence and Protestant Winds: The Spanish Armada of 1688," in Jonathan I. Israel, ed., *The Anglo-Dutch Moment: Essays on the Glorious Revolution and its World Impact* (Cambridge: Cambridge University Press, 1991), 335–63.

6 In 1607, Simon Harward wrote that God had stirred up "the inhabitants of the said town, as us all, to feare him, & to give us some taste of his judgments, to summon us all to true repentance." Simon Harward, *A Discourse of the Several Kind and Causes of Lightnings* (London: John Windet, 1607), 3. The Dutch Presbyterian Georg Nuber considered lightning to be the retribution for the sins of "impenitence, incredulity, neglect of the repair of churches, fraud in the payment of the tithes to clergy, and oppression of subordinates," quoted in White, *A History of the Warfare of Science with Theology*, 333–4. Catholic practices of baptizing bells were not welcomed by Luther and Protestants; devils were far too powerful to be frightened by their clang. The English clergy shared this opinion and declared the rite of bell-baptism unlawful and ineffective. Exegeses on lightning also include John Hilliards's *Fire From Heaven* (1613), and Robert Dingley's *Vox Coeli* (1658). Other early modern texts that touch upon the subject are Stephen Batman, *The Doome Warning All Men to the Iudgmente* (1581), Thomas Beard, *The Theatre of Gods Judgments* (1597), *A Divine Tragedie lately acted* (1636), and an anti-Royalist collection *Mirabilis Annus* (1661) are the most representative of the genre. The extent to which clergy thought it crucial to collect the instances of God's intervention is suggested by the *Designe for Registring of Illustrious Providence*, a scheme which Matthew Poole modeled on Bacon's natural history project. See Thomas, *Religion and the Decline of Magic*, 94. For a mythopoetic meaning of thunder of Giambattista Vico, see Michael Mooney, *Vico in the Tradition of Rhetoric* (Princeton: Princeton University Press, 1985), 211–15.

7 *Looke Up and See Wonders*, Preface to the Reader (unpaginated).

8 Blair Worden, "Providence and Politics in Cromwellian England," *Past and Present* 109 (1985): 55; Redwood, *Reason, Ridicule and Religion*, 145 ff. The arguments seeking to synthesize general and special providence defined the "unofficial apologetic position of the Royal Society" during the period he called the "first" Royal Society (1660–1741). See James E. Force, "Hume and the Relation of Science to Religion Among Certain Members of the Royal Society," in John W. Yolton, ed., *Philosophy, Religion and Science in the Seventeenth and Eighteenth Centuries* (Rochester: Rochester University Press, 1990), 228–47, 231.

9 Jacob Viner, *The Role of Providence in the Social Order* (Philadelphia: American Philosophical Society, 1972), 12 ff; J. W. Packer, *The Transformation of Anglicanism* (Manchester: Manchester University Press, 1969), 45–87; Margaret Jacob, *The Newtonians and the English Revolution 1689–1720* (Ithaca: Cornell University Press, 1976), 34–5.

10 Redwood, *Reason, Religion and Ridicule*, 135, who discusses Dr Samuel Bradford, *The Credibility of the Christian Revelation, from its intrinsick evidence* (London, 1699).

11 *A Strange and Wonderful Relation of a Clap of Thunder which lately set Fire to the Dwelling-House of one Widdow Rosingrean living in the town of Ewloe, in the parish of Howerden in the County of Flint* (London: W. Harris, 1677); *A Brief Relation of a Wonderful Account of a Wonderful Accident a Dissolution of the Earth in the Forest of Charnwood, about two miles from Loughborough in Leicestershire; Lately done, and discovered and resorted to by many people, both Old, and Young* (London: Two Lovers of Art, I. C. and I. W., 1679); *A Strange and True Account from Shresbury of a Dreadful Storm, which happened on the 4th of May last* (London, 1681); *Strange News from Oxfordshire: being a True and Faithful Account of a Wonderful and Dreadful Earthquake that happened in those Parts on Monday the 17th of this Present September 1683; Strange and Terrible News from Alton in Hampshire: Being a full Account of a Dreadful Tempest which happened there by Thunder and Lightning, December 19th, 1686* (London: for S. M., 1687); *A True Account of the damages done by the Late Storm which happened between the Hours of 12 and 4 of the Clock on Sunday Morning, January 12, 1689* (1689); *God's Marvellous Wonders in England in 1694* (London: P. Brocksby, 1694); *Strange and Wonderful News from Ireland, Giving a Dreadful Relation of a Prodigious Motion of the Earth near Charleville, Limerick, June 7, 1697* (London: J. Wilkins, 1697).

12 *Memento to the World; Or An Historical Collection of Divers Wonderful Comets and Prodigious Signs in Heaven* (London: T. Haly, 1680).

13 Reverend Divine, *A Practical Discourse on the Late Earthquakes with An Historical Account of Prodigies and their Various Effects* (London: J. Dunton, 1692), 7.

14 Ibid., 14. The tract argued that the ancients *and* moderns sanctioned the divine function of these phenomena: Church Fathers, philosophers, poets, statesmen, lawyers, historians, "yea Men of the most different religions, Jews, Pagans, Christians, Protestants, Papists agree that the things above named portend great changes to the Publick." Did not those flaming Torches appearing in 1664 usher in the "Raging Pestilence, and Devouring Fire? And now of late we have been alarmed with a Prodigy of another nature, the *Trembling and Shaking of the Earth.*" Ibid., 21, 22.

15 Worden, "Providence and Politics," 61. For the role of providentialism in Oliver Cromwell's thought see Christopher Hill, *God's Englishman* (New York: Penguin Books, 1970), chapter 9.

16 Alex Walsham suggests that English compilations of special providences published since the Elizabethan period largely plagiarized pre-Reformation Continental sources and were a continuation of a Catholic homiletic tradition. Walsham, "Aspects of Providentialism," 70–1. See also John Spurr, "'Virtue, Religion and Government': the Anglican Uses of Providence," in Tim Harris, Paul Seaward and Mark Goldie, eds, *The Politics of Religion in Restoration England* (London: Basil Blackwell, 1990), 29–47.

17 According to a contemporary, in Holland during these tense weeks "the common thing every morning . . . was first to go and see how the wind sate, and if there were any probability of a change. When any person came unto a house, in the heart of their city, concerning any manner of businesse, the very first question of all was, Sir, I pray how is the wind to day? Are we likely to get an easterly wind ere long? Pray God, send it, and such like. The ministers themselves pray'd that God would be pleas's for to grant an East Wind." Quoted in Jonathan I. Israel and Geoffrey Parker, "Of Providence and Protestant Winds: the Spanish Armada of 1588 and the Dutch Armada of 1688," in Jonathan I. Israel, ed., *The Anglo-Dutch Moment,* 357–9.

18 Gilbert Burnet, *History of My Time* (London: J. Hopkins, 1723), vol. 1, 789–90. At the centennial commemoration of the Protestant Wind, Samuel Stennet, the minister of the Little Wilde Street Chapel in London, reminded the audience that "the great storm that blew from the West, immediately upon the Prince's landing, which prevented the King's fleet from continuing their pursuits, and so shattered them that they were no more fit for service that year; was a providential Circumstance in favor of the Revolution." Samuel Stennet, DD, *A Sermon in Commemoration of the Great Storm of Wind on November 27, 1703 and of the More Dreadful Storm Which Threatened the Destruction of British Freedom at the Eve of the Revolution, Preached in Little-Wilde Street, Nov. 27, 1788* (London: J. Buckland, 1788).

19 "An Account of what happened from Thunder in Carmarthenshire; . . . communicated to the Royal Society, By John Eames, FRS as he received it in a Letter from Mr Evan Davies," *Phil. Trans.* 36 (1729–30): 444–8.

20 George Garden, "An Account of the Effects of a Very Extraordinary Thunder," *Phil. Trans.* 19 (1695–97): 311–12, 313. Garden seems to have suffered under William and Mary's reign: in 1692 he was ejected from the Ministry of St Nicholas of Aberdeen and during the 1690s wrote many protest articles announcing "the great end of Christianity." *DNB.*

21 Robert Mawgridge, "A True and Exact Relation of the Dismal and Surprizing Effects of a Terrible and Unusual Clap of Thunder with Lightning," *Phil. Trans.* 19 (1695–97): 783. In 1712, Thoresby inquired into the "most remarkable Particulars" left after the storm in Yorkshire. Hailstones, which killed several pigeons, were 3–5 inches wide, but the chief damage was done to the glass windows; "but the Damage in the Corn was severe upon the poorer sort of Inhabitants. I shall conclude this, as the good old Minister (who was a Sufferer of this Calamity) does his Letter: When thy Judgments, O Lord, are in the Earth, the Inhabitants of the World should learn Righteousness. *Isai.* 26. 9," Ralph Thoresby, "An Account of the Damage done by a Storm of Hail, which happened near Rotherham in Yorkshire on June 7, 1711." *Phil. Trans.* 27 (1710–12): 514. A particularly well-documented report on the damage and victims of a thunderstorm was drawn up by John Wallis in his letter to Dr Sloane: "The Effects of a Great Storm of Thunder and Lightning at Everdon in Northamptonshire," *Phil. Trans.* 20 (1698): 5. Thomas Molyneux of the Philosophical Society at Dublin used the testimony of Mrs. Close to reconstruct the "Extraordinary Accident" in which her whole house was "shook and inflamed" by a sudden thunderclap, "A Relation of the strange Effects of Thunder and Lightning," *Phil. Trans.* 26 (1708–09): 36. "Our waterspout", wrote Zachary Mayne in 1694, "happened Tuesday last; it was then very near, if not quite low Woter [sic], which is lookt [sic] on as a special Providence, since had it been High Water, 'tis concluded its strength would have been much greater," "Mr Zachary's Letter, 1694," *Phil. Trans.* 18 (1694): 28–31. Meteorological "semiotics" had a hold over European men of letters in general; the Sicilian Nobleman Vincentius Bonajutus wrote to Malpighi: "The Great Secrets of Nature are so inscrutable, that we find them far out reach whenever we go about to form a true and nice Judgment of them; nor is the Task less difficult to discover the Effects of Portentous Events, divers Accidents being often confounded with the Effects themselves." *Phil. Trans.* 17 (1693): 2.

22 Hunter, *Before Novels*, 180.

23 Geoffrey Holmes, "Religion and Party in Late Stuart England," in idem., *Politics, Religion and Society in England* (London: Hambledon Press, 1986), 185, 201–2. For the uses and interpretations of Newtonianism in the religious debates during Queen Anne's reign, see Larry Stewart, "Samuel Clark, Newtonianism, and the Factions of Post-Revolutionary England," *Journal of the History of Ideas* 42 (1981): 53–72.

24 On the sources of vitality of Jacobitism up until the mid-eighteenth century see Paul Kleber Monod, *Jacobitism and the English People 1688–1788* (Cambridge: Cambridge University Press, 1989); Murray G. H. Pittock, *Poetry and Jacobite Politics in Eighteenth-Century Britain and Ireland* (Cambridge: Cambridge University Press, 1994); Daniel Szechi, *The Jacobites. Britain and Europe 1688–1788* (Manchester: Manchester University Press, 1994); G. V. Bennett, *The Tory Crisis in Church and State* (Oxford: Oxford University Press, 1975), 206–41; N. Rogers, "Riot and Popular Jacobitism in Early Hanoverian England," in Eveline Cruickshanks, ed., *Ideology and Conspiracy: Aspects of Jacobitism 1689–1759* (Edinburgh: John Donald Publishers, 1982); Geoffrey Holmes, "Catholicism and Jacobitism, 1689–1746," in Geoffrey Holmes and Daniel Szechi, *The Age of Oligarchy* (London: Longman, 1993), 89–100. Issues relating to monarchical succession, the nature of royal legitimacy and its relevance in the discussion of natural order are discussed in Steven Shapin, "Of Gods and Kings: Natural Philosophy and Politics in the Leibniz-Clarke Dispute," *Isis* 72 (1981): 187–215.

25 Daniel Defoe, *The Storm: or, a Collection of the most remarkable Casualties and Disasters Which happened in the Late Dreadful Tempest, both by Sea and Land,* (London: G. Sawbridge, 1704) in *The Novels and Miscellaneous Works of Daniel Defoe* (London: George Bell and Son, 1911), vol. v, 284. In the general discussion of the storm, Defoe used "A Letter from the Reverend

William Derham, FRS, Containing his Observations concerning the late Storm," *Phil. Trans.* 24 (1704–05): 1530–34; other reports included: "Part of a Letter from Mr Anthony van Leuwenhoek, FRS giving his Observations on the late Storm," *Phil. Trans.* 24 (1704–05): 1535–37, and "Part of the Letter from John Fuller of Sussex, Esq; concerning a strange effect of the late Great Storm in that Country," *Phil. Trans.* 24 (1704–05): 1530.

26 Modern meteorological analysis of the storm is in Hubert Lamb and Knud Frydenhal, *Historic Storms of the North Sea, British Isles and Northwest Europe* (Cambridge: Cambridge University Press, 1992).

27 Derham, "The Storm of 1704," 1532. On Derham's various scientific interests see A. D. Atkinson, "William Derham, FRS (1657–1735)," *Annals of Science* 8 (1952): 368–92.

28 Defoe, *Storm*, 276–7.

29 A. Gifford, *A Sermon in Commemoration of the Great Storm, Commonly called the High Wind, in the year 1703, preached at the CHAPEL in Little Wilde Street, London, November 27, 1733, with an Account of the Damage done by it* (London: Aaron Ward), 1733; Samuel Stennet, DD, *A Sermon in Commemoration of the Great Storm of Wind on November 27*; Thomas Short, *A General Chronological History of the Air, Seasons, Meteors, etc.* (London: T. Longman and A. Millar, 1749), vol. 1, 428; *London Gazette*, no. 3917 (1703). A more recent account of the damage is in George Macaulay Trevelyan, *Blenheim. England under the Queen Anne* (London: Longman, Green and Company, 1959), 307–11.

30 *The Terrible Stormy Wind and Tempest November 27th 1703 Consider'd, Improved, and Collected to be had in Everlasting Remembrance* (London: W. Freeman, 1705).

31 William Hone, *The Table Book* (London: Hunt and Clarke, 1827), 1512–21.

32 Richard Chapman, *The Necessity of Repentance Asserted: In Order to Avert Those Judgments which the Present War and Strange Unseasonableness of the Weather At Present seem to threaten this Nation with in Sermon preached on Wednesday the 26th of May 1703 being the Fast-Day appointed by Her Majesty's Proclamation* (London: M. Wotton, 1704). Chapman relied on *Jeremiah* 18, 7–8: "At what instant I shall speak concerning a nation, and concerning a kingdom, to pluck up, and to pull down, and to destroy it; If that nation, against whom I have pronounced, turn from their evil, I will repent of the evil that I thought to do unto them."

33 Thomas Bradbury, *God's Empire over the Wind, Consider'd in a Sermon on the Fast-Day, January 19, 1703/4* (London: Jonathan Robinson, 1704). Joseph Hussey, Pastor of the Congregation Church at Cambridge, *A Warning From The Winds. A Sermon preach'd upon Wednesday, January XIX, 1703/4. Being the Day of Public Humiliation for the Late Terrible and Awakening Storm of Wind, sent in Great Rebuke upon this Kingdom* (London: William and Joseph Marshal, 1704).

34 Bradbury, "God's Empire," 14; *The Terrible Stormy Wind*, 8.

35 "And the most cruel Efforts of War, its concluding and most merciless Stroke, borrows its name hence, the Storming of a Place, Town or City; being frequently, the preceding Act to utter Desolation and Destruction of its inhabitants." *The Terrible Stormy Wind*, 22.

36 Defoe, *The Storm*, 256. The conclusions Defoe draws are: (1) winds refer us to God's power more than other parts of nature, (2) wind is more expressive of God's immediate power than other phenomena, and (3) it is more frequently used as "the executioner of judgments in the world," ibid., 265. In 1733 Andrew Gifford argued that *John* III, 8, "The Wind loweth where it listeth," implied that "The original and causes of the Wind are much unknown."

37 Gifford, *A Sermon in Commemoration of the Great Storm*, 9.

38 Defoe, *The Storm*, 259.

39 "To preserve the Rememberance of the late dreadful Tempest, an exact and faithful Collection is preparing of the most remarkable Disasters which happened on that occasion with the Places where and Persons concerned, whether at Sea or Shore. For the perfecting so good a Work, it is humbly recommended by the Author to all Gentlemen of the Clergy, or Others, who have made any Observations of this Calamity, that they would transmit as distinct an Account as possible, of what they have observed, to the Undertaker, directed to John Nutt near Stationers Hall, London. All Gentlemen that are pleas'd to send any such Accounts are desired to write no Particulars but what they are well satisfied to be true, and to set their Names and Observations

they send, which the Undertaker of this Work promise shall be faithfully recorded, and the Favour publicly acknowledged." *The London Gazette*, 6 December 1703.

40 Ilse Vickers, *Defoe and the New Sciences* (Cambridge: Cambridge University Press, 1996), 30, 66, 106–10.

41 "From whence we now inform the people / The danger of the church is from the steeple . . . / The time will come when all the town, / To save the church, will put the steeple down. / Two tempests are blown over, now prepare / For storms of treason and interstine war. / For high-church fury to the north extends, / In haste to ruin all their friends. / Occasional conform-ity led the way, / And now occasional rebellion comes in play / To let the wond'ring nation know, / That high-church honesty's an empty show," Defoe, *Storm*, 420–1.

42 *To begin Harvest Three Days too soon rather than Two Days too late, or Sentences of the Dissenters containing, relations of pretended Judgments, Prodigies and Apparitions in Behalf of the Non-conformists, in Opposition to the Established Church. With timely remarks* (London: B. Bragg, 1708). "The following Treatise, containing a Collection of many Prodigies, Signs, Apparitions, Accidents and Judgments (either pretended or real) being published by a Club of Dissenters (as is apparent from their Preface and Relations) in the year 1661, under the title of *Mirabilis Annus*, at the juncture when the Tottering Cause of the Non-Conformist Party, after the Restauration, stood in need in of all their strength and Policy, to keep it from sinking quite under Ground; the said collection was thought to be made Public one more, but with such usual Remarks, as were judged sufficient to convince an Unbiased reader, both of the weakness of their Design and Arguments," *Harvest*, 1.

43 *Harvest*, 49. As late as in 1788, Samuel Stennet, Dissenting minister at Little Wilde Street Chapel, (at which annual commemorations of the storm were regularly observed), turned in his sermon from the theological to a "civil or political description" of the storms of 1688 and 1703. It amounted to a detailed historical narration of political and social disabilities Dissenters suffered from the Corporation Act in 1662 to the times of the Glorious Deliverer, William of Orange.

44 It should be added that in the moralizing pamphlets the claims to empirical fidelity derived sometimes from the "guide-book" directions given to readers to locations where they could "autopsically" inspect the providential sight and wonders described in the text. On this, see Walsham, "Aspects of Providentialism," 34.

45 Simon Schaffer, "Making Certain," *Social Studies of Science* 14 (1984): 137–52, 146.

46 Some of these publications are reproduced in Hyder Edward Rollins, ed., *The Pack of Autolycus* (Cambridge: Cambridge University Press, 1927). Not mentioned in the volume are *More Warning yet. Being a True Relation of A Strange and most Dreadful Apparition Which was Seen in the Air By several Persons at Hull, the third day of this present Sept, 1654* (London: J. Cottrel, 1654); *Strange News from BERKSHIRE, of an Apparition of Several Ships and Men in the Air* (n.d., c. 1695–1710); *The Age of Wonders: or, a Particular Descripttion [sic] of the Remarkable, and Fiery Appartiion [sic] that was Seen in the Air, on Thursday, May the 11th 1710* (London, 1715); *O-YES from the Court of Heaven to the Northern Nations* (London: the Author, 1741).

47 *Looke Up and See Wonders*, 2–5.

48 Godfridus, *The New Book of Knowledge Shewing the Effects of Planetary and other Astronomical Constellations* (London: A. Wilde, 1758), 85; *Mirabilis Annus*, 49; *Strange and Wonderful News from Chippingnorton* (1716), 3; *A Full and True Relation of the Strange and Wonderful Appari-tions which were Seen in the Clouds upon Tuesday Evening at Seven of the Clock . . . the 6th and 7th of March 1715/16 (n.d.)*, 4.

49 The presentist dichotomy between the popular "beliefs" and scientific "certainties" have recently been maintained in Asgeir Brekke and Alv Egeland, *The Northern Light: From Mytho-logy to Space Research* (Berlin: Springer Verlag, 1983). Sections of this monograph bear titles such as: "The Northern Light – A Vengeful Being," "The Northern Light – Flames from the Realm of Dead," or "The Northern Light as an Omen of War, Disaster and Plagues." The works and chronicles from antiquity onwards featured both "military" and "neutral" descrip-tions: Aristotle and followers spoke of "gapings," Pliny reported heavenly fights. It would be historically amiss to claim they referred to the same phenomenon.

50 "Superstition," the term traditionally used by those who rejected the authenticity of these accounts, stood for everything illicit and spurious, everything in contrast to the canonical and

authorized. To condemn practices as superstitious was to deny their entry into canonical forms of knowledge. Mary R. O'Neil, "Superstition," in Mircea Elliade, ed., *Encyclopedia of Religion* (New York: Macmillan, 1982), vol. 5, 225. Stuart Clark has written on the concept of superstition as an instrument used by the early modern clergy to denounce routine material and spiritual practices of the rural and plebeian populace. See his "The Rational Witchfinder: Conscience, Demonological Naturalism and Popular Superstitions," in Stephen Pumfrey, Paolo Rossi, and Maurice Slawinski, eds, *Science, Culture and Popular Belief in Renaissance Europe* (Manchester: Manchester University Press, 1991), 222–49.

51 Stuart Clark, *Thinking with Demons: The Idea of Witchcraft in Early Modern Europe* (Oxford: Clarendon Press, 1997), 5.

52 J. F. C. Harrison, *The Common People* (London: Croom Helm, 1984), 174: "Signs and wonders were a staple of this type of belief. Any instance in which the laws of nature appeared to have been set aside, any abnormal or inexplicable happening, any unusual behaviour in man or beast aroused a widespread interest and speculation." It created wonder in itself, and prompted thoughts as to what it might portend as a "sign of the times." Ibid., 182.

53 E. P. Thompson, "Eighteenth-Century English Society: Class Struggle without Class?" *Social History* 3 (1978): 133–65, 157.

54 J. A. Sharpe interprets "death speeches" of condemned criminals as a form of ideological control: "'Last Dying Speeches': Religion, Ideology, and Public Execution in Seventeenth-Century England," *Past and Present* 107 (1985): 144–67, 148. Manipulation of the "common opinion" by printed means was also discussed in Mervyn James, *English Politics and the Concept of Honour, 1482–1647* (Oxford: Oxford University Press, 1978): "Civil order depended to a much greater extent than in the bureaucratized societies of a later age, on the effective internalization of obedience, the external sanctions, being so often unreliable" ibid., 44. The dialectic between the spiritual and titillating aspects of genre are addressed in Peter Lake, "Popular Form, Puritan Content? Two Puritan appropriations of the murder pamphlet from mid-seventeenth-century London," in Anthony Fletcher and Peter Roberts, eds, *Religion, Culture and Society in Early Modern Britain* (Cambridge: Cambridge University Press, 1994), 313–34. For extra-legal dimension of providential anecdotes, see Walsham, "Aspects of Providentialism," 98. In relation to "subversive expressive symbolism," Hans Medick has written: "Plebeian culture appears to constitute a peculiar type of public. It possesses that sensuous physical character which Basil Bernstein has termed 'expressive symbolism' and distinguishes itself both from the 'public realm of reason' of the educated bourgeoisie and from the seigneurial, ceremonial self-display of the aristocracy, 'the official public.' [In this culture and its public] symbol and experience that comprehends it are linked not by intellect but by ritual customary action." See Hans Medick, "Plebeian Culture in the Transition to Capitalism," in Raphael Samuel and Gareth Stedman Jones, eds, *Culture, Ideology and Politics* (London: Routlege and Kegan Paul, 1982), 84–113, 86.

55 Christopher Hill argues that "when a natural catastrophe affected them, [common people] had recourse to the intercession, by prayers and ceremonies, of the priest or the magician: the relationship [with nature] was vicarious, second hand." See his *God's Englishman*, 233.

56 *A Strange and Wonderful Relation of a Clap of Thunder which lately set Fire to the Dwelling-House of one Widdow Rosingrean living in the town of Ewloe, in the parish of Howerden in the County of Flint* (London: W. Harris), 1677. *Strange and Wonderful News from Chippingnorton in the County of Oxon*, c. 1700, 3.

57 *Looke Up and See Wonders*, 4; *A True Relation of the Terrible Earthquake at West-Brummidge in Staffordshire And the places adjacent on Tuesday the 4th of this instant January 1675/6* (London: D. M., 1676); *A True Relation of the Wonderful Apparition of a Cross in the Moon*, 1688; William Whiston, *An Account of a Surprizing Meteor, Seen in the Air, March 1715/16 at Night* (London: J. Senex, 1716).

58 Daniel Defoe, *History of the Plague in London*, 1665, in *The Novels and Miscellaneous Works of Daniel Defoe*, 16.

59 *Looke up and See Wonders*, 5.

60 Baruch de Spinoza, "On Miracles," *Tractatus Theologico-Politicus* (Leiden: E. J. Brill, 1989), 67.

61 John Morton, *Natural History of Northamptonshire* (London: R. Knaplock, 1712), 7.

62 Henry More, *Enthusiasmus Triumphatus* (Reprint; Los Angeles: William Andrews Clark Memorial Library, 1966), 12–13; Meric Causabon used the theory in his *A Treatise Concerning Enthusiasm* (London, 1655). Rob Illife discusses More in this context in his "The Puzzleing Problem: Isaac Newton and the Political Physiology of Self," *Medical History* 39, 4 (1995): 433–58.

63 *Freethinker*, 24 October 1718, No. 62 (London 1711–21), 34 (London 1722–3), 246–7. See also No. 34: 247–8: "The heads of such a creed are never for stinting mankind of pretended mysteries and miracles, suitable for every age and complexion; which, tough they be of weight sufficient to sink a good cause, with people of common sense; yet with minds duly prepared they pass as undeniable arguments in favour of the superstitions they are forged to support," quoted in Redwood, *Reason, Ridicule and Religion*, 139.

64 *Strange News from Oxfordshire: Being a True and Faithful Account of a Wonderful and Dreadful Earthquake* (Oxford, 1683), 2.

65 *Strange News from Berkshire*, 1679, 3.

66 John Howie, *An Alarm Unto a Secure Generation* (Kilmarnock: M. & S. Crawford, 1809), 29.

67 Reverend Divine, *A Practical Discourse*, 19.

68 "[K]nowing there was such a strict Tye betwixt that Church and State, that the ruin of one (especially at that Conjucture, [the Restoration]) could not be as much as attempted without imminent Danger of other; they thought fit to level their prodigies and strange apparitions against the government, as they had their judgments against the church." *To Begin Harvest Three Days Too Soon Than Two Days Too Late*, 5.

69 John Flamsteed, "Hecker," quoted in Patrick Curry, "Astrology in Early modern England: the Making of a Vulgar Knowledge," in Stephen Pumfrey et al., eds, *Science, Culture and Popular Belief in Renaissance Europe* (Manchester: Manchester University Press, 1989), 274–91, 281.

70 John Gadbury, *De Cometis* (London: L. Chapman, 1665). The explanation for the credulity of the English was in the wider cultural context grounded in the time-hallowed belief in their notorious inconstancy. The same year that Flamsteed made his charges against astrology, Sir Thomas Baines blamed "the mutability of air in an island" for an Englishman's gullibility and predisposition for novelties. See Sara Warneke, "A Taste for Newfangledness," 881–96, 887. One of the most disturbing effects of such fickleness was thought to be civic disloyalty and rebellion against political regime. The *Harvest* 5, comments: "Baker in his Chronicles among many other Excellent Observations concerning the different inclinations of various nations, gives the English the character of being soon surprised and led away by Prodigies, Apparitions and such like Strange Accidents. I know not whether the compilers of this treatise [*Mirabilis Annus*, 1661/2], had so much veneration for this Great Man, as to rely absolutely upon his Authority and Judgment. Or, whether, being convinced by their own Experience (especially among their Ignorant Brethren) of the Truth of this Assertion, *they though it fit to improve their weakness to their advantage.*" My italics.

71 *Historical Register*, March 7, 1716.

72 *Flying Post*, March 8, 1716.

73 Whiston, *An Account, 1715/6*, 27, 49. Halley supplied an illustration being aware "how insufficient such a verbal Description of a thing so extraordinary and unknown may be to most readers," Edmund Halley, "An Account of the Late Surprizing Appearance of the Lights Seen in the Air, on the sixth of March last; with an Attempt to explain the Principal Phenomena thereof," *Phil. Trans.* 29 (1714–16): 406–28, 414.

74 William Derham, "Observations on the Lumen Boreale," *Phil. Trans.* 35 (1727–28): 245–8.

75 Halley, "An Account of the Late Surprizing Appearance," 428–9.

76 Whiston, *An Account, 1715/16*, 29, 38. William Maunder wrote three years later that the Aurora of which he was a witness was "represented here in all sorts of Appearances, Armies, Battles, etc, and has put abundance of People in dismal Frights: But I had not an Imagination strong enough for it." [William Maunder], "A Relation of [the Luminous Appearance], seen at Cruwys Morchard in Devonshire," *Phil. Trans.* 30 (1717–19): 1101–04.

77 "A letter from Elston, near Newark, March 7," in Whiston, *An Account, 1715/16*, 48. Another observer described "sudden Agitations of small Whitish Clouds, that seemed to Dart, and shoot over and across, and oppose each other, and shoot along in Streaks and Sudden drifts," ibid., 36.

78 Ibid., 27, 49, 26.

79 Diary of the Countess Cowper, quoted in Ralph Arnold, *Northern Lights, the Story of Lord Derwentwater* (London: Constable, 1969), 13.

80 *A Full and True Relation of the Strange and Wonderful Apparitions which were Seen in the Clouds upon Tuesday Evening at Seven of the Clock . . . the 6th and 7th of March 1715/6*, (n.d.), 4–5. The battle was followed by the apparition of two flaming swords, and those replaced by daggers, pikes, spears, and other weaponry. Individual reports on flaming swords were common throughout the period. It is worth noting with what speed pamphleteers seized on the new topic; *The Flying Post* of March 10 informed readers of the newly published *Their Vision at Deal; being an Account of most Wonderful Apparitions of Ships, Castles and Other Figures in the Air, to which is added an Account of the late strange Appearance in the Sky over City of London* (London: S. Popping, 1716).

81 "Throughout these two decades the political temperature at the grass roots of politics . . . was feverishly high." Geoffrey Holmes, *British Politics in the Age of Anne* (New York: Macmillan, 1967), 218.

82 John Stevenson, *Popular Disturbances in England, 1700–1832* (New York and London: Longman, 1992), 23–30.

83 Holmes and Szechi, *The Age of Oligarchy*, 97, 98.

84 Arnold, *Northern Lights*, 14.

85 I haven't located such usages, but Angus Armitage refers to it in his *Edmund Halley* (London: Nelson, 1966), 180. "This illumination received from superstition the name of 'Lord Derwentwater's Lights,' occurring as it did within a fortnight of the execution of the Jacobite nobleman."

86 Arnold, *Northern Lights*, 14.

87 The paper added: "But they would do well to remember, that the great Eclipse in April last, which they expected would issue in their Favour, if it has had any Influence at all, has been fatal only to themselves and their Friends at home and abroad; and since by a Continued series of Providences, all their wicked and rebellious Attempts have been hitherto blasted, they have all the Reason in the World to believe that this last Prodigy . . . portends a due Chastisement for their Obstinacy, in carrying on Design against their King, their country and the Protestant Religion." *Flying Post*, 6 March 1716, 2.

88 *A Full and True Relation of the Strange and Wonderful Apparitions*, 2. The author cites Dr Flamsteed as interpreter of the apparitions; the rest of the pamphlet offers the interpretation of the Great Ball of Fire (religious passion, fury and rage), two blazing stars (two great enemies), two armies fighting until 11 o'clock at night (symbolizing an undecided war), two flaming swords (two kings). "Thus I gave you a full and true Relation of all the wonderful, strange and amazing Things which were seen. I have not here related a Fable or Factious Matter, but what is attested by several Thousands and I myself was a Spectator," *Full and True Relation*, 7.

89 *Daily Courant*, 7 March 1716.

90 Among many who have discussed the present problematics are Patricia Fara, " 'A Treasure of Hidden Vertues': the Attraction of Magnetic Marketing," *British Journal for the History of Science* 28 (1995): 5–35, 9. For other uses of natural philosophy in the public domain see Simon Schaffer, "Natural Philosophy and Public Spectacle in the Eighteenth Century," *History of Science* 21 (1983): 1–43; Steven Shapin, "Social Uses of Science," in George Rousseau and Roy Porter, eds, *The Ferment of Knowledge: Studies in the Historiography of Eighteenth Century Science* (Cambridge: Cambridge University Press, 1980), 93–139; Larry Stewart, *The Rise of Public Science*, Cambridge: Cambridge University Press, 1992); Jan Golinski, *Science as Public Culture: Chemistry and Enlightenment in Britain, 1760–1820* (Cambridge: Cambridge University Press, 1992); Shapin, "Of Gods and Kings", 187–215; Arnold Thackray, "The Business of Experimental Philosophy: The Early Newtonian Group at the Royal Society," *Actes du XIIe Congrès International d'Histoire des Sciences*, 1970–71, 3B: 155–9; Larry Stewart, "Samuel Clarke, Newtonianism, and the Factions of Post-Revolutionary England," *Journal of the History of Ideas*

42 (1981): 53–72. On the relationship between the Newtonian vision of the natural world and the public order, see Margaret Jacob, *The Newtonians and the English Revolution, 1679–1720* (Ithaca: Cornell University Press, 1976), 269. For the provincial context of the English Moderate Enlightenment, see Roy Porter "Science, Provincial Culture and Public Opinion in Enlightenment England," *British Journal for Eighteenth-century Studies* 3 (1980): 20–46, 27.

91 Schaffer, "Natural Philosophy and Public Spectacle," 16.

92 Valuable discussion with regard to the 1716 lights is Patricia Fara, "Lord Derwentwater's Light: Prediction and the Aurora Polaris," *Journal for the History of Astronomy* 27 (1996): 239–58, 244.

93 Frederick B. Burnham, "The More-Vaughan Controversy: The Revolt Against Philosophical Enthusiasm," *Journal of the History of Ideas* 25 (1974): 33–49; P. B. Wood, "Methodology and Apologetics: Thomas Sprat's History of the Royal Society," *The British Journal for the History of Science* 13 (1980): 1–26. More recently, Michael Heyd has shown how both critics and apologists of the new science mutually accused themselves of intellectual enthusiasm, "The New Experimental Philosophy: A Manifestation of "Enthusiasm" or an Antidote to it?" *Minerva* 25 (1987): 423–40; idem., *'Be Sober and Reasonable': The Critique of Enthusiasm in the Seventeenth and Early Eighteenth Centuries* (Leiden: E. J. Brill, 1995).

94 Sprat, *History of the Royal Society*, 364.

95 Ibid., 362. Even lesser authors adopted this position. A pamphlet from 1711 argued that, because thunder and lightning were results of nitrous and sulfurous matter, "'tis natural to conclude how much they are wrong, who look upon these kinds of Deaths as extraordinary and immediate judgments of those who suffer 'em; for there is nothing in all this which supposes or implies any immediate Interposition of God." *A True and Particular Account of a Storm of Thunder and Lightning which fell at Richmond, Surrey, . . . May 20th, 1711* (London: John Morphew, 1711), 15.

96 Geoffrey Holmes, "The Sacheverell Riots: the Crowd and the Church in Early Eighteenth-Century London," *Past and Present* 72 (1976): 42.

97 Redwood, *Reason, Religion and Ridicule*, 167.

98 Reverend Divine, *A Practical Discourse*, 12.

99 Whiston, *An Account, 1715/16*; *An Account of a Surprizing Meteor, seen in the Air March 19 1718/9 at Night* (London: W. Taylor, 1719). It is interesting to note that in the British Library, both tracts appear in the volume entitled *Astrological Tracts, 1711–1792*. Whiston was something of an intellectual puzzle: son of a Presbyterian minister, Cambridge graduate, writer of a Mosaic cosmogony, Newton's successor at Trinity, and the reputable Boyle lecturer, he became an Arian after 1700. This heterodoxy lost him a chair at Cambridge in 1710 and between 1711–14 the Tory Convocation even intended a prosecution which, delayed at first, came to nil soon after Queen Anne's death. Support from friends and royals as well as "eclipses, comets, and lectures," gave him "such a competency as greatly contented him." "William Whiston," *DNB*.

100 Whiston, *An Account, 1715/16*, 78.

101 Whiston, *An Account, 1715/16*, 57, 67. For other theories see J. Morton Briggs Jr., "Aurora and Enlightenment: Eighteenth-century Explanations of Aurora Borealis," *Isis* 58 (1967): 491–503.

102 In context of public lecturing, Fara observes that "the conservators of the social hierarchy needed to maintain a careful distinction between the consumers of polite philosophy and those who enjoyed the more popular forms of entertainment excluded from the domain of cultural validity," and adds that in the rhetoric of legitimation "natural philosophers defined new public sciences like chemistry by contrasting them with magical and superstitious practices," Fara, "'A Treasure of Hidden Vertues'," 32, 28. On this aspect of distancing in an earlier period, William Eamon writes: "The argument for the social utility of science did not die out . . . Increasingly, however, it was framed in the context of the growing split between elite and popular cultures, and by the fear and paranoia that elite culture exhibited toward the recalcitrant masses." William Eamon, *Science and the Secrets of Nature: Books of Secrets in Medieval and Early Modern Culture* (Princeton: Princeton University Press, 1994), 265. It becomes increasingly possible to link pronouncements regarding the intellectual distancing with those referring to the social and political sphere; the intellectual withdrawal of gentry in the late Stuart era meant an enforcement of secular thought. According to Fletcher and

Stevenson, secular thought at the time enjoyed a greater social respectability. See Anthony Fletcher and John Stevenson, eds, *Order and Disorder in Early Modern England* (Cambridge: Cambridge University Press, 1985), 12–13. Michael Macdonald sees this "respectability" as a political tool: new medical theories particularly, lacked the subversive political implications that religious psychology and therapy had acquired during the seventeenth century." See Michael Macdonald, *Mystical Bedlam* (Cambridge: Cambridge University Press, 1981), 8. The polarization between patrician and plebeian cultures which Keith Thomas and Patrick Curry (among others) used to account for the supposed decline of magic and astrology in the late seventeenth century is a central theme of K. Wrightson, *English Society 1580–1680* (New Brunswick, N. J.: Rutgers University Press, 1982), and L. Stone and J. C. F. Stone, *An Open Elite? England 1540–1880* (Oxford: Oxford University Press, 1984).

103 Whiston, *An Account, 1715/16*, 62. The "unthinking," for their part, could with some reason wonder as to how anything might resemble "Fermenting Matter." Whiston continued: "Hence we readily leave the Origin of our common stories, concurring the pretended Openings of the Heavens, the Air being on Fire, the exhibition of Weapons, Armies, Navies, and Battels in the Air, and the like [as arising from] the Fears, and Fancies, and Superstitions, and Prejudiced of Vulgar and Injudicious Spectators; as we accordingly find the like Chimerical Representations of this Appearance so current at this day among us." Whiston, *An Account, 1715/16*, 73.

104 Edmund Halley, "An Account of the Late Surprising Appearance," 416–28; Halley, "An Account of the Extraordinary Meteor seen all over England, on the 19th of March 1718–9," *Phil. Trans.* 30 (1717–19): 978–92. Morton, *Natural History of Northamptonshire*, 351. "I omit several others [instances of aerial battles]," wrote Godfridus in 1758, "for Melancholy Heads, by the Strength of Imagination and Conceit often they see things that really are not," Godfridus, *The New Book of Knowledge*, 85.

105 *Flying Post*, March 8, 1716. So blatantly hegemonic could these "empirical" arguments appear to the public, that the poet Edward Ward sympathized with the indicted ignoramuses and shrewdly exposed the practice in the poem of "Hanover-induced" prodigies:
"But men of art, who proudly aim / at universal praise and fame, / must, true or false, their judgment show, / in matters they professed to know, / or fools would think the learn'd but muddy / proficients in arts they study. / *Thus most mens excellency lies / In puzzling those they find less wise. / Be that alone the Gown and Band, / gain, of the crowd, the upper hand, / in things that neither understand.*" [Edward Ward] *British Wonders: or, A Poetical Description of the several Prodigies and most Remarkable Accidents that have happened in Britain since the Death of Queen Anne* (London: John Morphew, 1717), 4. Later writers used similar words: "Aurora Borealis, vulgarly called Streamers, or Merry Dancers;" "Strange are the Conjectures of the Unlearned concerning this Appearance in the Heavens: Some imagine they see Armies of Men, Horses and Chariots fighting in the Air which they take to be presages of War. But the real cause is Natural: the Sun rarifies lower Airy Region in the Day time, in the Night particles are raised into etherial Region and produce Pyramidal Glades, Streams etc." Campbell, *Time's Telescope*, 129.

106 "We hence learn . . . the weakness of those Vulgar Prognostications and Omens which are Perpetually drawn from all such uncommon Phenomena of Nature; and that, I observe, such is the Guiltiness of Mankind, for Judgments, and Deaths, and Wars, and Miseries, though usually they apply them not to themselves, or their own Party, . . . but to others and to Other Parties, for the gratification of their own Passions, and support of their own interest in the World." Whiston, *An Account, 1715/16*, 75.

107 Pointer, *Rational Account*, 196–7.

108 "Desiring to secure the peaceful enjoyment of [the values of English Moderate Enlightenment], the educated and propertied elites were also committed to public harmony and order, a stability which would partly flow from individual progress, and which party was to be imposed by the exercise of rationality, moderation, politeness, and humanitarianism. For many provincials, science could play a large part in both of these quests." Roy Porter, "Science, Provincial Culture and Public Opinion in Enlightenment England," 27.

109 Patrick Curry, *Prophecy and Power: Astrology in Early Modern England* (Cambridge: Polity Press, 1989), p. 161. John Rule wrote that – rather than a survival of traditional forms – the eighteenth-century popular culture was the creation of working people who were "defensive against repression and reactive against constraints." John Rule, *The Labouring Classes in Early Industrial England, 1750–1850* (London: Longman, 1986), p. 226, quoted in Curry, *Prophecy and Power*, 114. Curry's insight is well-illustrated by Simon Schaffer's discussion on Jethro Tull's farming schemes and Roy Porter's on popular medical beliefs. Tull had no qualms in directing his mathematical rules at "disciplining treacherous employees." He wrote how "the Ancients" had absurdly told us that "without all these Accidents meet . . . we must abstain from Ploughing. Our Ploughmen would be glad their Masters were as superstitious, for then the Plough might keep holidays enough." Simon Schaffer, "A Social History of Plausibility: Country, City and Calculation in Augustan Britain," in Adrian Wilson, ed., *Rethinking Social History: English Society 1750–1920 and its Interpretation* (Manchester: Manchester University Press, 1993), pp. 128–57, 147. Illustrating how "[t]he polite deplored the medical follies of their inferiors," Porter tells about the case of two maids, one ill of smallpox, the other healthy but convinced to be ill too. "This has provoked me," wrote her mistress, "& done me more harm then anything else, as she wou'd sit like a dead thing, and no reason had any effect on her . . . so strongly was she prepossessed she would have the small Pox." Roy Porter, "The People's Health in Georgian England," in Tim Harris, ed., *Popular Culture in England, c. 1500–1850* (New York: St Martin's Press, 1995), pp. 124–42, 126. One can wonder what Tull's opinion would have been on the behavior of the "stupid abigail" to whom "no reason" would have made her work at the time when the sick maid enjoyed a small-pox holiday from housework. In any case, as the cloak of superstition conveniently hid the interests of the working people, "folklore can provide a means of either actually or vicariously doing what the folk would like to do [or *not* do]." Alan Dundes, *The Study of Folklore* (New Jersey, 1965), p. 278, quoted in Bob Bushaway, *By Rite: Custom, Ceremony, and Community in England 1700–1880* (London: Junction Books, 1982), p. 12.

110 Rogers, "Popular Protest," p. 100.

111 Wrightson, *English Society*, 14. This divide entailed disappearance of judicial astrology from the discourse of higher classes in the seventeenth century. See Curry, *Prophecy and Power*, 153–8.

112 For the eighteenth-century polarizations see E. P. Thompson, "Patrician Society, Plebeian Culture," *Journal of Social History* 7 (1973): 382–405 397; E. P. Thompson, "Anthropology and the Discipline of Historical Context," *Midland History* 1 (1972): 52. Other supernatural readings of the 1716 lights included *An Essay concerning an Apparition in the Heavens, on the 7th of March, 1715/6. Proving by Mathematical, Logical, and Moral Arguments, that it cou'd not have been produc'd merely by the Ordinary Course of Nature, but must by necessity be a Prodigy.* Quoted in Pointer, *Rational Account*, 194–204; J. W., *O-Yes from the Court of Heaven to the Northern Nations, by the Streaming Lights that have appeared of the Late Years in the Air; or Mathematical Reasons shewing, that the said Lights &c, are no less than Supernatural. Presented to the Royal Society and others* (London: the Author, 1741). On the Lisbon earthquake, see Thomas Gibbons, *A Sermon Produced in November* [Lisbon Earthquake] (London, 1755), and Kendrick, *The Lisbon Earthquake*. For other phenomena, see William Lux, *Poems on several occasions: viz. The garden. The phaenomenon, a poem on the late surprising meteor, seen in the sky March 19th, 1719* (Oxford: Leon Lichfield, 1719). Icedore Frostiface of Freesland, *A True and Wonderful Narrative of Two Particular Phenomena, which were seen in the sky in Germany* (Philadelphia: Anthony Armbruster, 1764); Philotheus, *A True and Particular History of Earthquakes containing A Relation of that dreadful Earthquake which happened at Lima and Callao in Peru, October 28, 1746* (London: for the Author, 1748); *A true and wonderful narrative of two particular phenomena, which were seen in the sky in Germany* (Philadelphia: Anthony Armbruster, 1764); Rev. Isaac Farrer, Curate of Eggleton, *A Sermon or the Occasional Discourse on the late Great Flood in the North of England, Nov. 16 1771* (Newcastle: T. Slack, 1772), etc.

113 "Meteorology from the Intelligence of the Learned Societies," *Philosophical Magazine* 1 (1798): 428.

Notes to chapter 4

1 Arthur Young, *Arthur Young's Tour in Ireland 1776–9* (London and New York: C. Bell and Sons, 1892), vol. ii, 7–9.

2 Geography, according to Peter Heylen, was a description of Earth "by her parts and their limits, situations, inhabitants, rivers, fertility, and observable matters, with all other things annexed thereunto." It was divided into Hydrographie (a study of seas, promontories, creeks, springs and rivers), Topographie (a description of a particular town, city or village) and Chorographie (a "deciphering of any whole region, kingdome or nation"). The relation of geography to chorography stood as the whole to its part, an image universally familiar to the learned of Heylen's time. Peter Heylen, *Microcosmus, or a Little Description of the Great World. A Treatise Historicall, Geographicall, etc.* (Oxford: J. Lichfield, 1621), 15. Seven editions appeared before 1639.

3 Stan A. E. Mendyk, *'Speculum Britanniae.' Regional Study, Antiquarianism, and Science in Britain to 1700* (Toronto: University of Toronto Press, 1989), 67. William Lambarde, *A Perambulation of Kent, Containing the Description, Hystories, and Customes of that Shyre* (London: Ralphe Newberrie, 1576) was a model William Camden admitted to have used for his monumental *Britannia* (1586) with five editions before Heylen's *Microcosmus*. For the history of chorography I have relied on Lesley McCormack, "'Good Fences Make Good Neighbors': Geography as Self-Definition in Early Modern England," *Isis* 82 (1991): 639–61; W. G. Hoskins, *Local History in England* (London: Longmans, 1959); Kate Tiller, *English Local History, An Introduction* (Wolfeboro Falls, N. H.: Alan Sutton, 1992); Stuart Piggott, *Ancient Britons and the Antiquarian Imagination* (London and New York: Thames and Hudson, 1989) and idem., *Ruins in the Landscape. Essay in Antiquarianism* (Edinburgh: Edinburgh University Press, 1976); D. R. Woolf, *The Idea of History in Early Stuart England* (Toronto: University of Toronto Press, 1990); Arthur B. Ferguson, *Clio Unbound: Perception of the Social and Cultural Past in Renaissance England* (Durham, N. C.: Duke University Press, 1979).

4 Tiller, *English Local History*, 12.

5 Richard Helgerson, *Forms of Nationhood. The Elizabethan Writing of England* (Chicago and London: The University of Chicago Press, 1992), 136.

6 E. W. Gilbert, "The Idea of the Region," *Geography* 45 (1960): 157–75, 158, my italics. For F. W. Morgan, the humans and their habitat should be viewed as a "geographic individuality . . . appearing through the '*physionomie*,' aspect or landscape." See his "Three Aspects of Regional Consciousness," *The Sociological Review* 31 (1939): 68–88, 86. The literature on definition and meaning of landscape is overwhelming. The eighteenth-century esthetics of landscape are discussed in John Barrell, *The Idea of Landscape and the Sense of Place 1730–1840* (Cambridge: Cambridge University Press, 1972), 1–64. A study of landscape as a nexus of political ideologies is Nigel Everett, *The Tory View of Landscape* (New Haven and London: Yale University Press, 1994). For the ways in which landscape is viewed as constitutive of national identity see James Turner, *The Politics of Landscape: Rural Scenery and Society in English Poetry 1630–1660* (Cambridge, Mass.: Harvard University Press, 1979) and Stephen Daniels, *Fields of Vision: Landscape Imagery and National Identity in England and United States* (Cambridge: Cambridge University Press, 1993). For a recent view, see Kenneth Robert Olwig, "Sexual Cosmology: Nation and Landscape at the Conceptual Interstices of Nature and Culture; or What does Landscape really Mean?" in Barbara Bender, ed., *Landscape, Politics and Perspectives* (Providence and Oxford: Berg, 1993). For the eighteenth-century construction of Britishness through literary landscapes, a useful study is Tim Fulford, *Landscape, Liberty and Authority* (Cambridge: Cambridge University Press, 1996), chapter 1.

7 The author appears unaware of the works of McCormack and Mendyk which deal with the same issue and offer similar theses. See also John R. Stilgoe in *Common Landscape of America, 1580–1845* (New Haven: Yale University Press, 1982), who suggests that the origin of the comparative evaluation of different topographical districts of the Stuart "perambulations" lay in their "traditional, strongly agricultural bias: beautiful, useful landscapes are those made and maintained by husbandmen," ibid., 25. Stilgoe briefly explains the process by which the economical *Landschaft* – a discrete collection of dwellings structured within a circle of pasture, meadow, and

planting fields – gradually acquired humanistic identity by the late Middle Ages, representing a spot of land invested with meaning and collective memories: "Each [landschaft/region] was for its inhabitants a representation of the world, because each was the world . . . Each was so uncritically accepted as the emblem of all order that each was thought natural." Ibid., 19.

8 Mendyk, '*Speculum Britanniae*,' 19.

9 White Kennet, *Parochial Antiquities attempted in the History of Ambrosden and Other Adjacent Parts* (1695), quoted in Hoskins, *Local History in England*, 27.

10 William Burton, *Description of Leicester Shire* (London, 1622); John Norden, *Speculum Britanniae, The First part; an Historicall and Chorographicall Description of Middlesex* (London, 1593); William Camden, *Britain, or Chorographical Description of the Most Flourishing Kingdomes, England, Scotland and Ireland* (London, 1610); quoted in Mendyk, '*Speculum Britanniae*'.

11 John Norden, *The Description of Hartfordshire* (1598), John Speed, *Theatre of the Empire of Great Britaine* (1611), Robert Reyce, *Breviary of Suffolk* (1618), William Burton, *Description of Leicester Shire* (1622), Tristram Rison, *Devon* (1630). All quoted in Mendyk, '*Speculum Britanniae*,' 66, 79, 85, 88, 91.

12 Lambarde, *Perambulations of Kent*, 7.

13 Richard Carew, *The Survey of Cornwall* (S[tafford]: J. Jaggard, 1602), 4–5.

14 Gerard Boate, *Ireland's Natural History* (London: John Wright, 1652).

15 Michael Hunter, *John Aubrey and The Realm of Learning* (London: Duckworth, 1975), 70. Mendyk '*Speculum Britanniae*,' 7. McCormack contends, however, that "these Baconian tendencies existed long before the Civil War and themselves helped to shape the ideology of both their Lord Chancellor and the Royal Society." McCormack, "Good Fences Make Good Neighbors," 656. I believe that the debate as to whether chorography possessed its own empirical character or whether it borrowed it from the Baconian program, could be resolved by looking at Bacon's understanding of the "history of cosmography." According to Bacon, the history of cosmography is "compounded of natural history, in respect of the regions themselves; of civil history in respect of the habitations, regiments and manners of people; and mathematics, in respect of the climates and configurations towards the heavens." Bacon's "history of cosmography" hence already included chorography *as* the natural history of regions, and whoever studied it was therefore necessarily following the path independently charted by pre-Baconian county chorographies.

16 Natural histories on particular subjects also made an appearance during this period, from Cesalpino's *De Plantis* (1583) to John Ray's *History of Fishes* (1685). For Arthur Hopton, chorography was an art of describing "any particular place, without relation unto the whole, delivering all things of note contained therein, as ports, villages, rivers, not omitting the smallest . . . and therefore a *Topographicall* description ought to express every particular, which caused me the rather to call this instrument the *Topographicall Glasse*." Arthur Hopton, *Speculum Topographicum: or the Topographicall Glasse* (London, 1611).

17 Gunther, *Early Science in Oxford*, vol. 12, preface. On Plot's assimilation of Boate's principles see Webster, *The Great Instauration*, 427–31.

18 Piggott, *Ancients Britons*, 29.

19 Webster, *The Great Instauration*, 431. The full title was *An Interrogatory Relating More Particularly to the Husbandry and Natural History of Ireland*. It was subsequently published and circulated independently from the *Legacy*. For the activities of the "Georgical Committee," see Mayling Stubbs, "John Beale, Philosophical Gardener of Herefordshire. part 2: The Improvement of Agriculture and Trade in the Royal Society," *Annals of Science* 46 (1989): 323–63, 328.

20 The meteorological portion of queries required information on the winds, degrees of heat and cold, dryness and moisture, the pressure of air, the face of the sky (clouds and any form of precipitation), the effects of lightning on beer, milk and silk-worms, extraordinary tides, etc. Boyle and Hooke made proposals on a uniform terminology and described instruments. See Sprat, *History*, 173–79. For John Hoskyns's scheme, see Hunter, *John Aubrey*, 113. Some twelve programs of research for travelers were published in the two first years of the *Philosophical Transactions* and were reprinted together in the book *General Heads for the Natural History of a Country* (1692). For a larger perspective on travel and its place in the early science see George B. Parks, "Travel as Education," in Richard Foster Jones, ed., *The Seventeenth Century: Studies*

in the History of English Thought and Literature from Bacon to Pope (Stanford: Stanford University Press, 1951), 264–91, 285.

21 Gunther, *Early Science in Oxford*, vol. 4, 34–5. Lister's scheme was made public in 1683 and was less exacting than Hooke's.

22 Gunther, *Early Science in Oxford*, vol. 4, 42–3, 79, 170. For the activities of William Molyneux and George Ashe at the Dublin Philosophical Society during 1683–87, see K. Theodore Hoppen, *The Common Scientist in the Seventeenth Century. A Study of the Dublin Philosophical Society 1683–1708* (Charlottesville: The University Press of Virginia, 1970), 131–6. Hoppen even suggests that the weather-diary craze represented an instance of early "'generalized' specialization" within natural philosophy. (134). In his topographic surveys, Plot was helped by Henry Oldenburg's provincial correspondents, while Oxford authorities came forward with a testimonial in Plot's itinerant quest for information. On this see Thomas Birch, *History of the Royal Society* (New York and London: Johnson Reprint Corporation, 1968), vol. 4, 144. The Royal Cosmographer John Ogilby, before he discontinued his project of county surveys, on hearing about Plot's project issued a letter of introduction to John Aubrey in which he was "requiring in His Majesty's name, all Justices of Peace, Mayors, Bailiffs, Sheriffs, Parsons, Vicars, Church Wardens, High Constables, Constables, and Head-borrows and all other His Majesty's Officers, Ministers, and Subjects whatever, to be aiding and assisting to my said Deputy, or his Agents, in the said actual survey." John Aubrey, *Natural History and Antiquities of the County of Surrey* (London: E. Curll, 1719), preface.

23 Michael Hunter, *Science and Society in Restoration England* (Cambridge: Cambridge University Press, 1981), 55, 142. Information on Ashmolean is according to R. G. Frank, *Harvey and the Oxford Physiologists* (Berkeley: University of California Press, 1980), 74. Plot was well-connected in the virtuosi circle, his close friends being Wallis, Willis, Boyle, Bathurst, Millington, J. Ward, and Mayow.

24 Aubrey, *Natural History*, preface.

25 Plot, *Oxfordshire*, 1.

26 Plot, *Oxfordshire*, 5. The storm was also reported by John Wallis in his "Relation of an Accident by Thunder and Lightning at Oxford," *Phil. Trans.* 1–2 (1666–67): 222.

27 Plot, *Oxfordshire*, 4.

28 Ann Blair, "Humanist Methods in Natural Philosophy: The Commonplace Book," *Journal of the History of Ideas* 20 (1992): 541–51.

29 Ibid., 547.

30 Robert Plot, *Natural History of Staffordshire* (Oxford: Theatre, 1686).

31 Carl Estabrook, *Urbane and Rustic England* (Manchester: Manchester University Press, 1998), 57.

32 Plot, *Staffordshire*, 23–4, 20–2. The physical explanation, Plot writes, is that when the light remains compact, it is because the matter of the storm is "no where spent," and the storm is impending. When the storm matter is weak or broke, the light appears on several places and is called Castor and Pollux. If on ship-masts it foretells calm weather.

33 Mendyk, '*Speculum Britanniae*', 170–84, Hunter, *John Aubrey*, chapters 1–2; the introduction to the reprint of *Aubrey's Natural History of Wiltshire* (Reprint; Ponting, New York: Augustus M. Kelley, 1969) also gives a detailed account of Aubrey's life and work. On John Ray's criticism of Wiltshire's inadequacy, see ibid., 7. Mendyk discovered that a late eighteenth-century issue of the *Gentleman's Magazine* called Aubrey a "gossiping anecdote-monger." Mendyk, "*Speculum Britaninae*," 171.

34 *Aubrey's Natural History*, 14–19. At the Royal Society meeting on February 5, 1673, Aubrey presented "some written observations concerning winds, their blowing down many hundreds of oaks at once." On April 20, 1681, he produced a letter of one Mr John Rodgers who wrote that: "The prodigious fog in the Temple was between three and four of the clock in the afternoon the 27th of November, 1674." Aubrey explicitly linked these reports to "a natural history of the weather." Birch, *History of the Royal Society*, vol. 3, 122; vol. 4, 81.

35 In connection to Plot's contribution to *Staffordshire*, Leigh noted "the unfairness of some Modern Authors in laying down Theorems and Experiments which were not their own and not acknowledging from whom they had them," Leigh, *Lancashire*, preface. One hundred and twelve coats of arms are engraved in the opening pages. The list of subscribers is alphabetical

only to a first approximation, as within the list under the same letter, subscribers of the higher rank are put first; Knights, higher clergy, landed gentry, MAs. Mendyk quotes the charges of vagueness from Richard Rawlinson's letter to Edward Lhwyd, in 1701. Mendyk, '*Speculum Britanniae*', 322.

36 Leigh, *Lancashire*, "Dedicatory Epistle."

37 Leigh, *Lancashire*, 6.

38 Ibid., 9, 17.

39 Quoted in Mendyk, '*Speculum Britanniae*', 225. Morton thanked the Lord Archbishop of Canterbury, the Lord Bishop of Carlisle, and the Lord Bishop of Ely, the last permitting him use of his library. Morton, *Northamptonshire*, iii.

40 Morton, *Northamptonshire*, 328. Morton's work exudes with pride for the county: the fertility of Northamptonshire's soil is praised, as is the abundance of grain (unequaled in other parts). As far as the air was concerned: "no one can accuse [it] of any extremes . . . it having all the advantages of a friendly and benign climate." Even in the foggy fenland, "inhabitants neither complain, nor have any need of complaining of more than ordinary coldness of it." Ibid., 325.

41 As Robert Mayer has recently remarked about the character of the Baconian historiography implicit in works of this nature, "knowing strange things was of the very essence of historical discourse in this period." Mayer, *History and the Early English Novel*, 136. On guidebooks, see Barbara Benedict, "The 'Curious Attitude' in Eighteenth-Century Britain: Observing and Owning," *Eighteenth-Century Life* 14 (1990): 59–8, 65.

42 Barbara Shapiro, *Probability and Certainty in Seventeenth-Century England* (Princeton: Princeton University Press, 1983), 135–9. Levine, *Dr. Woodward's Shield*, 118; Michael Hunter, *Establishing the New Science: The Experience of the Early Royal Society* (Woodbridge: The Boydell Press, 1989), 135.

43 Barbara Shapiro, "History and Natural History in 16th- and 17th-century England: An Essay on the Relationship Between Humanities and Science," in idem., ed., *English Scientific Virtuosi in the 16th and 17th Centuries* (Los Angeles: William Andrews Clarke Memorial Library, 1979), 3–55, 37. But see also Joseph M. Levine, *Humanism and History: Origins of Modern English Historiography* (Ithaca and London: Cornell University Press, 1987), 123–54. On the effects of the literary technology used in establishing "matters of fact," see Schaffer, "Making Certain," 143.

44 Shapiro, "History and Natural History," 38.

45 Aubrey, *Surrey*, preface.

46 Barbara Shapiro, "The Concept of 'Fact': Legal Origins and Cultural Diffusion," *Albion* 26 (1994): 1–26, 11.

47 David P. Miller, "'Into the Valley of Darkness': Reflections on the Royal Society in the Eighteenth Century," *History of Science* 27 (1989): 155–66, 160, n. 24. "The eighteenth century was to produce its own great contribution to the antiquarian description of the counties of England, and in the sphere of local history the break between the two epochs was less than elsewhere." David C. Douglas, *English Scholars* (London: Jonathan Cape, 1939), 357. Shapiro, "The Concept of 'Fact'," 11; D. R. Woolf, "The 'Common Voice': History, Folklore and Oral Tradition in Early Modern England," *Past and Present* 120 (1988): 26–52.

48 Levine, *Humanism and History*, 94, n. 108.

49 Morton, *Northamptonshire*, ii–iii.

50 Andrew Wear, "William Harvey and the 'Way of Anatomists,'" *History of Science* 21 (1983): 220–49, 221.

51 Ibid., 238.

52 Plot, *Oxfordshire*, Preface, unpaginated.

53 Derick Thomson, *Edward Lhuyd in the Scottish Highlands 1699–1700* (Oxford: Oxford University Press, 1963), 33; [Edward Lhwyd], "Extracts of Several Letters from Mr Edward Lhwyd Containing Observations Made on His Travels thro't Wales and Scotland," *Phil. Trans.* 28 (1713): 99–105; Lhywd to Ralph Thoresby, May 4, 1703 in *Letters to Eminent Men addressed to Ralph Thoresby*, FRS (London, 1832), II, 2. Sloane's collection is the subject of the study of Arthur MacGregor, ed., *Sir Hans Sloane: Collector, Scientist, Antiquary, Founding Father of the British Museum* (London: British Museum Press, 1994), 180–98. Other European authors

discussing thunderbolts as atmospheric stones included Nicolas Lemery (*Cours de Chimie*, 1675), Paul M. Terzago (*Musaeum Septalianum*, 1664), G. E. Rumphius (*Amboinsche Rariter Kamer*, 1705), Charles N. Lang (*Histoire de pierres figureés de la Suisse*, 1708).

54 Benjamin Cook, "A Letter . . . Concerning a Ball of Sulphur Supposed to be Generated in the Air," *Phil. Trans.* 40 (1737–38): 427–8; During the 1730s, a common practice was to seek for the "Jellies which are supposed to owe their Beings to [fiery] Meteors." [Mr Crocker], "An Account of a Meteor Seen in the Air in the Day-Time on Dec. 8, 1733," *Phil. Trans.* 41 (1739–41): 346–7.

55 Crocker, "a Meteor Seen in the Air." Personal presence was important for authenticity of a report, but it was also important because it confirmed that some *person* was present: "No person could have a better opportunity of discerning this awful meteor than myself," begins a communication of Archdeacon of York to Sir Joseph Banks. "Observation on a Remarkable Meteor Seen On the 18th of August 1783, . . . by William Cooper, DD, FRS," *Phil. Trans.* 73 (1783): 116–17.

56 Thoresby was a prominent historian of Leeds and a collector of curiosities. See the Rev. Joseph Hunter, *The Diary of Ralph Thoresby, FRS, Author of the Topography of Leeds* (London: Colburn and Bentley, 1830), Preface, i. Another source for biographical information is in the *Biographia Britannica*, a biography written by his son Ralph Thoresby; and the "Life of Thoresby" prefixed to *Ducatus Leodiensis* Edited by J. D. Whitaker (Leeds, 1816). In 1677 and 1681, Thoresby toured Scotland and did research in local genealogy, inscriptions, architecture, numismatics, and heraldry. The surviving diary illuminates Thoresby's routine in transcribing epitaphs, benefactions, and mottos from churches, but also his interest in unusual weather phenomena observed on travels: "September 12, 1681. Morning, from Berwick over the Moors, where we found the proverb verified, that a Scotch mist, for I cannot say it rained, wets the Englishman to the skin, to Hayton, a country town, seated upon the river Hay." Thomas Kirk and Ralph Thoresby, *Tours in Scotland 1677, 1681*, edited by P. Hume Brown (Edinburgh: David Douglas, 1892), 50.

57 Ralph Thoresby, "An Account of a Young Man Slain with Thunder and Lightning," *Phil. Trans.* 21 (1699): 51–4.

58 Ibid., 52.

59 Ralph Thoresby, "Of an Accident by Thunder and Lightning at Leeds," *Phil. Trans.* 22 (1700–01): 577–9.

60 Ralph Thoresby, "A Letter . . . Giving an Account of a Lunar Rainbow seen in Darbishire, and of a Storm of Thunder and Lightning which happened near Leeds in Yorkshire," *Phil. Trans.* 27 (1710–12): 320–1.

61 Krzystof Pomian, *Collectors and Curiosities: Paris and Venice, 1500–1800.* (Cambridge: Cambridge University Press, 1990), 24.

62 Joseph Wasse, "Two Letters on the Effects of Lightning," *Phil. Trans.* 33 (1724–25): 366–70.

63 Ibid., 367.

64 Ibid., 368. Similar reports were common throughout the eighteenth century. In 1718, Henry Barham, surgeon-major of the military forces in Jamaica, sent a report to Hans Sloane about a ball of fire that fell three miles west from St Jago de la Vega. When he reached the site of the fall, he observed Holes in the Ground and the Grass which was "perfectly burnt near the Holes." "A Letter of That Curious Naturalist Mr Henry Barham, R.S.S. to the Publisher, giving a Relation of a Fiery Meteor Seen by Him, in Jamaica, to Strike Into the Earth," *Phil. Trans.* 30 (1717–19): 837–8. In the same letter information were given on two lightning victims, on the extraordinary loudness of Jamaican thunder-claps, and on the influence of the rainy season on the silk-worm population.

65 Dr Cookson, "A Further Account of the Extraordinary Effects of the Lightning at Wakefield," *Phil. Trans.* 38 (1733–34): 74–5.

66 "Part of a Letter from Sir John Clark to Roger Gale, Esq.," *Phil. Trans.* 41 (1739–41): 235.

67 Stephen Fuller, "A Letter . . . concerning a Violent Hurricane," *Phil. Trans.* 41 (1739–41): 851–3.

68 William Hamilton, "An Account of the Effects of a Thunder-Storm, on the 15th of March 1773, upon the House of Lord Tylney at Naples," *Phil. Trans.* 63 (1773): 324–32.

69 Levine, *Dr Woodward's Shield*, 114–33; Richard Olson, "Tory-High Church Opposition to Science and Scientism in the Eighteenth Century: The Works of John Arbuthnot, Jonathan Swift, and Samuel Johnson," in John G. Burke, ed., *The Uses of Science in the Age of Newton* (Berkeley: University of California Press, 1983).

70 Peter Dear, "'*Totius in Verba*:' Rhetoric and Authority in the Early Royal Society," *Isis* 76 (1985): 145–61, 154. The appearance of stylistically-similar weather reports in the *Philosophical Transactions* and eighteenth century periodicals (such as *Gentleman's Magazine*) suggest that Shapin's "virtual witnessing" (literary transference of a reader into laboratory) was not exclusive to the Royal Society. Newspaper accounts, sermons, and books of wonders all indiscriminately utilized "virtual witnessing" to establish not the mere "truth" of an experiment or an event, but to incite spiritual empathy, a sense of wonder or attraction.

71 Dear, "Totius in verba," 154.

72 Steven Shapin and Simon Schaffer, *Leviathan and the Air Pump: Hobbes, Boyle and the Experimental Life* (Princeton: Princeton University Press, 1985), 62.

73 Simon Schaffer, "Making Certain," 143.

74 My premise here is, however, that the meteoric tradition did not distinguish between the meteorologically and publicly relevant; I treat it as a hybrid of the socially conspicuous and ontologically extreme.

75 Shapin and Schaffer, *Leviathan*; Steven Shapin, *A Social History of Truth: Civility and Science in Seventeenth-Century England* (Chicago and London: The University of Chicago Press, 1994).

76 Martin Folkes, "An Account of the Aurora Borealis, seen at London, the 30th of March Last," *Phil. Trans.* 30 (1717–19): 586–7; Philip Percival, "An Account of a Luminous Appearance in the Air at Dublin," *Phil. Trans.* 31–2 (1720–23): 21–2; John Pringle, "Several Accounts of the Fiery Meteors," *Phil. Trans.* 51 (1759–60): 265–81; John Dorby, "Of a Terrible Whirlwind," *Phil. Trans.* 41 (1739–41): 230–2; Fuller, "Concerning a Violent Hurricane," 854–6; Rev. William Gostling, "Concerning the Meteor Seen in Kent," *Phil. Trans.* 41 (1739–41): 202; Stephen Hales, "A Collection of Various Papers Presented to the Royal Society Concerning Several Earthquakes," *Phil. Trans.* 46 (1749–50): 602–17; Thomas Lord Lovell, "Of a Meteor Seen Near Holkam in Norfolk," *Phil. Trans.* 42 (1742–43): 184.

77 Edward Bayley, "A Narrative of the Earthquake," *Phil. Trans.* 39 (1735–36): 362–7.

78 D. R. Woolf, "The 'Common Voice'," 30.

79 Ibid., 29.

80 Boate, *Ireland's Natural History*, 170.

81 *Aubrey's Natural History of Wiltshire*, 17.

82 Childrey, *Britannia Baconica*, 176.

83 "Mark Browell's Diary" in *North Country Diaries* (Durham: Surtees Society, 1915), 184; "Jacob Bee's Chronicle," in ibid., 62; Robert Latham and William Matthews, eds, *The Diary of Samuel Pepys*. 2 vols. (Berkeley: University of California Press, 1970), vol. 2, 86; *The Diary of Abraham de la Pryme* (Durham: Surtees Society, 1869), 68; "The Diary of John Hobson," in *Yorkshire Diaries and Autobiographies in the Seventeenth and Eighteenth Centuries* (Durham: Surtees Society, 1877), 375; Guy de la Bedoyère, ed., *The Diary of John Evelyn* (Woodbridge: The Boydell Press, 1995), 108.

84 Robert Boyle, "The General History of the Air," (1695); *The Works*, iv, 642.

85 Robert Taylor, "Account of a Great Hail-Storm in Herefordshire," *Phil. Trans.* 19 (1695–97): 577–9.

86 "A Copy of a Letter from R. P., Vicar of Kildwick in Yorkshire, to a Friend of his in those parts, wherein he gives an Account of an Extraordinary Eruption of Water, which happened in June, 1698," *Phil. Trans.* 20 (1698): 382–5; Taylor, "Account of a great Hailstorm", 577; Fuller, "Concerning a Violent Hurricane," 851; Lord Petre "Effects of Lightning," 136–40; "A Letter of Mr Perry on the Earthquake in Sumatra," *Phil. Trans.* 50 (1757–58): 491–3.

87 Serge Soupel, "Science and Medicine and the Mid-Eighteenth-Century Novel: Literature and the Language of Science," in idem., *Literature and Science and Medicine* (Los Angeles: William Andrews Clark Memorial Library, 1982), 31. For the ways in which the epistemology and the narrative of natural-historical pursuits endorsed aristocratic values, see Robert James Merret,

"Natural History and the Eighteenth-Century Novel," *Eighteenth Century Studies* 25 (1991): 145–70.

88 Soupel, "Science and Medicine," 8.

89 [John Whiston], "The 1703 Weather Diary," MS., National Meteorological Archives, Bracknell, Berkshire. I am grateful to Jan Golinski for letting me see his unpublished "Notes" on this diary in which he has presented strong circumstantial evidence that the diarist was not John Whiston, but Thomas Appletree from Dodington, Oxfordshire.

90 Samuel Cruwys, "A Relation of the [aurora borealis], seen at Cruwys Morchard in Devonshire," *Phil. Trans.* 30 (1717–19): 1101–02. My italics.

91 Dr Richard Richardson, "An Account of a Wonderful Fall of Water from a Spout in Lancashire," *Phil. Trans.* 30 (1717–19): 1097–1104; "A Further Relation of the same Appearance as seen at Dublin, Communicated to the Publisher by an Unknown Hand," *Phil. Trans.* 30 (1717–19): 1104–05; John Huxham, "[Aurora borealis] Described in a Letter to the Publisher," *Phil. Trans.* 34 (1726–27): 137–9; John Hattet, "A Letter to Dr Henry Pemberton, FRS. On the Same Subject," *Phil. Trans.* 34 (1726–27): 143–4; Abraham De la Pryme, "Concerning a Waterspout Observed by him in Yorkshire." *Phil. Trans.* 23 (1702–03): 1248–51; John Swinton, "An Account of Two Aurorae Boreales Observed in Oxford," *Phil. Trans.* 59 (1769): 367–70. My italics.

92 "Tuesday, 3 November 1761," *Gentleman's Magazine* 31 (1761): 532.

93 John V. Pickstone, "Ways of Knowing: Towards a Historical Sociology of Science, Technology and Medicine," *British Journal for the History of Science* 26 (1993): 439. See also idem., *Ways of Knowing* (Manchester: Manchester University Press, 2000).

Notes to chapter 5

1 Quoted in Richard Mabey, *Gilbert White A Biography of the Author of The Natural History of Selborne* (Vermont: Century, 1986), 49.

2 Sir James Balfour Paul, ed., *Diary of George Ridpath, Minister of Stitchel, 1755–1761* (Edinburgh: A. Constable, 1922).

3 Ralph Thoresby, *Ducatus Leodiensis: Or the Topography of the Ancient and Populous Town and Parish of Leeds, and Parts Adjacent in the West-Riding of the County of York* (London: Maurice Atkins, 1715), preface. On Thoresby's and Nicolson's friendship, see the Rev. Joseph Hunter, *The Diary of Ralph Thoresby, FRS* (London: Henry Colburn, and Richard Bentley, 1830), 7. A century after Thoresby's pronouncement, another chorographer of Cornwall, C. S. Gilbert, wrote that "'local attachment ranks among the best feelings of our nature,' and it was an irresistible impulse of this kind, which led to the publication of [this book]." C. S. Gilbert, *An Historical Survey of the County of Cornwall* (London: J. Congdon, 1817).

4 Porter, "Science, Provincial Culture and Public Opinion in Enlightenment England," 27. On the aspects of the influence of London on provincial society, see Borsay, *English Urban Renaissance*, chapter five; in the European context, the motivation for, and features of, the provincial urge to publish is ably researched by Anne Goldgar in *Impolite Learning: Conduct and Community in the Republic of Letters, 1680–1750* (New Haven: Yale University Press, 1995). For the ways in which the changes in the countryside affected the practice of natural history, see David E. Allen, "Natural History in Britain in the Eighteenth Century," *Archives of Natural History* 20 (1993): 333–47.

5 Margaret Jacob, *Scientific Culture and the Making of the Industrial West* (Oxford: Oxford University Press, 1997), 89. Uses of natural philosophy in Hanoverian Britain are discussed by Larry Stewart, "Public Lectures and Private Patronage in Newtonian England," *Isis* 77 (1986): 47–58, idem., *The Rise of Public Science*, Jan Golinski, "Utility and Audience in Eighteenth-Century Chemistry: Case Studies of William Cullen and Joseph Priestley," *British Journal for the History of Science* 68 (1988): 1–32; Margaret C. Jacob, "Scientific Culture in the Early English Enlightenment: Mechanisms, Industry, and Gentlemanly Facts," in Alan Charles Kors and Paul J. Korshin, eds, *Anticipations of the Enlightenment in England, France, and Germany* (Philadelphia: University of Pennsylvania Press, 1987), 134–65; Simon Schaffer, "The Consuming

Flame: Electrical Showmen and Tory Mystics in the World of Goods," in John Brewer and Roy Porter, eds, *Consumption and the World of Goods* (London: Routledge, 1996), 489–526; Patricia Fara, *Sympathetic Attractions* (Princeton: Princeton University Press, 1996); on the commercialization of instruments, see Alice Walters, "Tools of Enlightenment: The Material Culture of Science in Eighteenth-Century England" (Ph.D. Dissertation, University of Berkeley, 1992). Recently, Charles Withers has put geography in the commercial context in his "Toward a History of Geography in the Public Sphere," *History of Science* 36 (1998): 45–78. For an overview of the presuppositions of the theories of "conspicuous consumption" and "emulative behavior" on which most of the above studies are founded, see Colin Campbell, "Understanding Traditional and Modern Patterns of Consumption in Eighteenth-century England: a Character-Action Approach," in Brewer and Porter, *Consumption and the World of Goods* (London: Routledge, 1994), 40–58.

6 Michael Reed, "The Cultural Role of Small Towns in England 1600–1800," in Peter Clark, ed., *Small Towns in Early Modern Europe* (Cambridge: Cambridge University Press, 1995), 121–47. Reed defines "villetes" as places with under 5000 inhabitants at the time of the 1801 census. For a more detailed account of the "distractions," in English provinces in the year 1771, see R. M. Wiles, "Provincial Culture in Early Georgian England," in Paul Fritz and David Williams, eds, *The Triumph of Culture: 18th Century Perspectives* (Toronto: A. M. Hakkert, 1972), 49–68.

7 "The literary and antiquarian societies of the time met the needs of an interested public which read the *Gentleman's Magazine*, and the investigation of local antiquities was continued in many of the country houses of England. The eighteenth century was to produce its own great contribution to the antiquarian description of the counties of England, and in the sphere of local history the break between the two epochs was less than elsewhere." Douglas, *English Scholars*, 357.

8 Thomas Short, *New Observations, Natural, Moral, Civil, Political and Medical on City, Town, and Country Bills of Mortality* (London: T. Longman, 1750), 457.

9 For the popular scientific pursuits see Keith Thomas, *Man and the Natural World* (London: Harmondsworth, 1984), 229; T. Fawcet, "Measuring the Provincial Enlightenment: The Case of Norwich," *Eighteenth-Century Life*, New Series, 18, part 1 (1982): 15; R. E. Duthie, "English Florists. Societies and Feasts in the Seventeenth and First Half of the Nineteenth Centuries," *Garden History* 10 (1982): 17–35;

10 Robert Atkyns, *The Ancient and Present State of Glostershire* (London: W. Bowyer, 1712), preface.

11 Stuart Piggott, *William Stukeley* (Oxford: Clarendon Press, 1950), 137.

12 For the treatment of domicentricity as synonymous to ethnic and psychological rootedness see David E. Sopher, "The Landscape of Home. Myth, Experience, Social Meaning," in D. W. Meinig, ed., *The Interpretation of Ordinary Landscapes* (Oxford: Oxford University Press, 1979), 129–49, 133. The intensity of human attachment to the native place and home as an anchor of security is comprehensively discussed by E. Relph, *Place and Placelessness* (London: Pion, 1976), 31, 37, and passim. More recently, the idea of a "nature of one's own" and its presence in the early modern economic and legal thought is discussed in Gerhardt Jaritz and Verena Winiwarter, "On the Perception of Nature in a Renaissance Society," in Mikulas Teich, Roy Porter and Bo Gustafson, eds, *Nature and Society in Historical Context* (Cambridge: Cambridge University Press, 1997), 91–112.

13 Yi-Fu Tuan, *Topophilia. A Study of Environmental Perception, Attitudes, and Values* (Englewood Cliffs, NJ: Prentice Hall, 1974), 92–129; idem., "Place: An Experiential Perspective," *Geographical Review* 65 (1975): 151–65; idem., "Geopiety: A Theme in Man's Attachment to Nature and to Place," in F. Lowenthal and Martyn Bowden, eds, *Geographies of the Mind* (New York: Oxford University Press, 1976); D. Geoffrey Hayward, "Home as an Environmental and Psychological Concept," *Landscape* 20 (1) (1975): 2–9; J. Douglas Porteous, "Home: The Territorial Core," *Geographical Review* 66 (4) (1976): 383–90. Roy Porter has seen *topophilia* as coordinating the scientific interests during the Enlightenment years: "*Topophilia* lay not just in its traditionally agreeable aspects [pastoral etc] but in new modes – in the Sublime, the Picturesque and then the Romantic. [The] naturalists unashamedly viewed the

environment through esthetic filters, such as order and disorder, regularity, symmetry, organic form. Furthermore, such value-laden notions stood as yardsticks of *scientific* truth," Roy Porter, "The Terraquaeous Globe," in Rousseau and Porter, eds, *The Ferment of Knowledge*, 300.

14 Quoted in Mabey, *Gilbert White*, 48.

15 Such is the conclusion of J. Oliver, "William Borlase's Contribution to Eighteenth Century Meteorology and Climatology," *Annals of Science* 25 (1969): 275–317. Stan Mendyk documented the separation of antiquarian and natural study during the first decades of the eighteenth century; he sees Borlase as "the odd regional writer," but this interpretation seems to overlook the size and profile of the enterprise Borlase shared with his correspondents and with the larger community of provincial clerical naturalists. Mendyk, '*Speculum Britanniae*', 242.

16 For example, John Laurence, *Gardening Improved: The Clergyman's Recreation Shewing the Pleasure and Profit of the Art of Gardening* (London, 1718) and John Mortimer, *The Whole Art of Husbandry. The Way of Managing and Improving the Land* (London: R. Robinson, 1721).

17 Borlases of Pendeen belonged to the second rank of Cornish landed gentry. The wealth of William's father, John Borlase, an influential Whig squire, at one time MP for St Ives borough, was based on ownership of land and of mineral rights, Cornwall being an important tin-mining region at the time.

18 Andrew to Borlase, 5 March 1736, quoted in P. A. S. Pool, *William Borlase* (Truro: The Royal Institution of Cornwall, 1987), 26, 80. Richard Carew's study on Cornwall (mentioned in chapter 2) was the only chorographic study of prominence. Other writers on natural history and antiquities were Thomas Tonkin (1678–1742), George Jago (died 1726), and Walter Moyle (1672–1721). John Ray and Edward Lhuyd visited the county in 1667 and 1700 respectively. See Pool, *William Borlase*, passim.

19 Borlase to John Andrew, 12 September 1737, quoted in Pool, *William Borlase*, 54. More information on Borlase's connections with fashionable society in and around Strawberry Hill is available in Morris R. Brownell, *Alexander Pope and the Arts of Georgian England* (Oxford: Clarendon Press, 1978), 259–64; see also Marjorie Nicholson and G. S. Rousseau, *"This Long Disease, My Life:" Alexander Pope and the Sciences* (Princeton: Princeton University Press, 1968).

20 John Woodward, *Brief Instructions For Making Observations in All Parts of the World* (London: Richard Wilkin, 1696).

21 Borlase to William Oliver, 1 November 1737. Of Cornwall he thought to be "a place of an healthy air and of plenty of things which a gentleman would desire to see at his table. I have had pleasure of seeing some of the most considerable places in England, and I think there is hardly any place where I could so willingly wish that my lot has fallen, as where it has." Letter to George Borlase, 18 December 1727. To Andrew he wrote on 12 September 1737: "The climate [is] so temperate, the soil so plentiful, the plants so many, the natural advantages of havens for trades so commodious, that no one country of equal extent in the world will probably afford so great a variety . . . And let me add, that 'tis a county whose Natural History, well executed, will be superior to any thing of that kind which has yet come to my notice." Pool, *William Borlase*, 88, 29, 84.

22 Thomas Pennant, *The Literary Life of the Late Thomas Pennant, Esq, by himself* (London: Benjamin and John White, 1793), 2. In his words, "I drew several accounts of Earthquakes on the request of J. Banks, in vol. 71," ibid., 7.

23 Charles Lyttelton had a considerable stature as a local historian and archeologist; he was FRS from 1742 and in 1765 was elected president of the Society of Antiquaries. He was cousin of the "patriot Whig" and minister, George Grenville. Jeremiah Milles, FRS and FSA, succeeded Lyttelton in the presidency of the Society of Antiquaries; his research includes, among articles for the Society, "Topographical Notes on Bath, Wells." He published "Meteorological Observations for 1768, made at Bridgewater, Somerset, and Ludgvan, Cornwall," *Phil. Trans.* 59 (1769): 234. John Hutchins undertook a series of enquiries into the history of Dorset during the 1730s; in 1739 he circulated a single-sheet folio of six queries about the county. The information was eventually collated in his *The History and Antiquities of the County of Dorset with Some Remarkable Particulars of Natural History*, 2 vols (W. Bowyer and J. Nichols: London, 1774).

24 There were 14 "Parochial Queries" from 1752; 8 asked about antiquities, seats of nobility, buildings and customs. Nine were devoted to natural history, including one to the nature of air: "Is your Parish reckoned healthy as to air, or otherwise? Any men or women now living of a great age, what their diet? Have any monstrous births human or brutal happened in your Parish, and when? Have any remarkable, non-descript land or water insects been observed by you, or any one you know, lately? Have you, or any neighbouring Gentleman you know, made any observations towards forming a Register of the Weather? Do you know any uncommon plants near you, and where to be found?" quoted in Pool, *William Borlase*, 300. A similar questionnaire was distributed in 1754 titled *Queries Proposed to Gentlemen in Several Parts of Great Britain*. Lhwyd's undated folio sheet bore the title: "Parochial Queries in order to a geographical dictionary, a natural history of Wales. By the undertaker E. L." *Bibliotheca Topographica Britannica* (London: J. Nichols, 1780–90), vol. 4: 156.

25 Another chorographic study was R. Bradley, *The Natural History of Cambridgeshire and Essex*, c. 1735. It is likely that such method of collecting information was used by Silas Taylor for his *History and Antiquities of Harwich and Dovercourt*, published in 1730 and by Dr Treadway Nash for his *Collections for a History of Worcestershire*, published in 1781–82. See Michael Reed, "The Cultural Role of Small Towns in England 1600–1800," in Clark, ed., *Small Towns*, 127, 140.

26 William Borlase, *Observations on the Antiquities, Historical and Monumental, of the County of Cornwall* (Oxford: W. Jackson, 1754). Idem., *The Natural History of Cornwall, the Air, Climate, Waters, Rivers, Lakes, Sea, and Tides etc.* (Oxford: W. Jackson, 1758).

27 Ibid., v, xi.

28 Most chorographic studies were assisted by a subscription from local gentry and aristocracy. Leigh's *Natural History of Lancashire, Cheshire, and the Peak in Derbyshire* (1700) was made possible by more than 400 subscribers, of whom by far the largest group were local nobles. When in the 1770s, the geographer Edward Hasted approached the Morrices of Betteshanger for information on their family history, Mary Morrice wrote to her son, the Rev. James Morrice, that Hasted's writings were well approved and "if I'm not misinformed he has got money by it, and those families who present him highest are the most favourably inserted; but whether it is by liberal subscription for his book about to be published, or by way of present in hand, or how, I can't tell that his favour is really gained; but I know Lady D. let fall that it cost her son M. 30 Pounds to have the affairs of Knowlton, etc., inserted in Hasted's *History*." Quoted in Everitt, "Edward Hasted (1732–1812), in Jack Simmons, ed., *English County Historians* (East Ardsley: E. P. Publishing, 1978), 199.

29 Cited in Robert Dough, "John Hutchins," in Simmons, ed., *English County Historians*, 125.

30 Borlase, *The Natural History of Cornwall*, iii. Borlase associated the sense of natural grandeur with the admiration able to elicit spontaneous religiosity: "Look where you will, Admiration seizes us." The instances of extreme weather were, in the words of one of Borlase's Cornwall contemporaries, "astonishing phenomena of nature [which] may perhaps terrify us into an apprehension of a superior power; but this is a proof which works upon us in the most sweet and agreeable, though at the same time, forcible and convincing manner." Theophilus Botanista, M.D., *Rural Beauties, or the Natural History of the four following Counties: Cornwall, Dorsetshire, Devonshire and Somersetshire* (London: William Fenner, 1757), ii. On the grandeur of meteorological phenomena, see Samuel Halt Monk, *The Sublime: a Study of Critical Theories in XVIII Century England* (New York: Modern Language Association, 1935); David Lowenthal and Hugh C. Prince, "English Landscape Tastes," in Paul Ward English and Robert C. Mayfield, eds, *Man, Space, and Environment* (New York: Oxford University Press, 1972), 81–114, 90; Fulford, *Landscape, Liberty and Authority*, 21. "There is something engaging to the Fancy, as well as to the Reason, in the treatises of Metals, Minerals, Plants and Meteors," claimed the *Spectator*, 2 July 1712, and in connection to this explained that the greatness of objects (the sea during a tempest in concrete case) terrified imagination and "produced the idea of an Almighty Being." *Spectator* 19 September 1712. On the religious connotation of fieldwork among the "Hutchinsonians", a group of natural philosophers, followers of William Hutchinson see Michael Neve and Roy Porter, "Alexander Catcott: Glory and Geology," *British Journal for the History of Science* 10 (1977): 37–60.

31 Borlase to da Costa, 21 May 1749, quoted in Pool, *William Borlase*, 123. Oliver, "William Borlase," 276 ff. Borlase's article includes, William Borlase, "An Account of a Storm of Thunder and Lightning, near Ludgvan in Cornwall," *Phil. Trans.* 48 (1753–54): 86–98; "Of the Earthquake in the West Parts of Cornwall, July 1757," *Phil. Trans.* 50 (1757): 499–503; "On the Late Mild Weather in Cornwall," *Phil. Trans.* 53 (1763): 27–9; "Of the Quantity of Rain fallen at Mount Bay in Cornwall and of the Weather in that Place," *Phil. Trans.* 54 (1764): 59–60; "Meteorological Observations at Ludgvan in Mount's Bay, Cornwall," *Phil. Trans.* 58 (1768): 89; "Meteorological Observations for 1769, made at Bridgewater, Somersetshire; and at Mount's Bay, Cornwall," *Phil. Trans.* 60 (1770): 228; "Meteorological Observations for 1770 at Ludgvan in Mount's Bay, Cornwall," *Phil. Trans.* 61 (1771): 195; "Meteorological Observations for 1771, at Ludgvan in Mount's Bay," *Phil. Trans.* 62 (1772): 365.

32 Borlase, "An Account of a Storm of Thunder," 93.

33 The director of a local tin mine informed Borlase that no sound coming from the surface could be heard at 60 fathoms depth (at which level several miners heard the rumbling sound during the earthquake) and Borlase concluded that the sound was caused by "a real tremor of the earth, attended with a noise, owing to a current of air and vapour proceeding upwards from the earth." Borlase, "An Account of the Earthquake," 505. The informants might have been coordinated through personal connections; such was the case with "An Account of the Effects of a Storm of Thunder and Lightning, in the Parishes of Looe and Lanreath, in the County of Cornwall, on the 27th Day of June, 1756," sent to Jeremiah Milles by the Reverend James Dyer, minister of Looe and the Reverend Milles (Jeremiah's brother?), vicar of Duloe, Cornwall. *Phil. Trans.* 50 (1757): 104–7.

34 H. R. French, "'Ingenious & Learned Gentlemen': Social Perceptions and Self-fashioning among Parish Elites in Essex, 1680–1740," *Social History* 25 (2000): 44–66, 56.

35 Pool, *William Borlase*, 166.

36 The journal exists in manuscript form, titled *Barometrical and Thermometrical and Ombrometrical Observations with an Account of Winds and Weather at Ludgvan*, quoted in Oliver, "William Borlase," 276. Oliver gives a detailed analysis of the record which he describes as evolving in both precision and extent as Borlase started using information published in the newspapers. See ibid. (307) for drought of 1762.

37 Meteorological schemes of the seventeenth century are discussed briefly in chapter 2. In the eighteenth century, the most notable example was James Jurin's, "Invitatio ad Observatione Meteorologicas Communi Consilio Instituendas," *Phil. Trans.* 31–2 (1720–23): 422–7. During 1720s, he established correspondence with several weather observers, including John Horsley, a schoolmaster in Northumberland, John Huxham, a physician at Plymouth, and Thomas Nettleton in Halifax. Jan Golinski, "Barometers of Change: Meteorological Instruments as Machines of Enlightenment," in William Clark, Jan Golinski, and Simon Schaffer, eds, *The Sciences in Enlightened Europe* (Chicago: The University of Chicago Press, 1999) 69–93. Jurin's agenda was medical.

38 Pool, *William Borlase*, 253.

39 Sources are William Borlase, "Of the Earthquake in the West Parts of Cornwall, July 1757," *Phil. Trans.* 50 (1757): 499; idem., "An Account of a Storm of Thunder and Lightning, near Ludgvan in Cornwall," *Phil. Trans.* 48 (1753–54): 86; Pool, *William Borlase*, 167.

40 Borlase to Henry Baker, 18 July 1763, quoted in Pool, *William Borlase*, 252–3. A decade earlier, Baker was interested in weather himself: "Abstract of Several Observations of Aurorae Boreales," *Phil. Trans.* 46 (1749–50): 499–502; "An Account of the [Meteor of July 22, 1750]," *Phil. Trans.* 47 (1751–52): 3. As no evidence exists to suggest Borlase's commitment to astrometeorology, the reference to heavenly bodies might have been either an allusion to the contemporary theories of the moon's influence on the weather, or a colloquialism without literal meaning.

41 Golinski, "Barometers of Change," 70–2.

42 For this information I am grateful to Mick Wood, Esq., Archivist to the Meteorological Archive and Library at Bracknell.

43 S. Horsley, "An Abridged State of the Weather at London in the Year 1774," *Phil. Trans.* 65 (1775): 167–8. An interesting comparison can be made with the suspicion voiced several years

later by the officials of the Cercle des Philadelphes, the "tropical equivalent of a French provincial academy:" "Very precise meteorological observations have been made with very exact instruments, but the utility of these observations has not yet been sufficiently demonstrated. [One never sees causes], one only sees outcomes. No power to control, no ability to prevent or to remedy anything results. . . . The Cercle asks what can be the utility of meteorological observations." Quoted in McClellan III, *Colonialism and Science*, 167. Golinski, in "Barometers of Change," 72, concludes that the barometer was "hailed as a triumph of contemporary science and as an object of superstitious faith." This was a result of wide popularity of astrometeorological prediction. For an interpretation which postpones the demise of astrology (proposed by Bernard Capp, Patrick Curry and Keith Thomas), see Maureen Perkins's *Visions of the Future: Almanacs, Time, and Cultural Change, 1775–1870* (Oxford: Oxford University Press, 1996).

44 Robert Boyle, "The General History of the Air," (1695), *Works* iv, 642, 643. Boyle referred to the *chorographically* based natural histories of which one of the most successful examples was an Irish cartographic project undertaken during 1680s. William Molyneux collected information for an English atlas envisioned by Moses Pitt and coordinated by Hooke. Molyneux mobilized a number of Protestant and Catholic Irish to collaborate, the most enthusiastic of whom was Roderick O'Flaherty, writer of *A Chorographical description of West or h-Iar Connaught, written* AD 1684. See J. G. Simms, *William Molyneux of Dublin, 1656–1695* (Blackrock: Irish Academy Press, 1982), 34–46.

45 On utilitarian trends, see Goldgar, *Impolite Learning*, 239. "Leisure indeed is all in life I have to boast of," wrote Borlase to Oliver in 1737, "and that leisure would by this time have degenerated into little better than perfect idleness, but a friend I have at Leyden has been rousing me up, and taking care to prevent the approaching lethargy." Borlase to Oliver, 1 November 1737, quoted in Pool, *William Borlase*, 86. The significance of the regular observation, as Golinski observes, was "in the routinization of observational practices [more] than in the achievement of a naturalistic understanding of weather . . . [T]he daily or twice-daily drill of tapping the barometer and taking a reading yielded some more immediate satisfaction, perhaps as a means of structuring experienced time by laying down the benchmarks of a reassuring ritual." Jan Golinski, "Barometers of Change," 86.

46 Borlase wrote regarding his will to continue the rain record despite an improbability of discovering weather laws: "[Keeping a weather register] has been a constant amusement to me ever since my correspondence on the subject increased." Letter to William Huddesford (Keeper of the Ashmolean Museum), 13 February 1767, Oliver, "William Borlase," 292.

47 See Appendix for more information.

48 William Henry, "An Account of an Extraordinary Stream of Wind, which Shot thro' Part of the Parishes of Termonomungan and Urney, in the County of Tyrone, on Wednesday October 11, 1752," *Phil. Trans.* 48 (1753–54): 1–4. Lord Cardogan, sometime MP for Reading and Isle of Wight, was governor of Gravesend in 1753. He was a FRS and trustee of the British Museum.

49 Francis Griffin Stokes ed., *The Bletchley Diary of the Rev. William Cole, 1765–67* (London: Constable, 1931), quoted in *Sunday Times* 24 January 1932. At approximately the same time, Samuel Johnson, who otherwise deplored the ubiquitousness of weather-talk among his contemporaries, came to regard the keeping of a journal as a valuable practice, primarily for its contribution to the spiritual life: "My general resolution to which I humbly implore the help of god is to methodise my life, to resist sloth and combat scruples. I hope from this time to keep a Journal." The quote and other important information on the social relevance of the diarizing among the Augustan elites is to be found in Robert A. Fothergill, *Private Chronicles: A Study of the English Diaries* (London: Oxford Universing Press, 1974), 25.

50 John Beresford, ed., *The Diary of a Country Parson, The Rev. James Woodforde, vol. 2., 1782–1787* (London: Humphrey Milford, 1926), 176. An ingenious method of measurement is mentioned by Borlase's correspondent, Henry Ellis, governor of Georgia: "I have frequently walked a hundred yards under an umbrella, with a thermometer suspended from it by a thread, to the height of my nostrils, when the mercury rose to 105; which is prodigious." His letter shows the extreme weather as a topic of personal and scientific correspondence as well as a

subject of diary writing; the sense of moment is conveyed when he writes: "It's now 3 o'clock; the sun bears nearly S.W. and I am writing in piazza, open at each end, on the N.E. side of my house, perfectly in the shade; a small breeze at S.W. blows freely through it; Bird's thermometer stands at 102." *The Annual Register* (London: J. Dodsley, 1767).

51 The study appeared posthumously as the *Border History of England and Scotland* (1776). Paul B. James, ed., *Diary of George Ridpath* (London, 1922).

52 Warren R. Dawson, ed., *The Banks Letters* (London: British Museum, 1958), 438. Benjamin Hutchinson [Vicar of Kimbolton, Lincolnshire], *A Calendar of the Weather for the year 1781, With an Introductory Discourse on the Moon's Influence at common Lunations in general, and on the Winds at the Eclipses in particular, founded on observations at Kimbolton* (London: J. Fielding, 1782).

53 Miles entered into history of science mainly as the editor and transcriber (poorly assisted by Thomas Birch) of Robert Boyle's manuscripts purchased from Boyle's apothecary. See Marie Boas Hall, "Henry Miles FRS (1698–1763) and Thomas Birch (1705–1766)", *Notes and Records of the Royal Society* 18 (1963): 39–44. Boas suggests that Miles's FRS title was largely due to the work on his edition of Boyle's works despite a good previous publishing record. Miles's meteorological papers are "A Representation of the Parhelia seen in Kent, December 19, 1741," *Phil. Trans.* 42 (1742–43): 16–17; "Of the Storm of Thunder which happened June 12, 1748," *Phil. Trans.* 45 (1748): 383–5; "On Thermometers and the Weather," *Phil. Trans.* 46 (1749–50): 1–6; "Concerning Aurora Borealis, seen February 16, 1749," *Phil. Trans.* 46 (1749–50): 319; "On the heat of the Weather at Tooting in July and September last," *Phil. Trans.* 46 (1749–50): 571–2.

54 Geoffrey Holmes, *Augustan England: Professions, State and Society* (London: George Allen and Unwin, 1982), 85.

55 John Pringle, "Several Accounts of the Fiery Meteor," *Phil. Trans.* 51 (1759–60): 218; "Of an Unusual Agitation of the Sea," *Phil. Trans.* 49 (1755–56): 642–4; Rev. George Costard, "Concerning a fiery Meteor seen in the Air on July 14, 1745," *Phil. Trans.* 43 (1744–45): 522–4; Rev. Anthony Williams, "Of a Remarkable Thunder-Storm," *Phil. Trans.* 61 (1771): 71–2.

56 Goldgar, *Impolite Learning*, 15.

57 Quoted in Pool, *William Borlase*, 247.

58 Fothergill, *Private Chronicles*, 29.

59 "The Calendar of Flora. 1. Swedish and 2. English; made in 1755. 3. Greek; selected from Theophrastus," in Benjamin Stillingfleet, *Miscellaneous Tracts Relating to Natural History, Husbandry, and Physick* (London: R. and J. Dodsley, 1762), 402.

60 In 1754 Stillingfleet personally adjusted the thermometer of the poet Thomas Gray, and the next year, by coincidence, both Gray and Stillingfleet independently drew up their calendar schemes. Gray kept up the observations of Cambridge plants, insects, and weather for the rest of his life, from 1767 in the printed form as the "Naturalist's Journal," issued by Benjamin White, Gilbert's brother and an eminent publisher in London. William Powell Jones, *Thomas Gray, Scholar* (Cambridge, Mass: Harvard University Press, 1937), 125–42. Also see C. E. Norton, *Thomas Gray as a Naturalist* (Boston, 1903), and R. W. Ketton-Cremer, *Thomas Gray, A Biography* (Cambridge: Cambridge University Press, 1955), 182. The popularity of naturalistic diaries reached its climax during these decades, when, for instance, one Josiah Ringsted published *A Diary for Gentlemen, Sportsmen, Gardeners, Grasiers, Game-keepers, Cow-keepers, Horse dealers, Carriers, etc., etc. with a Coloumn [sic] for Observations on the Weather* (London: J. Dixwell, ca. 1780); see also, Reuben Burrow, *The Lady's and Gentleman's Diary* (London: T. Carnan and G. Robinson, 1755–8). Other examples of the Linnean/Georgic climatology is T[imothy] Sheldrake, *The Cause of Heat and Cold in the several Climates and Situations of this Globe* (London: the Author, 1756). Sheldrake's interest in climate began with his attempt to grow exotic plants in England. The work was mainly based on Hales's chemistry of plants, Boyle's and Ray's study of cold, and Boerhaave's iatrochemistry. The same year he published *The Gardener's Best Companion in the Greenhouse, or tables showing the greatest Heat and Cold of all Countries.* The more popular scheme was *The Scots Gardiner [sic] for the Climate of Scotland; Together with the Gardiner's Calendar, and Observations on the Weather* (Edinburgh: James

Reid, 1755). The compiler inserted *The Shepherd of Banbury's Rules to Judge the Changes of Weather*, of which see below, chapter 6. Thomas Ignatius Maria Forster informs us in 1827 that his father, the naturalist Thomas Forster, kept a season calendar and the diary of weather on a daily basis from 1780 to his death in 1825. T. Forster, *Pocket Encyclopedia* (London: John Nichols), 1827, x.

61 White switched to the "Naturalist's Journal" in 1768. On the first page he wrote: "The Gift of the Honourable Mr Barrington, the Inventer." Rashleigh Holt-White, *The Life and Letters of Gilbert White of Selbourne* (London: John Murray, 1901), I, 156.

62 Introduction to Gilbert White, *Natural History of Selborne* (1790, Harmondsworth: Penguin, 1977), v, and "Letter IX" to Barrington, quoted in Keith, *Rural Tradition*, 48–9. On the epistemological and literary evolution of the parish record see Mabey, *Gilbert White*, and Lansing V. Hammond, "Gilbert White, Poetizer of the Commonplace," in Frederick W. Hilles, ed., *The Age of Johnson. Essays Presented to Chauncey Brewster Tinker* (New Haven: Yale University Press, 1949), 377–83. On White's political conservatism and the trust in aristocratic ideology see Merrett, "Natural History and the Eighteenth-Century English Novel," 148, and Lucy B. Maddox, "Gilbert White and the Politics of Natural History," *Eighteenth Century Life* 10 (1986): 45–57.

63 William Whiston, "An Account of two Mock-suns, and an Arc of a Rainbow inverted, with a Halo," *Phil. Trans.* 31 (1720–23): 212.

64 John Kington, *The Weather Journals of a Rutland Squire: Thomas Barker of Lyndon Hall* (Rutland Record Society: Oakham, 1988).

65 Evidence appears in White's letter to Pennant in the *Natural History of Selborne*, and the confirmation can be found in Thomas Barker's jotting down young Gilbert's observations when in Rutland: "1736. *March 31*. A flock of wild Geese flew N. – G[ilbert]. W[hite]. *April 6*. The cuckow heard. – G. W." Quoted in Holt-White, *Letters*, 31.

66 Thomas Barker, "An Account of an Extraordinary Meteor seen in the County of Rutland, which resembled a Water-Spout," *Phil. Trans.* 46 (1749–50): 248–50; "An Account of a remarkable Halo," *Phil. Trans.* 52 (1761–62): 3; "[C]oncerning Observations of the Quantities of Rain at [Lyndon] for several years," *Phil. Trans.* 61 (1771): 221; "Extract of Mr Barker's Meteorological Register at Lyndon in Rutland [1771]," *Phil. Trans.* 62 (1772): 42; "Extract of a Register of the Barometer, Thermometer, and Rain, at Lyndon in Rutland, 1772," *Phil. Trans.* 63 (1773): 221; "Extract . . . for 1773," *Phil. Trans.* 64 (1774): 202; the *Transactions* annually published Barker's meteorological extracts and abstracts until 1800. Barker's work on comets was partially inspired by the imposing presence of his grandfather: Whiston's controversial reputation rested in no small part on the millenarian context which, from the earliest period, had informed his study of eclipses, comets, and meteors. These and other issues are discussed in Force, *William Whiston*. Simon Schaffer, "Newton's Comets and the Transformation of Astrology," in Patrick Curry, ed., *Astrology, Science, and Society: Historical Essays* (Woodbridge: The Boydell Press, 1987), 219–45, 233; Barker's theological works give strong support to the materiality of Whiston's influence as do his occasional reflections in his *Meteorological Journal 1733–1795*. Thomas Barker, *A Treatise on the Duty of Baptism* (London 1771); *On Prophesies Relating to the Messiah* (London, 1780); *On the Nature and Circumstances of the Demoniacks in the Gospels* (London, 1783).

67 Robert Marsham, "Naturalist Journal, 1736–1788," *Phil. Trans.* 79 (1789): 154.

68 John corresponded with Linnaeus and compiled *Fauna Calpensis, the Natural History of Gibraltar*. See the Rev. Robert Clutterbuck, *Notes on the Parishes of Fyfield, Kimpton, Penton Mewsey, Weyhill and Wherwell in the County of Hampshire* (Salisbury: Bennet Brothers, 1898), 5 ff. In fact, Gilbert's other brother Thomas as well as his two nephews, Charles Henry and Sampson Henry also kept naturalistic journals. Here is a typical entry from Henry White's journal: "1784, May 17th. Perfect settled dry weather continues, with summer warmth beyond ye season; thin light gray clouds with hot sun between. Went to serve Kimpton before 10 a.m. etc." Clutterbuck, *Notes*, 10. Some of these records were published: the Rev. B. Hutchinson [Vicar of Kimbolton, Lincol.] *A Kalendar of the Weather for the year 1781, with an Introductory Discourse on the Moon's Influence at common Lunations in general, and on the Winds at the Eclipses in particular, founded on observations at Kimbolton* (London: J. Fielding, 1782).

69 It is difficult not to perceive the study of natural history as a phenomenon of interpenetration, in Allen's formulation, of the "hard and rational on the one hand, the soft and sentimental on the other." Allen, "Natural History in Britain in the Eighteenth Century," 342. James Thomson's heritage was vital throughout the century: *The Seasons: Autumn and Winter. A Poem containing, a Short Review of each Season; but more particularly Winter, and its Severity. etc. etc.* (London: J. Wakelin, 1778). On the other hand, the specialness of the place owed significantly to providential considerations, as suggested by Christopher Hamlin in his "Chemistry, Medicine, and the Legitimization of English Spas, 1740–1840," *Medical History* Supplement 10 (1990): 67–81, 68.

70 Both quotes are from the advertisement to the first edition of *Natural History of Selborne*, quoted in Keith, *Rural Tradition*, 42.

71 Holt-White, *Letters*, I, 202.

72 White, *Selborne*, 272. My italics.

73 White, *Selborne*, 281. Italics added. Needless to say, White continued to arrange for observing "this strange severity of the weather," by placing an Adam's thermometer at a more elevated location in Newton, Hampshire and "expecting wonderful phenomena." He was "disturbed" to find that Newton's weather proved warmer. The subspecies of the meteoric genre focusing on unusual spells of heat and cold was introduced to the pages of the *Philosophical Transactions* by Henry Miles in 1750. James Six was, by 1784, performing a number of thermometrical measurements on Canterbury cathedral and John Cullum wrote for the journal an "Account of extraordinary Frost, 23 June 1783," *Phil. Trans.* 74 (1784): 416. Cullum was a divine pursuing the study of antiquities and botany with a wide network of correspondents, including Pennnant, William Cole, and possibly Thomas Gray. He kept a naturalist's diary and worked on "The History and Antiquities of Hawsteed and Hardwick in the County of Suffolk," published in the *Bibliotheca Topographica Britannica* in 1784. "John Cullum," *DNB*.

74 White, *Selborne*, 286. Gilbert's brother Henry recorded in his Fyfield diary, "ye superstitious in town and country have abounded with ye most direful presages and prognostication," quoted in Mabey, *Gilbert White*, 190.

75 Mr Nicholson, "An Account of a Storm of Lightning observed March 1, 1776," *Phil. Trans.* 54 (1764): 350–1. Even for physicians, unusual meteorological events had special powers: Thomas Short contributed an essay to *Phil. Trans.* with a description of the appearance of the blood-like streamers, resembling, but not identical, to an aurora borealis. This meteor, claimed Short, "put an End to the Remains of both Catarrh, and watery Diarrhea; and restored general Health, till the next epidemic Catarrh among Infants in February 1738." "An Account of several Meteors, communicated in a Letter from Thomas Short," *Phil. Trans.* 41 (1739–41): 625–30.

76 John Swinton, "An Account of a Remarkable Meteor seen at Oxford," *Phil. Trans.* 52 (1761): 99; Swinton's other arresting descriptions are "An Account of an Anthelion Observed near Oxford," *Phil. Trans.* 52 (1761): 94; "An Account of a Remarkable Meteor seen at Oxford, March 5, 1764," *Phil. Trans.* 54 (1764): 326–9. Swinton was Rector in St Peter-le-Bailey, in Oxford and a FRS from 1728. His later career was connected with Christ Church at Oxford, where he was appointed keeper of the archives of the university. Between 1739 and 1750 he published six book-length studies on etymology, numismatics and inscriptions and numerous articles in *Philosophical Transactions*.

77 [Hayman Rooke], *A Continuation of the Register of Weather, 1801–2* (Nottingham: S. Tupman), 1802.

78 [Hayman Rooke], *A Meteorological Register kept at Mansfield Woodhouse. With Probable Indications of Weather* (Nottingham: S. Tupman, 1795). Another example of a "non-stationary" diary was Richard Townley, *A Journal kept in the Isle of Man. Giving an Account of the Wind, Weather, and Daily Occurrences for Upwards of 11 Months.* 2 vols. (Oxford: J. and J. Merrill, 1791). Alain Corbin discusses Townley in *The Lure of the Sea. The Discovery of the Seaside in the Western World, 1750–1840* (London: Polity Press, 1994), 90. In 1736 Henry Forth wrote to William Derham: "I have for some time pursued [observations of the weather], for my own private satisfaction, upon your ingenious Model," "An Account of the Storm, Jan. 8, 1734–5," *Phil. Trans.* 39 (1735–36): 285–7.

79 J. C. Loudon, quoted in D. E. Allen, *The Naturalist in Britain*: A Social History (Princeton, Princeton University Press, 1994), 18, 19.

80 "In Country writings, philosophy and history were blended into an ideology which has been described as the 'politics of nostalgia.' [Isaac Kramnick's term]." In William Speck, "Whigs and Tories Dim Their Glories: English Political Parties Under the First Two Georges," in John Cannon, ed., *The Whig Ascendancy: Colloquies on Hanoverian England* (New York: St Martin's Press, 1981), 51–71, 64.

81 Estabrook, *Urbane and Rustic England*, 276. This divide was analogous to the esthetic perception of the countryside. As Archibald Alison observed in 1790, the man who lived in the countryside regarded it differently from the way in which it has was seen by the tourist. "To the latter, the streams were known only by their gentleness or their majesty, the woods by their solemnity, the rocks by their awfulnes or terror." But to the former "they serve as distinctions of different properties, or of different divisions of the country. They become boundaries or landmarks, by which the knowledge of the neighbourhood is ascertained. Even a circumstance so trifling as the assignation of particular names contributes in a great degree to produce this effect, because the use of such names in marking the particular situation or place of such objects naturally leads him to consider the objects themselves in no other light than that of their place or situation" Archibald Alison, *Essay on the Nature and Principles of Taste* (Edinburgh, 1790), 312–13, quoted in Thomas, *Man and the Natural World*, 264.

82 Alan Everitt, *Landscape and Community in England* (London and Ronceverte: The Hambledon Press, 1985), 8. Everitt locates the source of provincial integration in the role that the English counties assumed during the seventeenth century, in which period they "ceased to be simply administrative units and in many cases became genuine self-conscious regions" (21). It would be interesting to pursue this conclusion by asking about the place of topography and surveying in this development. A step in this direction is Helgerson, *Forms of Nationhood*.

83 J. C. D. Clark, *Revolution and Rebellion. State and Society in England in the Seventeenth and Eighteenth Centuries* (Cambridge: Cambridge University Press, 1986), chapter 4. See also his *English Society 1688–1832*, chapter 2. For a most recent revaluation of the role of the Established Church, see W. M. Jacob, *Lay People and Religion in the Early Eighteenth Century* (Cambridge: Cambridge University Press, 1996), 20–56. Like Clark, Jacob rejects the thesis that the "old world" of the popular religiosity ended during the reign of Queen Anne: "[t]he evidence suggests that this 'old world' continued for much of the eighteenth century," ibid., 4.

84 Paul Langford, *Public Life and the Propertied Englishman, 1689–1789* (Oxford: Clarendon Press, 1991), 367.

85 Dror Wahrman, "National Society, Communal Culture: an Argument about the Recent Historiography of Eighteenth-Century Britain," *Social History* 17 (1992): 43–72, 45.

86 Norma Landau, *The Justices of Peace, 1679–1760* (Berkeley: University of California Press, 1984), 3–4, 360. See also D. Rollinson, "Property, Ideology and Popular Culture in a Gloucestershire Village 1660–1740," *Past and Present* 93 (1981): 70–97.

87 But the regional study could also be politically motivated: Phillipson shows that post-union elites of the Scottish province reasserted their challenged political authority by an outburst of literary activity concerned with local antiquarianism. N. T. Phillipson, "Culture and Society in the 18th Century Province: the Case of Edinburgh and the Scottish Enlightenment," in Lawrence Stone, *The University in Society* (Princeton: Princeton University Press, 1974), 407–48. The antiquarian circle in Canterbury presents another example, even though it was less politically defined. Alan Everitt, "Country, County and Town: Patterns of Regional Evolution in England," in Peter Borsay, ed., *The Eighteenth-Century Town* (London: Longman, 1990), 29, n. 41. Borsay has emphasized that local history was a particularly cultivated field in the publishing activities of provincial presses. See Borsay, *English Urban Renaissance*, 128.

88 "Queries, addressed to the Gentlemen and Clergy of North-Britain, respecting the Antiquities and Natural History of their respective Parishes, with a view to favour the World with a fuller and more satisfactory Account of their Country, than it is a Power of a Stranger and transient Visitant to give." Quoted in Charles Withers, "Geography, Natural History and the Eighteenth-Century Enlightenment: Putting the World in Place," *History Workshop Journal* 39 (1995): 137–64, 152.

89 J. C. D. Clark, *Samuel Johnson. Literature, Religion and English Cultural Politics from Restoration to Romanticism* (Cambridge: Cambridge University Press, 1994), 1–43.

90 Gerald Newman, *The Rise of English Nationalism 1740–1830* (London: Weidenfield and Nicolson, 1987), 111. On the politicization of georgics and the significance of the countryside and the farm as traditional sources of criticism of social and political life, see Richard Feingold, *Nature and Society: Later Eighteenth-Century Uses of the Pastoral and Georgics* (New Brunswick, N. J.: Rutgers University Press, 1978), 121–55.

91 Merrett, "Natural History and the Eighteenth-Century English Novel," 147. The escapist retreat to an illusion of a stable regional world was underscored in regional fiction which is ably discussed in W. J. Keith, *Regions of the Imagination: The Development of British Rural Fiction* (Toronto: University of Toronto Press, 1988), 9. For Oliver Goldsmith, for example, the greatest pleasure was "to grow old in the spot in which one was born. Native walks and fields have, by association, beauties beyond the most delightful scenes 'that ever art improved' or 'fancy painted.' To observe *familiar objects*, and to measure the years by the growth of trees planted by *one's own* hands, are the basis of personal contentment, of loyalty to individuals and communities, and ultimately of love of country." Quoted in Everett, *The Tory View of Landscape*, 63, my italics.

92 Speck, "Whigs and Tories Dim Their Glories," 51–71, 64. The court/country polarization has been the subject of much discussion. The main lines of thought are summarized in Geoffrey Holmes and Szechi, *The Age of Oligarchy*, 39–40.

93 Simmons, ed., *English County Historians*. I am grateful to Newton Key for pointing out a similarity between my pairing of country and chorography and his connection between country and county feast sermons. These sermons, preached in London in celebration of counties, were inventories of "local topography, buildings and institutions, and worthies." In the eighteenth century, one feast sermon even celebrated Oxfordshire's natural history, reminding natives of "the rareness of its Minerals; the richness of its Soil," and mentioning Robert Plot's geological work, *The Natural History of Oxfordshire*. Newton Key, "The Localism of the County Feast in Late Stuart Political Culture," *Huntington Library Quarterly* 58 (1996): 211–37, 227.

94 Beginning with the middle decades of the century, the impulse for the study of provinces was sometimes being blended with the various forms of English cultural nationalism. It has been proposed that the consolidation of English nationalistic sentiments at this time owed some of its academic content to jealousy of the intellectual achievement of the neighboring French. On the popular side, xenophobia targeting the French, served as a pretext for condemnation of the metropolis and the court, both accused of being conduits for alien influences. Jeremy Black, *Natural and Necessary Enemies: Anglo-French Relations in the Eighteenth Century* (Athens: University of Georgia Press, 1986), 178 and passim. The early "romantic" effort to validate national virtues in terms of its intellectual heritage resulted in "the proliferation of historical, philological, ethnological, socio-demographic, art-historical, musicological, and other forms of historicist enquiries." Newman, *The Rise of English Nationalism*, 111. As Newman observes, the production of this kind was essentially ethnic in its concerns, antiquarian in content, educational in its impact, and ideological in its overall significance. The chartering of the Society of Antiquaries (1751), the opening of the British Museum (1759), the preparation of *Biographia Britannica* (1747–66) are all taken as symptoms of a budding national self-study and self-promotion. Ibid., 111.

Notes to chapter 6

1 Nathaniel Kent [of Fulham], *Hints to Gentlemen of Landed Property* (London: J. Dodsley, 1773, 1793), 3.

2 Reed, *Romantic Weather*, 23. In 1749, a report from Rome said: "Wednesday, about two hours after midnight, we had a tempest here, which threatened the return of all the elements into their first chaos," *Gentleman's Magazine* 19 (1749): 273.

3 Edward Barlow, *Meteorological Essays Concerning the Origin of Springs, Generation of Rain and Production of Wind* (London: J. Hooke, 1715). Preface, 2. The second edition appeared in 1722.

Barlow was a Catholic priest educated at the English College at Lisbon in 1660s, practicing later in Parkhall in Lancashire. The explanatory reticence was notorious in respect to wind. To the question "Is the cause of Wind in Motions of the Planets?" editors of *The Athenian Oracle* answered: "Our Saviour who knew Nature well enough, has told us '*We know not whence it comes, nor whither it goes.*'" *The Athenian Oracle*, vol. II (London, 1718).

4 Barlow, *Meteorological Essays*, 72, 57. It should be noted that Barlow uses an abundance of military metaphors to describe this elemental disorder.

5 John Smith, *A Compleat Discourse of the Nature, Use, and Right Managing of that Wonderful Instrument, the Baroscope, or Quick-silver weather-glass.* (No publisher cited, c. 1700), 91. Henry Miles wrote in 1749: "It has been often complained that the theories of the air and weather are so imperfect, and that an unfinished one of Mr Boyle, published since his death, is the best we have yet. Perhaps there is equal reason for complaint, that the thermometer first introduced into use in England, by the same excellent philosopher, has been so little improved for more than half a century, that it serves for little more than amusement." "On Thermometers and the Weather," *Phil. Trans.* 46 (1749–50): 223.

6 Edward Saul, *An Historical and Philosophical Account of the Barometer, or Weather Glass* (London: A. Bettesworth and C. Hitch, 1730), 75. Saul wrote that aqueous meteors proceeded from the "collection of Streams, Mists and Vapours, raised from the surface, confusedly driven together and accumulated by the Winds," ibid., 23.

7 Samuel Johnson, *The History of Rasselas, Prince of Abissinia* (London: Oxford University Press, 1971, 1759), 112. See also David O. Ross, Jr., *Virgil's Elements: Physics and Poetry in the Georgics* (Princeton, N. J.: Princeton University Press, 1987), 82.

8 Rev. William Derham, "The History of the Great Frost in the Last Winter, 1708 and 1708/9," *Phil. Trans.* 26 (1708–09): 477.

9 Pasqual R. Pedini, "An Account of the Earthquakes felt in Leghorn," *Phil. Trans.* 42 (1742–43): 77.

10 "Meteorology," *Encyclopedia [Britannica]* (Philadelphia, 1798).

11 Barlow, *Meteorological Essays*, 45.

12 Harris, *Lexicon Technicum*, I, 245.

13 "Meteorology," in Chambers, *Cyclopaedia: or An Universal Dictionary of Arts and Sciences* (London, 1728). Chambers's entries stayed unchanged as late as 1784 edition. Chambers listed three kinds of meteors: Igneous (made up of sulfurous smoke set on Fire), Aerial (made up of flatulent and spirituous Exhalations), and Aqueous (of Vapors, or watery Particles). A scrupulous discussion of the nature of "Mixed" bodies is given in Charles Morton, *Compendium Physicae*, a textbook used by Harvard College students during the period 1687–1728. Morton founded a dissenting academy in London (attended by Defoe and Wesley) but left England in 1685. I did not find evidence that his book was used in England. See S. E. Morison's preface to the *Compendium* in the *Publications of the Colonial Society of Massachusetts, Boston* 33 (1940): 3–237. For Morton's relation to Defoe see Ilse Vickers, *Defoe and the New Sciences* (Cambridge: Cambridge University Press, 1996). Even as late as 1771, the meteor was defined as "an imperfect, changeable and mixt body . . . formed by the action of the heavenly bodies, out of the common elements." "Meteorology," M. Hinde, *New Royal and Universal Dictionary of Arts and Sciences* (London: J. Cooke, 1771).

14 We may recall that in his *Godly Gallerye*, Fulke argued that no one knew *what* those impressions might be: no earlier definition was of service, and indeed, Aristotle's own was derived from "doutfullnes" and one "of that whereof there is no name" (which is defined). The ambiguity of the meaning of "meteors" was further enhanced by the frequent use of the term "affections" known in the sixteenth century to signify "a temporary or non-essential state, condition, or relation of anything" (*OED*), as in the construction "The coldness or other affection of Air."

15 Morton, *Natural History of Northamptonshire*, 1712.

16 George Costard, "Concerning a Fiery Meteor seen in the Air on July 14, 1745," *Phil. Trans.* 43 (1744–45): 522–4.

17 John Pringle, "Some Remarks upon the Several Accounts of the Fiery Meteor," *Phil. Trans.* 51 (1759–60): 259–60. The uncertainty was sometimes a part of the narratives: William Gordon

wrote about a fireball that "I really at first took it for some artificial Fire-Work, but was soon undeceived by the different Forms it appeared in." Capt. William Gordon, "An Account of the Fireball," *Phil. Trans.* 42 (1742–43): 58.

18 John Hedley Brooke, *Science and Religion* (Cambridge: Cambridge University Press, 1991), 213. For natural-theological origins of a geographical concept, see Yi-Fu Tuan, *The Hydrologic Cycle and the Wisdom of God* (Toronto: University of Toronto Press, 1968).

19 See Charles Coulston Gillispie, *Genesis and Geology* (Cambridge, Mass.: Harvard University Press, 1951), 1–40, 9. For Darwin see Frank M. Turner, *Contesting Cultural Authority: Essays in Victorian Intellectual Life* (Cambridge: Cambridge University Press, 1993), 111.

20 On Derham's scientific activities see A. D. Atkinson, "William Derham, FRS (1657–1735)," *Annals of Science* 8 (1952): 368–92.

21 William Derham, *Physico-Theology* (London: W. Innys, 1716). The British Museum General Catalogue of Printed Books notes thirteen English editions by 1768, three new editions by the end of the century, and translations into Dutch, French, Swedish and German, John J. Dahm, "Science and Apologetics in the Early Boyle Lectures," *Church History* 39 (June 1970): 176. On the numerous prominent readers of the work, see Atkinson, "William Derham, FRS (1657–1735)," 368. Derham's Cartesian explanation of watery meteors assumes only moist vapours and rejects the hot and dry exhalations, being "really no other than the humid parts of bodies respectively dry." Derham, *Physico-Theology*, 28. The purpose of the work made it a compilation of textual and empirical evidence gleaned from an array of sources. Derham borrowed from Seneca's *Naturales Quaestiones*, Pliny's *Historia Naturalis*, Cicero's *De Natura Deorum*, and Lucretius's *De Natura Rerum*, but there was also a mention of Drebbel's submarine "aerology," the pneumatic experiments of Marcello Malpighi and Nehemiah Grew, Ralph Bohun's treatise on wind, and even the weather journal of Samuel Clarke's grandfather.

22 George Cheyne, *Philosophical Principles of Religion Natural and Revealed*, 4th edition (London: George Strahan, 1734), 289. An example of the contrast between the popular versions of natural theology and those of polemical homiletics is given by *A Short Essay upon the Cause and Usefulness of the W. and S.W. Wind's frequent Blowing in England*, by way of solution to a passage in Mr Ray's Wisdom of the Creator, in the works of creation: with some observations upon the weather. Also a diary during the wet season which succeeded the drought in 1737; intermixed with verses (Liverpool: the Author, 1742).

23 John Claridge, *The Shepherd of Banbury's Rules to Judge the Changes of Weather* (London: T. Waller, 1723), 45. My italics.

24 "Never rob other countries of rain to pour it on thine own. For us the Nile is sufficient." Johnson, *Rasselas*, 113.

25 In *Goodly Gallerye*, Fulke had a short chapter on the signs of rain concluding with: "But these kynde of signes perteine not so properly to Meteorologie, as to maryners and husbandrie, which haue a great many more than these. And Virgil in his first boke of Georgikes, hath a great number for them," pp. 92–3. According to Heninger this was the first instance of the word "meteorology" in English writings, Heninger, *Handbook of Renaissance Meteorology*, 4n.

26 L. P. Wilkinson, "The Intention of Virgil's *Georgics*," *Greece and Rome* 55 (1950): 19–28; L. A. S. Jermyn, "Virgil's Agricultural Lore," *Greece and Rome* 53 (1949): 49–69; idem., "Weather-Signs in Virgil," *Greece and Rome* 58 (1951): 28–59.

27 The equation in question was: "METEOROLOGY or, Certain Rules to know the Weather," Thomas Todd, Philomath, *Perpetuum Kalendarum Atronomicum: or, a Perpetual Astronomical Kalendar* (Edinburgh: Thomas Lumisden and John Robertson, 1738), 70.

28 Richard Kirwan, *A Comparative View of Meteorological Observations Made in Ireland Since the year MDCCLXXXVIIII (1789) with some Hints toward Forming Prognostics of the Weather* (Dublin: George Bonham, 1794), 3–4. The writings of these authors coincided with the renewed interest in natural theology, culminating in William Paley's *Natural Theology* (1802). As Frank Turner has emphasized, the vindication of utility and commerce was intrinsic to natural theological argument, which makes it reasonable to expect that the Derhamesque emphasis on the uses of atmosphere would resurface in late eighteenth-century works. This time around, however, the majority of meteorologists would claim theoretical and social priority over the meteoric tradition, as we will discuss in chapter seven. Turner, *Contesting Cultural*

Authority, 101–9. See also Donald Worster, *Nature's Economy: A History of Ecological Ideas* (Cambridge: Cambridge University Press, 1977), 3–51.

29 Anthony Low, *The Georgic Revolution* (Princeton, N. J.: Princeton University Press, 1985). For other practical manuals see John Donaldson, *Agricultural Biography* (London: the Author), 1854. An illuminating discussion of the Georgic tradition is Frans de Bruyn, "From Virgilian Georgic to Agricultural Science: An Instance in the Transvaluation of Literature in Eighteenth-Century Britain," in Albert J. Rivero, ed., *Augustan Subjects: Essays in Honor of Martin C. Battestin* (Delaware, 1997), 47–67.

30 In 1697, John Dryden published his translation of Virgil's *Georgics*, inaugurating the eighteenth-century "astonishing vogue for georgic poetry." Other poems were John Philips, *Cyder* (1706), Somerville, *The Chase* (1735), Armstrong, *The Art of Preserving Health* (1744), Smart, *The Hop-Garden* (1752), Dodsley, *Agriculture* (1754), John Dyer, *The Fleece* (1757), Grainger, *The Sugar Cane* (1763), Mason, *The English Garden* (1772–82) and William Cowper, *Task* (1785). See L. P. Wilkinson, *The Georgics of Virgil* (Cambridge: Cambridge University Press, 1969), 299. See also William Powell Jones, *The Rhetoric of Science: A Study of Scientific Ideas and Imagery in Eighteenth-Century English Poetry* (London: Routledge and Kegan Paul, 1966), 200–12.

31 John Martyn, FRS, *The Bucolicks of Virgil, with an English Translation and Notes* (London, 1741), vi. Martyn was a botany professor at Cambridge and his botanical reading of Virgil is discussed in terms of the tension between "moderate Neoclassicism and scientific positivism" by Annabel Patterson, *Pastoral and Ideology. Virgil to Valéry.* (Berkeley: University of California Press), 1987, 238–41. For a contemporary publication on classical agriculture, in England and in Europe, see G. E. Fussell, *The Classical Tradition in West European Farming* (Rutherford: Fairleigh Dickinson University Press, 1972), 138–74. A more detailed discussion of Bradley appears in his *The Old English Farming Books from Fitzherbert to Tull 1523–1730* (London: Crosby Lockwood and Son, 1947), 94–125.

32 Quoted in Fussell, *The Classical Tradition*, 142, 244.

33 Pointer, *Rational Account*; Pointer graduated from Oxford in 1691, took orders in 1693, and held a small rectory in Northamptonshire. In the 1710s he wrote a piece on the Roman Pavement of Stunsfield, disputing, with Thomas Hearne on iconographic issues, in which effort he gained support from John Morton, the "meteorological" antiquary and chorographer of Northamptonshire. Pointer collected "Rarities" for his cabinet in the Rectory, contributed to the two-volume *Chronological History of England* (1714), and compiled a guide for classical authors with instructions on pagan mythology. The meteorological treatise appeared in 1723, enlarged and corrected in 1738. See *DNB*. See also Gunther, *Early Science in Oxford*, III, 454 ff.; Levine, *Humanism and History*, 116–17.

34 Pointer, *Rational Account*, 9.

35 Ibid., 12.

36 John Claridge's prognostics appeared as *The Shepherds Prophecy or John Claridges Forty Years of Experience of the Weather* (London, 1670); during the mid-eighteenth century, coinciding with the revival of georgics, there was a flurry of reprints of Claridge in 1744, 1748, 1755, 1781. The edition of 1748 followed Pointer's work; a reviewer from *Gentleman's Magazine* (1748), 255 wrote that: "I find almost all [Claridge's] observations to be transcrib'd *verbatim* from it." The "shepherd" thus wrote in that "the Air is composed of Exhalations of all earthly Bodies, as well solid as fluid, as also of Fire, whether of the Sun or the Stars, or of earthly Bodies burnt, or of Fire breaking out from the entrails of the Earth . . . and though it be this compounded, . . . yet we find it perfectly wholesome," Claridge, *Shepherd*, 46. A similar compilation was John Reid, *The Scots Gardener for the Climate of Scotland* [including Banbury's Rules], (Edinburgh, 1756, 1766). This publication, like that of Claridge, was based on the earlier "template," *The Scots's Gardener* (Edinburgh, 1683, 1721). See also James Justice, *The British Gardener's Calendar, chiefly Adapted to the Climate of North-Britain etc.* (Edinburgh: R. Fleming, 1759).

37 James Capper, *Meteorological and Miscellaneous Tracts, Applicable to Navigation, Gardening, and Farming with Calendars of Flora for Greece, France, England, and Sweden* (Cardiff: J. D. Bird, 1810), 29.

38 Claridge, *Shepherd*, iv.

39 Ibid., 41. Pointer explained that the use of instruments might in principle be justified as prognosticating tools by the fact that "natural causes do naturally (according to the settled order and nature of things) produce natural effects, as a dry air will naturally produce fair weather." Pointer, *Rational Account*, vii.

40 Claridge, *Shepherd*, viii.

41 Ibid., iv.

42 Ibid., v.

43 Ibid., v.

44 *Stars and Planets the best Barometers and Truest Interpreters of all Airy Vicissitudes. With some Brief Rules of Knowledg [sic] of the Weather at all Times. By an humble Adorer of God in his Word and Works* (London: John Nutt, 1701).

45 Pointer, *Rational Account*, v–vii.

46 Leigh, *Natural History*, 6. A century later the *Cambrian Register* recorded that the Welsh could with a great certainty predict a storm by the roaring of the sea and that such an accuracy was explicable by causes as natural as the forebodings of shepherds, "for which they have rules and data as well known to themselves, and perhaps, as little liable to error, as any of those established by the more enlightened philosophers of the present day," quoted in John Brand, *Observations on Popular Antiquities* (London: Vernon *et al.*, 1810), vol. 3, 247.

47 Claridge, *Shepherd*, vi–vii.

48 In reality, however, a considerable demand for the instrument existed in the countryside and was indicated by a blooming of the "black market" for cheap weather glasses "hawk'd about the Country, by needy Foreigners or peddling Philosophers." Saul, *A Historical and Philosophical Account of the Barometer*, 98. Saul warned against these "Cheats and Impositions," recommending instruments of better quality and higher price. Claridge's downplaying the barometer as an instrument of useful knowledge could also be seen as a ploy to detract its potential buyers into the shepherd's flock.

49 Claridge, *Shepherd*, vii.

50 Thomas Short, *New Observations, Natural, Moral, Civil, Political and Medical on City, Town, and Country Bills of Mortality* (London: T. Longman, 1750), 457. Short also published *A General Chronological History of the Air, Weather, Seasons, Meteors etc. in Sundry Places and Different Times; with Some of their Most Remarkable Effects on Animal (especially Human) Bodies, and Vegetables*, 2 vols (London: T. Longman and A. Millar, 1749). His medico-meteorological observations are abridged in *A Comparative History of the Increase and Decrease of Mankind in England and Several Countries Abroad* (London: W. Nicoll, 1767), 125–53, 455–6.

51 Capper, *Meteorological and Miscellaneous Tracts*, 30. Claridge, *Shepherd*, 34, also pointed out the connection: "[W]e live much within doors, by which [the alterations of the weather] are less obvious to us, and it is for this reason that Husbandmen, Seamen, Fishermen, but above all Shepherds, who are more in the open Air than other Men, are better" judges of the weather "than those who live altogether within doors." A contemporary German bureaucrat was cited by Lowood as asserting that "[a]lert country folk understand signs of the sky, and are better weather prophets than we city folk." Lowood, *Patriotism, Profit and the Promotion of Science in the German Enlightenment*, 260.

52 Pointer, *Rational Account*, 40.

53 For seaweed, see W. Carew Hazlitt, *Faiths and Folklore of the British Isles* (London: Murray, 1846), 625; for leeches, see a letter quoted in William Hone, *The Table Book*: "A phial of water, containing a leech, I kept on the frame of my lower sash window so that when I looke in the morning I could know what would be the weather of the following day . . . What reasons may be assigned for [its predictive properties] I must leave philosophers to determine, though one thing is evident by everybody, that it must be affected in the same way as that of the mercury and spirits in the weather glass. This is a weather glass which may be purchased at a very trifling expense, and which will last I do not know how many years." A similar "experiment" with prognostic leeches is reported by a "lover of the sciences", *A Succint Treatise of Popular Astronomy to which are subjoined Prognostics of the Weather* (Edinburgh [1780?]), 47–8. For a surprising hygrometer invented by a barber, see an essay by Richard Steele in the *Spectator*, July 30, 1712. For the sap of trees, see Dr Ezekial Tonge, "Some Observations

Concerning the Variety of the Running of Sap in Trees," *Phil. Trans.* 5–6 (1670–71): 2070–71. In 1772, Louis Cotte discussed whether the human sensation or thermometers were better indicators of the heat. Louis Cotte, *Traité de météorologie* (Paris, 1772), 248–62.

54 On "mundane means," see Steven Shapin's "Rarely Pure and Never Simple: Talking About Truth," *Configurations* 7 (Winter 1999), 1–14. It should be noted that the rustic apologists resided in cities, too. Estabrook explains this flight from urbanisation as a reaction to the mid-century grievances of city dwellers and to the confluence of the urban and rustic environments into a distinct suburbian mentality. Estabrook, *Urbane and Rustic England*, 253–75. For the political undertones of rustication, see Isaac Kramnick, *Bolingbroke and his Circle: The Politics of Nostalgia in the Age of Walpole* (Cambridge, Mass., Harvard University Press, 1968), 223–30. The intellectualization of shepherds is discussed by Leo Marx, *The Machine in the Garden* (London: Oxford University Press, 1964), 102; on the esthetics of rural space, see Simon Schama, *Landscape and Memory* (London: Fontana Press, 1996), 526–46.

55 Benjamin Stillingfleet *Literary Life and Selected Works of Benjamin Stillingfleet*, 2 vols (London: J. Nichols, 1811), II, 371.

56 Thomas Short, *New Observations*, 457.

57 The idea is elaboration upon *Georgics* 1: 353–5, and 1: 373: "The Father himself decreed what warning the monthly moon should give, what should signal the fall of the wind, and what sight, oft seen, should prompt the farmer to keep his cattle nearer to their stalls," and "Never has rain brought ill to men unwarned." *Virgil*, translated by H. Rushton Fair-clough (Cambridge, Mass.: Harvard University Press, 1940), 105, 107. Nature is inherently prophetic, Brooks Otis noted, and "the voice of gods speaks in the heavens and in the dumb animals and in the sputtering lamp wicks as it foretells through them the approach of bad and good weather." Brooks Otis, *Virgil: A Study in Civilized Poetry* (Oxford: Clarendon Press, 1963), 160.

58 Virgil explains the quality of this connection in following words: "Not, methinks, that they [birds] have wisdom from on high, or from Fate of a larger foreknowledge of things to be; but that when the weather and fitful vapours of the sky have turned their course, and Jove, wet with the south winds, thickens what just now was rare, and makes rare what just now was thick, the phases of their minds change, and their breasts now conceive impulses, other than they felt, when the wind was chasing the clouds." Virgil, *Georgics* 1: 415–22. A "lover of the Sciences" wrote the following in his popular astronomical manual: "In consequence of the observation, that such creatures as live in the open air have a quick sensation of what changes are about to happen in that region, the following rules have been laid down by various naturalists who have applied to themselves signs for good or bad weather," *A Succint Treatise of Popular Astronomy*, 33. The agricultural writer William Marshall discussed this issue in his *Minutes of Agriculture* (London: J. Dodsley, 1783). He asked if "Animals, Vegetables, and Fossils [can] be the Cause of Rain," and whether they "can spend forth the aquaeous particles, assembled them in the Atmosphere, or call them down in Rain." Answering negatively, he thought that the reason they are useful for meteorological prediction was because they were "*actuated* by the Causes."

59 Quoted in Turner, *Contesting Cultural Authority*, 113.

60 M. Toaldo, "On the Signs exhibited by Animals which indicate Changes of the Weather, with Remarks on other Prognostications," *Philosophical Magazine* 4 (1799): 367–75.

61 Daniell, *Meteorological Essays and Observations*, 1. Italics mine. Also: "To [the mariner, fisherman, and husbandman], the necessity of their condition prescribes the observance of the face of the sky, and of the changes of the fluctuating deep, *as part of their education*; they become weather-wise by tradition and experience; and are often able to communicate the results of a certain local knowledge, without being prepared to assign a *reason* for any thing they say," Luke Howard, *Seven Lectures on Meteorology* (London: Harvey and Darton, 1837, 1843), 1–2; or: "It cannot escape the notice of those who are conversant either in maritime or rural affairs that experienced seamen and well-informed husbandmen are seldom mistaken in the diurnal pre-dictions of the weather," Capper, *Meteorological and Miscellaneous Tracts*, 4: "It is by sort of *intuition* that [country folks and seamen] seize as it were the precursors of the coming pheno-mena, and for the most part, they merely obey an *instinct* which compels them to prognosticate

the impending shock of the gaseous ocean which surrounds us." Andrew Steinmetz, *Sunshine and Showers* (London: Reeve and Co., 1867), 5. Italics mine.

62 In the natural philosophic discourse, the common attitude toward the ancients was the opposite: "I would give to Aristotle the electrical shock: I would carry Alexander to see the experiments upon the Warren at Woolwich . . . I would shew to Julius Caesar, the invader of Britain, an English man of war; to Archimedes a fire engine, and a reflecting telescope." William Jones, *Physiological Disquisitions, or Discourses on the Natural Philosophy of the Elements* (London: J. Rivington, 1781), xiv.

63 Marshall, *Minutes of Agriculture*, 143. Paying respect to ancient wisdom occurs with greater frequency toward the end of the century. "[T]he more wise and sagacious have, in all Ages, and several Countries, found Matter to make their Observations from, and carefully handed them down to us in a Cloud of no despicable Authors." Short, *New Observations*, 457. William Jones protested that "Philosophers, who speculate in their closets, treat the notion of subterraneous winds with contempt, as nothing more than a fable of the Heathen poets; whose representation, though fabulous, is yet built upon the real history of nature, and confirmed by many accounts both ancient and modern." Jones, *Physiological Disquisitions*, 572. Capper noted that "these traditions must not be considered as the vain dreams and idle fancies of illiterate peasants; on the contrary, they are the result of the judicious observations by men of acknowledged wisdom and virtue. The sentiments of such men on the operations and appearances of nature merit our attention and respect." Capper, *Meteorological and Miscellaneous Tracts*, 4. See also David Purdie Thomson, *Introduction to Meteorology* (Edinburgh and London: W. Blackwood, 1849), 1: "The observations of the ancients were directed chiefly to changes in the weather; and by personal assiduity, they were enabled to prognosticate often with considerable certainty."

64 Claridge, *Shepherd*, ix. Writing in 1827, the London physician Thomas I. M. Forster, was struck "by agreement between the prognostics of Theophrastus, Aratus, Virgil, Columella, and other ancient writers, and the proverbial prognosticology of more recent times," *Pocket Encyclopedia* (London: John Nichols, 1827), vi.

65 Heninger, *A Handbook of Renaissance Meteorology*, 219. In an entry for August 1777, Marshall explained the comparative procedure he hoped would discriminate between false and probable signs: "A whirlwind (a); the sky beautifully mottled with shell-like Clouds and with deep-blue ground (b) – the Clouds high (c), but the barometer kept getting up (d) and the Sun set foul (e)." Marshall, *Minutes of Agriculture*, 127. It is important to note that the barometer reading represents only *one* of the natural *signs*.

66 In Jacobean England the following sources were available for empirical weather rules: Theophrastus's *De Signis aquarum et ventorum*, Aratus's *Diosemea*, Virgil's *Georgics*, Pliny's *Historia Naturalis*, and the more recently issued prognostications of Leonard Digges, Philip Moore, Godfridus and others. Despite the fact that the lampooning of "empiric" and astrological forecasts made in such works was rampant since Elizabethan times, they continued to circulate throughout the seventeenth and eighteenth centuries. See Carroll Camden, Jr. "Elizabethan Almanacs and Prognostications," *The Library* 12 (1931), 100–8; Don Cameron Allen, *The Star-Crossed Renaissance* (Durham, N. C.: Duke University Press, 1941), 190–246.

67 Feingold, *Nature and Society*, 26, 16.

68 Claridge, *Shepherd*, 41.

69 Information on Mills exists in *DNB* and Donaldson's *Agricultural Biography*, 51. See also André Bourde, *The Influence of England on the French Agronomes, 1750–1789* (London, 1953).

70 John Mills, Esq. FRS, *An Essay on the Weather; with Remarks on the Shepherd of Banbury Rules for Judging of it's Changes: Intended Chiefly for the Use of Husbandmen* (London: S. Hooper, 1770), 1–2.

71 Mills, *An Essay on the Weather*, x–xi. The French analogue to Mills was the agro-meteorological writer Louis Cotte who, like Mills, began his studies after reading Duhamel du Monceau's agronomic science. Cotte believed that the goal of meteorology was in the "perfection of the sciences of agriculture and medicine," and he wished that "all farmeres would become observers," because they had better opportunity of knowing the land and sky than "savants

who can only watch from a distance." These and similar insinuations about the superiority of a "farmer's epistemology" of science occurred in Cotte's manuscripts and his influential *Traité de météorologie* (Paris, 1772). Quoted in Feldman, "The History of Meteorology," 230–41.

72 Mills, *An Essay on the Weather*, 72. Despite the common view of the ancient naturalists as inventors of astrometeorology, Bishop Samuel Horsley argued that neither all ancients nor husbandmen would believe in the physical influence of the moon on the weather, because they "derive their prognostics from circumstance, which neither argue any real influence of the moon as a cause, nor any belief of such an influence, but are merely indications of the state of the air at the time of observation." Samuel Horsley, "An Abridged State of the Weather at London in the Year 1774," *Phil. Trans.* 65 (1775): 167.

73 Allen Hall, *Observations on the Weather* (London: Drury Office, 1788). Similar claims to establish local regularities were made by Henry Adams, a teacher of mathematics at Royston, Herts, in his *A Royal Almanac and Meteorological Diary for the Year of Our Lord 1778* (London: T. Carnan, 1778). Robert Marsham, "Indications of Spring," *Phil. Trans.* 79 (1789): 154–6. Samuel Hopkinson, *Causes of the Scarcity Investigated: also an Account of the Most Striking Variations in the Weather, from October 1789 to September, 1800* (Stamford: R. Newcomb, 1800). One of the essays that John Dalton included in his *Meteorological Observations and Essays* (Manchester: Harrison, 1793) was titled "General Rules for judging of the Weather." One of the most reprinted practical texts was *The Newest, Best, and Very Much Esteemed Book of Knowledge. Together with the Husbandman's practice, Also a Brief Discourse of the Natural Causes of Meteors, and Observations on the Weather* (London: A. Wilde, 1764). In the early nineteenth century, there is a continuing interest in the subject witnessed by works of Murdo Downie, *Observations upon the Nature and Properties of the Atmosphere, to which are Added Observations on the Moon's Influence upon the Atmosphere, and the Rise and Fall in the Barometrical Tube* (Aberdeen: J. Chalmers and Co., 1801); Henry Robertson, *A General View of the Natural History of the Atmosphere and of its Connection with the Sciences of Medicine and Agriculture.* (Edinburgh, 1808); Joseph Taylor, *The Complete Weather Guide: A Collection of Practical Observations for Prognosticating the Weather, Drawn from Plants, Animals, Inanimate Bodies, and also by Means of Philosophical Instruments; Including Shepherd and Banbury Rules* (London: John Harding, 1812); John Adams, *Extract from a Meteorological Journal kept at Edmonton, Middlesex, together with a Collection of such Observations as have been Considered the Surest Guides whereby to Foretell the Weather* (London: the Author, 1814); T. Forster, *The Pocket Encyclopaedia of Natural Phenomena; being a Compendium of Prognostication of the Weather, Signs of the Seasons* (London: John Nichols, 1827). "Soon after my attention was directed to atmospheric science I observed that mariners, shepherds, husbandmen, and others whose employment kept them constantly out of doors, could foretell with more certainty what sorts of weather were coming than the more scientific meteorologist could do; they seemed to me to have a sort of code of prognostics of their own, founded partly on tradition and partly on experience: they used numberless trite sayings and proverbial adages respecting the weather which were handed down from the remotest antiquity, but which, in the long run, seldom failed to be right," ibid., vi.

74 George Adams, *Lectures on Natural and Experimental Philosophy* (London: J. Dillon & Co., 1798), 264. Thomas Garnett of the Manchester Literary and Philosophical Society prefixed Adams's opinion to an article in the Society's *Memoirs* in which he wrote: "By the weather the traveller endeavours to regulate his journies, and the farmer his operations; by it plenty and famine are dispensed, and millions are furnished with the necessities of life." "Meteorological Observations, Collected and Arranged by Thomas Garnett," *Memoirs of the Literary and Philosophical Society of Manchester* 4 (1793): 517. Kirwan wrote in 1787: "If from the present state of the atmosphere, the tiller of the soil, or of the main, could foresee, or have a table presented to him of the changes it would undergo for six, or even three months, with what confidence and security would not each of them be enabled to direct, and pursue their respective operations?" Kirwan, *Estimate of the Temperature*, v. In his research on the influence of the moon on the cycles of weather, printed in the *Philosophical Magazine* for 1799, Toaldo related an anecdote that upon crowning, Mexican emperors followed tradition in promising fair weather to their subjects and commented that "the multitude imagine that the meteorologist enters into an obligation of a like kind; but all that can be expected from him are conjectural

rules respecting changes of the weather." "An Account of Toaldo's System respecting the Probability of a Change of Weather at the Different Changes of the Moon," *Philosophical Magazine* 3 (1799): 120–7, 121.

75 G. Gregory, *The Economy of Nature: Explained and Illustrated on the Principles of Modern Philosophy* (London: J. Johnson, 1796), 519.

76 "Hence, though the sky be fitful, we can foretell the weather changes, hence the harvest time and the sowing time." Virgil, *Georgics*, 99. In John Dryden's translation of 1697: "From hence uncertain seasons we may know: and when to reap the grain, and when to sow," *The Works of Virgil* (Philadelphia: Claxton, 1870), 73. In Capper's opinion, the poem represented an unquestionable "epitome of practical prognostics." Capper, *Meteorological and Miscellaneous Essays*, 50. Capper used local lore: "A mackarel sky, and mares' tail (say the seamen) make tall ships carry low sails. Those fleckered and fleecy clouds, which are fringed and light towards the edges and dark in the middle, generally portend sudden and violent showers." Ibid., 51.

77 Ibid., 52. Capper concludes the discussion of seasonal changes and the agricultural calendar with Pope's theodicy: "All Nature is but art unknown to thee / All chance directions, which thou cans't see./ All discord, harmony not understood, / All partial evil, universals good." Ibid., 185. Capper could have had in mind someone like Mr Alexander Copland from Dumfries who collated his observations in "Meteorological Observations and Remarks on the weather at Dumfries," first published in the *Dumfries Weekly Journal* and later in the *Manchester Literary and Philosophical Society Memoirs*. Copland combined instrumental and visual information, as in rule 21, for instance: "That the longer a fall has been indicated by streamers [Aurora Borealis] and a low barometer, not accompanied or followed by cold, without its taking place, the heavier and more continued it will be when it once commences." Dr Thomas Garnett called Copland's collection the "most rational rules for judging the weather" but warned that their applicability to other parts of England ought to be ascertained by observations in those locations. "Meteorological Observations and Remarks on the Weather at Dumfries," *Manchester Literary and Philosophical Society Memoirs* 4 (1793): 243, 248. Garnett is quoted in "Meteorological Observations made on different Parts of the Western Coast of Great Britain, arranged by T. Garnett," *Memoirs of the Literary and Philosophical Society of Manchester* 4 (1793): 235.

78 Marshall, *Minutes of Agriculture*, 143. Marshall's earlier work was entitled *Experiments and Observations Concerning Agriculture and the Weather* (London: J. Dodsley, 1779).

79 Richard Kirwan observed in relation to the signs of wet and dry weather: "If meteorological observations were taken at proper distances all over the globe, and with tolerable accuracy, they probably would in a few years disclose that connexion which all the phenomena of the atmosphere have with each other, and particular *species* of weather might be *foreseen* either to a certainty or to a high degree of probability, but until this happens, the only use of meteorological tables, as far as regards the art of forming *prognostics*, is to exhibit a view of the sort of weather that most usually precedes wet, dry, hot, or cold season (these being modifications most interesting to agriculture and medicine), and tracing their recurrency by the laws of probability." Kirwan, *Meteorological Observations*, 19. Italics mine.

80 John Ruskin, "Remarks on the Present State of Meteorological Science," *Transactions of the Meteorological Society*, vol. i (London, 1839), 56–7.

Notes to chapter 7

1 Jones, *Physiological Disquisitions*, 553.

2 On the editorial sieve at the Royal Society, see Rusnock, "Correspondence Networks," 159–69, 162 where she illustrates the procedure of rejection in the case of Benjamin Langwith, a rector at Petworth in Sussex, one of the meteorological authors during James Jurin's secretaryship. Langwith sent three articles on aurorae borealis in the period 1723–27.

3 George Reuben Potter, "The Significance to the History of English Natural Science of John Hill's Review of the Works of the Royal Society," *University of California Publications in English* 14 (1943): 157–80.

4 Hans Sloane, Preface to *Philosophical Transactions* (1699), cited in Potter, "The Significance to the History of English Natural Science," 175.

5 Miller, "Into the Valley of Darkness', 155–66; Schaffer, "Natural Philosophy and Public Spectacle."

6 Miller, "'Into the Valley of Darkness'," 162.

7 Ibid., 162 Miller argues that Banks had specific reasons for choosing specific individuals (or constituencies) for the Council, the Society's ruling body.

8 Harold B. Carter, *Sir Joseph Banks, 1743–1820* (London: British Museum, 1988), 573.

9 For instance, the editors Charles Hutton, George Shaw and Richard Pearson (editors of the 1809 abridgment of the *Philosophical Transactions*), divided the eighteenth-century contributions into "Pneumatics" and "Meteorology." The latter category covered the interests overlapping with the "meteoric tradition" writers and the weather observers proper, while the former category covered research conducted by Priestley, Maskelyne, Deluc, etc.

10 Thomas L. Hankins, *Science and the Enlightenment* (Cambridge: Cambridge University Press, 1985), 50–67.

11 Benjamin Franklin, "Concerning the Effects of Lightning," *Phil. Trans.* 47 (1751–52): 289–97; for an earlier instance of the electrical theory of lightning by J. H. Winkler (1746), see Karl Schneider-Carius, *Weather Science*, 100. For Franklin's later meteorological speculations, see J. M. Walker, "A Meteorological Curio," *Weather* 25 (1970): 30–2.

12 "Since the invention of gunpowder," wrote John Mills, "[thunder and lightning] have been generally ascribed to a mixture of nitrous and sulphureous vapours by some means set on fire in the air and exploding like that powder. [But] Franklin's soaring genius has realized the fable of Prometheus bringing fire down from heaven and furnished us with a better theory." Mills, *An Essay on the Weather*, 17.

13 [Albrect von Haller], "An historical account of the wonderful discoveries made in Germany concerning electricity," *Gentleman's Magazine* 15 (1745): 193–7. See also J. L. Heilbron, "Franklin, Haller and Franklinist History," *Isis* 68 (1977): 539–49.

14 Simon Schaffer, "The Consuming Flame: Electrical Showmen and Tory Mystics in the World of Goods," 491. See also his "Natural Philosophy Spectacle" as well as Patricia Fara, *Sympathetic Attractions*.

15 William Stukeley, in "A Collection of Various Papers Concerning several Earthquakes," *Phil. Trans.* 46 (1749–50): 602. Stukeley proposed an elaborate hypothesis of the electrical nature of earthquakes in his *Philosophy of Earthquakes* (London, 1751). See also Kendrick, *Lisbon Earthquake*, and Schaffer, "Natural Philosophy and Spectacle."

16 Quoted in Feldman, "The History of Meteorology," 167.

17 The letter is from Franklin's correspondence in 1753, quoted in Burke, *Cosmic Debris*, 10. In 1783, Charles Blagden used the electrical hypothesis to explain the balls of fire. See below.

18 Henry Eeles, "Letter Concerning the Cause of the Ascent of Vapour and Exhalation," *Phil. Trans.* 49 (1755–56): 124–9; Eeles early subscribed to Franklin's suggestion in his "On the Cause of Thunder," *Phil. Trans.* 47 (1751–52): 324–7. This notion was elaborated in James Capper's *Meteorological and Miscellaneous Tracts*. In 1757, Erasmus Darwin published a study in the *Phil. Trans.* in which he rejected Eeles hypothesis. He held that the rise of exhalations, the suspension of clouds and the rainfall could be accounted for without the introduction of electricity. The electrical theories of rain persisted in the work of Italian experimenter Giovanni Beccaria (1758) and as late as 1801, when Richard Kirwan supported the idea in his *Of the Variation of the Atmosphere* (Dublin, 1801), 220. See W. E. Knowles Middleton, *A History of the Theories of Rain* (London: Oldbourne Book Co. Ltd, 1965), 111–15.

19 Thomas Henry "An Account of the Earthquake which was felt at Manchester and other Places, on the 14th day of September, 1777," *Phil. Trans.* 67 (1777): 221–5.

20 John Williams, *The Climate of Great Britain; or Remarks on the Change it has Undergone, Particularly within the Last Fifty Years* (London: C. and R. Baldwin, 1806), 318. Other studies on the electrical subject include John Read, *A Summary View of the Spontaneous Electricity of the Earth and Atmosphere* (London: the Author, 1795); Abraham Bennet, [Curate of Wirksworth], *New Experiments on Electricity* (Derby: John Drewry, 1789); Thomas Kirby, *An Analysis of the Electrical Fire* (Chatham: the Author, 1777). These and other similar documents discussed the electrification of the atmosphere and the circumstances necessary for the production of lightning.

The protection of buildings from lightning was one of the major concerns among many electricians. There were some dissenting voices too. John Lyon wrote an *Account of Several New and Interesting Phenomena, Discovered in [bodies] Killed by Lightning. With Remarks on the Insufficiency of Electricity to Explain Them* (London: James Phillips, 1796).

21 John Gerard William de Brahm, *The Levelling Balance and Counter Balance* (London: T. Spilsbury, 1774), 2–4.

22 Walpole to Mann, 19 May 1750, in P. Cunningham, ed., *Letters of Horace Walpole* (London: Richard Bentley, 1857), 2, 207.

23 Charles Blagden, "An Account of Some Late Meteors," *Phil. Trans.* 74 (1784): 201–31, 222.

24 Ibid., 223, 204.`As support, Blagden adduced the observed high velocity of the meteoric fall, the fact that various electrical phenomena attend meteors, e.g. lambent flames emanating with sparks, and of the damage inflicted by lightning and those by meteors. The hissing sound attending the fall is explicable as the streams of electric matter issuing from the meteor. A certain Mr Robinson of Hickley, Leicestershire, made observations on the sparkling meteor following the storm, which gave "the hissing sound reminiscent of the one from the electric machine when the electric matter is running away." Additional confirmation was the similarity of these meteors to the northern lights "positively electrical in nature." Blagden however believed that the hardest evidence was the direction of meteor's movement, i.e., to or from North or Northwest, that is, along the magnetic meridian.

25 "Here below we have thunder and lightning, from the unequal distribution of the electric fluid among the clouds, in the loftier regions, whither the clouds never reach, we have the various gradations of falling stars; till beyond the limits of our crepuscular atmosphere the fluid is put into motion in sufficient masses to hold a determined course, and exhibit the different appearances of what we call fire-balls and finally, above all, aurora borealis," Ibid, 231.

26 Burke, *Cosmic Debris*, 13, quoting the seventeenth-century naturalist Anselm Boece de Boodt.

27 Ibid., 15.

28 "Several accounts of the Fiery Meteor, which appeared on Sunday Nov 26, 1758 between 8 and 9 at Night. Collected by John Pringle, MD, FRS." *Phil. Trans.* 51 (1759–60): 218–59; idem., "Some Remarks upon the several Accounts of the Fiery Meteor," *Phil. Trans.* 51 (1759–60): 265–81.

29 Pringle, "Remarks," 267. Pringle adduced two chief reasons for this conclusion: the first was the otherwise inexplicable velocity of the meteor (30 miles per second), and the second was the report published in the Memoirs of the Academy of Bologna, describing a meteor appearing in Italy in 1719 in which "several chasms were distinguished, each emitting smoke." Ibid., 267. Pringle also denied the electrical nature of the meteor.

30 Ibid., 269.

31 Ibid., 269.

32 Ibid., 271. A disavowal of the old theory occurs in the 1809 abridgment of the *Philosophical Transactions*. In a long footnote to Halley's 1718 article on the fiery meteor, the editors explain that even though Halley's "mind fixes on nothing but vapours and exhalations, to solve the appearance . . . [l]ater observations however have induced a belief that these luminous appearances are allied to, if not the same as the stones which have frequently been known to fall from the atmosphere." *Phil. Trans.* (abr.) 6 (1809): 100.

33 Blagden, who in 1783 had opted for the electrical explanation of the same phenomena believed Pringle "took his ideas from [Professor Winthrop of Cambridge, New England], which Maskelyne is now going to hash up warm." Blagden to Joseph Banks, 21 October 1783, in Warren R. Dawson, ed., *The Banks Letters* (London: British Museum, 1958), 60. Blagden believed he had an ally in Erasmus Darwin.

34 Another early contribution to this notion was the treatise by Thomas Clap, the president of Yale College 1739–66, titled *Conjectures upon the Nature and Motion of Meteors which are Above the Atmosphere* (Norwich, Connect.: John Trumbull, 1781) (posthumously). Clap conceived fireballs as earth-orbiting comets. Burke notes that only Pringle and the American David Rittenhouse (in 1780) seriously considered the extraterrestrial origins of fireballs. Burke, *Cosmic Debris*, 24. His inclusion of Halley seems to me to be problematic on the grounds of internal evidence, because Halley entertained the idea of meteors as combinations of "atoms"

above the atmosphere only as a conjecture, whereas his "official" position was strictly that of the mineral meteorology of exhalations. Clap, however, should be included in the list.

35 It is interesting to note that for Pringle this ontological displacement implied nomological order: the phrasing in the last paragraph (i.e. if these bodies are of extraterrestrial origins, "surely we are not to consider them as indifferent to us, much less as fortuitous masses") implies that as soon as they "leave" their sublunary province, natural bodies acquire a place in the regulated beneficence of the Creation.

36 Ron Westrum, "Science and Social Intelligence about Anomalies. The Case of Meteorites," *Social Studies of Science* 8 (1978): 461–93. I agree with Burke that Westrum's insistence on the workings of the "social intelligence system" (the system of report-transmission) overlooks the theoretical reasons scientists had for disbelieving that stones fell.

37 Burke, *Cosmic Debris*, 37. Thomas Birch, President of the Royal Society was soliciting information on fiery meteors during the 1750s. *The Philosophical Transactions* published a large number of the reports: Mr Chalmers, "Of an Extraordinary Fireball Bursting at Sea," *Phil. Trans.* 46 (1749–50): 366; William Hirst, "Of a Fireball Seen at Hornsey," *Phil. Trans.* 48 (1753–54): 773. Hirst subscribed to the exhalation theory. Rev. Forster, "Of a Meteor Seen at Shefford in Berkshire," *Phil. Trans.* 51 (1759–60): 299; Mr Josiah Colebrooke, "On the Same meteor," ibid., 301; William Dutton, "Of the same Meteor seen at Chigwell Row, in Essex," ibid., 302; John Winthorp, "Of a Meteor Seen in New England," *Phil. Trans.* 52 (1761–62): 6; idem., "Of Several Fiery Meteors," *Phil. Trans.* 54 (1764): 185; John Swinton, "Of a Remarkable Meteor seen at Oxford," *Phil. Trans.* 52 (1761–62): 99; Peter Gabry, "Observation of a Fiery Meteor made at Hague," *Phil. Trans.* 52 (1761–62): 300; Samuel Dunn, "Of a Remarkable Meteor," *Phil. Trans.* 53 (1763): 351; Patrick Brydone, "Of a Fiery Meteor," *Phil. Trans.* 63 (1773): 163–5; Tiberius Cavallo, "Description of a Meteor, Aug. 18, 1783," *Phil. Trans.* 73 (1783): 108–15; Alexander Aubert, "An Account of Meteors," *Phil. Trans.* 73 (1783): 112, etc.

38 Stephen Hales, *Vegetable Staticks* (London, 1727).

39 For more detailed account of these developments see Hankins, *Science and the Enlightenment*, 81–113; Arthur L. Donovan, *Philosophical Chemistry in the Scottish Enlightenment: the Doctrines and Discoveries of William Cullen and Joseph Black* (Edinburgh, 1975); Maurice P. Crosland, "The Development of Chemistry in the Eighteenth Century," *Studies on Voltaire and the Eighteenth Century* 24 (1963): 369–441. The connection between pneumatic chemistry, doctrines of imponderable fluids and research on atmospheric processes is dealt with in Feldman, "The History of Meteorology," 48–100; Middleton, *Theories of Rain*, 20–43; and idem., "Chemistry and Meteorology," *Annals of Science* 20 (1965): 125–41.

40 A difference should be observed between this idea and the earlier notion of air as the mixture of gross substances produced in the earth or ocean and transmitted into the atmosphere. Oliver Goldsmith wrote in 1774 that air was "one of the most compounded bodies in all nature . . . A thousand substances that escape all our senses we know to be there." In his *A History of the Earth and Animated Nature* (1774. London: Blackie and Son, 1852), 109. Hinde's *Dictionary* went as far to claim that the "atmosphere was a perfect chaos of different effluvia, consisting of all kinds of corpuscles, confusedly jumbled together." *A New Royal and Universal Dictionary of Arts and Sciences* (London, 1771). In his widely cited *System of Familiar Philosophy* (London, 1802) the lecturer Adam Walker wrote about the atmosphere as "a compound of every kind of body capable of gaseous state by means of heat or fermentation," ibid., 254.

41 Jean Andre Deluc, *Idées sur la météorologie* (London, 1786), quoted in Feldman, "The History of Meteorology," 86. Hygrometry, the science about atmospheric humidity, had by this time become a large "subfield" of physiology (science of nature). Johann Lambert and de Saussure were the most prominent of the early theorists whose theories Deluc discussed and contested. See Feldman, "The History of Meteorology," 48–100.

42 Deluc, *Idées*, quoted in Feldman, "The History of Meteorology," 85. Deluc's treatment of "vapors" as composed of a "ponderable and deferent fluid" provides the basis for a 459-pages-long discussion of the topic in the *Idées*. Middleton calls the work a "chemical manifesto." See Middleton, "Chemistry and Meteorology," 135. In the early 1790s, Deluc's theories became politicized because of his opposition to the French Lavoisierians, whom he dismissed as

chemical "néologues." The German intellectual Georg Lichtenberg used Deluc's theory of rain to substantiate his attacks on French chemistry, and in 1795, Johann D. O. Zylius wrote that thanks to Deluc, "at least in chemistry the counter revolution has succeeded," quoted in Middleton, "Chemistry and Meteorology," 139. For a discussion on Deluc's Neptunist recreation of the Mosaic narrative and his engagement with James Hutton, see Gillispie, *Genesis and Geology*, 56–66.

43 Horace Benedict de Saussure, *Essais sur l'hygrométrie* (Neuchatel, 1783), quoted in Middleton, *Theories of Rain*, 34.

44 Quoted in Schneider-Carius, *Weather Science*, 123.

45 Henry Robertson, *A General View of the Natural History of the Atmosphere* (Edinburgh: Abernethy and Walker, 1808), 1. Before this time, Erasmus Darwin had applied the premises of this program in his "Frigorific Experiments on the Mechanical Expansion of Air," *Phil. Trans.* 78 (1788): 654. Another important contribution to the genre was the work of the Rev. Hugh Hamilton, Professor of Philosophy at Aberdeen University: "A Dissertation on the Nature of Evaporation and Several Phenomena of Air, Water, and Boiling Vapors," *Phil. Trans.* 55 (1765): 146–52. The essay was reprinted in a collection *Philosophical Essays on [several subjects]* (London: J. Nourse, 1772). Hamilton subscribed to solution theory of evaporation and thought the *Aurora* and the cometary tail to result from the electrical matter. On the solution theory, see Middleton, *Theories of Rain*, 28–36. Walker wrote that the streams and exhalations "are kept in a fluid state by their union with fire or light." Walker, *A System of Familiar Philosophy*, 254.

46 Deluc also published "Barometrical Observations on the Depth of the Mines in the Hartz," [transl. from the French], *Phil. Trans.* 67 (1777): 388–91; "An Essay on Pyrometry and Aerometry, and on Physical Measures in General," *Phil. Trans.* 67 (1777): 419; "A Second Paper on Hygrometry," *Phil. Trans.* 81 (1791): 1; "On Evaporation," *Phil. Trans.* 82 (1792): 283; In Shaw's 1809 abridgment of the *Phil. Trans.* these contributions are listed under "Pneumatics" rather than "Meteorology." In the late 1770s, the journal published other pneumatic studies by the Italian, Abbe Fontana, the Dutch, John Ingenhousz, and Cavendish.

47 Adams, *Lectures*, iv, 473n.

48 Adams, *Lectures*, 474–5. Similar claim was made by Horace Benedict de Saussure, Deluc's main meteorological critic and the designer of an alternative hygrometer, who in 1783 praised chemistry's potential to explain phenomena of "different branches of physics." Saussure, *Essais sur la hygrométrie*, x. Quoted in Feldman, "The History of Meteorology," 90.

49 A. F. Fourcroy, *Système des connaissances chymiques* (Paris, 1800), vol. 1, 7, quoted in Middleton, "Chemistry and Meteorology," 137. Identification of meteorology with the chemistry of atmosphere continued in the early nineteenth century, as chemists appropriated the field. See Thomas Thomson, *History of the Royal Society* (London: Murray, 1812), 505.

50 Le Roy, 1771, quoted in Burke, *Cosmic Debris*, 22.

51 Deluc, *Idées*, para. 535, quoted in Middleton, "Chemistry and Meteorology," 127.

52 Kirwan, *Estimate*, v.

53 Adams, *Lectures*, iv, 473–4.

54 Kirwan, *Estimate*, v. Quoting Kirwan's observation regarding the lack of control over atmospheric processes, John Williams nevertheless believed that "with the advance of knowledge it would be possible to change this; we may even at some time in future attempt amelioration of the weather." *The Climate of Great Britain* (London: C. and R. Baldwin, 1806), 314.

55 Hinde et al., *New Royal and Universal Dictionary*; E. Chambers and Abraham Rees, *Encyclopaedia, or an Universal Dictionary of Arts and Sciences* (London, 1784). This meaning was accepted by European scientists generally. Leonard Euler described meteorology as "the science of meteors, that is, of bodies floating in the air and quickly passing away," David Brewster, ed., *Letters of Euler on Different Subjects in Natural Philosophy* (New York: J. J. Harper, 1833), 431.

56 William Nicholson, *British Encyclopedia or the Dictionary of Arts and Sciences* (London: C. Whittingham, 1809).

57 Abraham Rees, *The Cyclopaedia; or Universal Dictionary of Arts, Sciences, and Literature* (London: Longman, 1819). Circumstantial evidence suggests John Dalton as the author of this

entry. It should be observed that the stipulation was less than universal; some nineteenth-century writers still thought of "fiery meteors" as part of meteorological science: "The meteors called shooting stars, and the stupendous masses of matter in combustion called fired balls, which cast down upon the earth immense blocks of red-hot iron, or showers of heated stones, – constitute another wide field of meteorological inquiry. E. W. Brayley, Jun., "Introductory Sketch of the Objects and Uses of Meteorological Science," *The Magazine of Natural History and Journal of Zoology, Botany, Mineralogy, Geology, and Meteorology* (London: Longman, 1836), 153–4.

58 Peter Barlow, *A New Mathematical Dictionary* (London: J. Nichols, 1814). Meteorology comprised both the classical and the emerging meaning: "the science which treats of meteors; the state of the weather etc."

59 *The Penny Encyclopedia of the Society for the Diffusion of Useful Knowledge* (London: Charles Knight, 1839). The contrast was more explicitly posited in the 1881 edition of *Chambers's*: "METEOROLOGY (Gr. *metéora*, meteors, or atmospheric phenomena) was originally applied to the considerations of all appearances in the sky, both astronomical and atmospherical; but the term is now confined to that department of natural philosophy which treats of the phenomena and modification of the atmosphere as regards weather and climate." Similarly, the American James Renwick, wrote that "[a]lthough the name of meteor is often applied so as to include the phenomena [like snow, wind, rain], we shall confine it to the phenomena which resemble the heavenly bodies in their appearance, but which are certainly nearer to us than the moon." *First Principles of Natural Philosophy*, (New York, 1856), 477.

60 Nicholson, *British Encyclopedia*.

61 In George Adams's view, the principal object of meteorological enquiry ought to disregard all but hydrometeors of which it ought to ask: "1. In what manner the atmosphere is supplied with humidity, 2. What causes and what prevents invisible humidity from being formed into clouds? 3. What occasions and prevents visible clouds from being precipitated into rain? That is, to know various ballancing of the clouds, and learn how such ponderous materials are suspended in the air; and how the waters are bound up in the thick clouds." Adams, *Lectures*, 477.

62 Comparable tendencies existed in other fields. In 1810, the chemist William Henry wrote that "mineralogy has been advanced from a confused assemblage of its objects, to the dignity of a well methodized and scientific system," Quoted in Golinski, *Science as Public Culture*, 272.

63 John Walker, *Lectures on Geology: Including Hydrography, Mineralogy, and Meteorology with an Introduction to Biology.* Edited with Notes and Introduction by Harold W. Scott (Chicago and London: The University of Chicago Press, 1982), 18, 49–118.

64 Walker, *Lectures*, 50. See also William Jones, *Physiological Disquisitions or, Discourses on the Natural Philosophy of the Elements* (London: J. Rivington and Sons, 1781). "By the weather, we mean the temperature of the air with respect to heat, cold, wind, rain, and other meteors," 552. Adam Walker did not think "Meteorology" was autonomous enough to be included in his *System of Familiar Philosophy* (1802). Lecture V is thus entitled "On the Atmosphere" and considers the mechanical and chemical properties of air, wind, air guns, sound, etc.

65 Murelo Downie, *Observations upon the Nature and Properties of the Atmosphere* (Aberdeen: J. Chalmers and Co., 1801). Thomas Thomson, a chemist and a historian of the Royal Society, made this clear in his definition of the object of meteorology as ascertaining "the changes which take place in the atmosphere, upon which the *weather* of every country, and of course the comforts of the inhabitants, entirely depends." Thomson, *History of the Royal Society*, 505. See also R. L. Denston, *A New Theory of the Atmosphere* (Birmingham: Knott and Lloyd, 1807). In 1816, Jeremy Bentham designed an entirely new, non-Baconian and non-d'Alembertian classification of sciences in which the following was said about meteorology: "No sooner does a substance break free from [earth's surface], than it enters into the province of Meteorology, and there continues, until by any of those revolutions of which the atmosphere is the constant theatre, it is again brought into immediate contact with . . . some one or more of those solid or liquid masses." Jeremy Bentham, *Chrestomathia*, edited by M. J. Smith and W. H. Burston (Oxford: Clarendon Press, 1983), 70.

66 Robertson, *A General View of the Natural History of the Atmosphere*. Thomas I. M. Forster, *Researches About Atmospheric Phenomena* (London: J. Moyes, 1813); John F. Daniell, *Meteorological Essays and Observations* (London: T. & G. Underwood, 1823). "Meteorology is that science, or branch of science, the object of which is the illustration of atmospheric phenomena." Forster, *Researches*, vii. In addition, it is symptomatic of the changing conceptual character of meteors that Robertson, for instance, thought them interesting only to the extent they "be of importance in the purposes of life," and this importance "in proportion to the knowledge we thereby obtain of predicting the various changes of the weather." In other words, knowledge about meteors was desirable only as far as it formed "our knowledge of the properties of the atmosphere," which in turn would lead to the prognostics of weather. Daniell, *Essays*, 379.

67 The value of the rare may be illustrated by the account produced by the Sussex antiquarian Benjamin Langwith who in 1726 wrote: "The Northern Lights have been so common in all Places of late Years, I did not think it worth while to write you about them; but those that appear'd on Saturday the 8th of the last Month, were too remarkable to be pass'd over in Silence". Rev. Dr Benjamin Langwith, "An Account of the Aurora Boeralis that appeared October 8, 1726," *Phil. Trans.* 34 (1726–27): 132. Or, consider the account sent in by Mr Samuel Dunn about a "remarkable Meteor," where his description opens with the statement that the appearance had been "not before noticed." Dunn, "of a Remarkable Meteor," *Phil. Trans.* 53 (1763): 351.

68 The fear of foreign climates could be found in nineteenth-century medical writings in which the tropics featured as the seeds of disease and bodily decay. In the 1850s Thomas Burgess went as far as to argue that nature adapted the human constitution to the climate of its ancestors and because this 'hereditary climate' defined pathological disposition, it also had a therapeutic power. It was thus inconsistent with nature's laws that a person born in England and attacked by consumption could be cured from it in a *foreign* climate, which was inimical to the hereditary constitution. In 1875, a Superannuation Act gave the Commissioners of the Treasury power to declare a country or 'place' unhealthy for the purposes of determining retiring allowances for civil servants. Thomas Burgess, "Inutility of Resorting to the Italian Climate for the Cure of Pulmonary Consumption," *Lancet* 1 (1850): 591. See also David Livingston, 'Human Acclimatization: Perspectives on a Contested Field of Inquiry in Science, Medicine and Geography,' *History of Science* 25 (1987): 359–94; L. J. Jordanova, 'Earth Sciences and Environmental Medicine,' in Jordanova and Roy Porter, eds, *Images of the Earth: Essays in the History of the Environmental Sciences* (Chalfont St Giles: British Society for the History of Science, 1979), 119–46. More generally on medical climatology see Frederick Sargent, *Hippocratic Heritage: a History of Ideas about Weather and Human Health* (New York and Oxford: Peryamon Press, 1982) and James Riley, *The Eighteenth-Century Campaign to Avoid Disease* (New York, St Martin's Press, 1987).

69 *Abstract of the Papers Printed in Philosophical Transactions 1800–1830* (London: Richard Taylor, 1851), vol. 1, 163, 239, 400; vol. 2, 183. On hypsometry, see Theodore Feldman, "Applied Mathematics and the Quantification of Experimental Physics: The Example of Barometric Hypsometry," *Historical Studies in the Physical Sciences* 15 (1985): 127–97.

70 Thomson, *History of the Royal Society*, 506.

71 Curiously enough, meteorological empiricism of the early nineteenth century does not fit well into the interpretation of the rise of geological empiricism, the staple of the early Geological Society (founded 1803). British meteorologists encouraged theorizing because meteorology (not *meteors*) had never entailed theological controversies such as were part of the geological repertoire from William Burnet to James Hutton. Furthermore, British meteorologists showed no signs of Francophobia as can be witnessed in the content of *Annals of Philosophy* and *Philosophical Magazine*. For a summary of the sources of geological empiricism, see Roy Porter, "Gentlemen and Geology: The Emergence of a Scientific Career, 1660–1920," *The Historical Journal* 21 (1978): 809–36, 822.

72 Luke Howard, "Meteorological Journals," *Annals of Philosophy* 1 (1813): 79; Dr James Clarke, "Meteorological Tables for the Year 1812, at Sidmouth," *Annals of Philosophy* 1 (1813): 265.

73 "Advertisement," *Annals of Science, or Magazine of Chemistry, Mineralogy, Mechanics, Natural History, Agriculture and Arts* 1 (1813): iii.

74 Kirwan, *Estimate*, iii.

75 It is worth noting that Kirwan's division implied that the local meteorological knowledge – the chorographic heritage – had become methodologically unacceptable.

76 Kirwan, *A Comparative View*, 4. Italics mine.

77 Ibid., iv.

78 Indeed, the search for regularities could not but find regularities. Some twenty years after Kirwan's pronouncements, an author in the first volume of William Nicholson's *Philosophical Magazine*, considered the correlations established between the Lunar phases and barometric data, and thought that with these correlations, meteorology had been "exalted into a science." "Cursory View of Some of the Late Discoveries in Science," *Philosophical Magazine* 1 (1798): 208. In 1800 the French meteorologist Louis Cotte announced that the temperature trends recurred every nineteenth years; regularities of other atmospheric parameters were given in the form of empirical "axioms" inferred from his own and foreign observations over the course of thirty years. L. Cotte, "Meteorological Axioms, reprinted from *Neues Journal der Physik*," *Philosophical Magazine* 6 (1800): 146–7.

79 Theodore M. Porter, *Trust in Numbers: The Pursuit of Objectivity in Science and Public Life* (Princeton, N. J.: Princeton University Press, 1995), 49–50.

80 Adams, *Lectures*, 472.

81 Daniell, *Essays*, x.

82 Arnold Thackray, "John Frederick Daniel," *Dictionary of Scientific Biography*, vol. 3, 556–8.

83 Ibid., xiv.

84 Ibid., xviii, 1.

85 William Nicholson pointed out the lack and inaccuracy of observations in his *Introduction to Natural Philosophy*, 63; Alexander Tilloch commented on 'unprofitability' of the existing observations, in *The Philosophical Magazine* 24 (1806): 272; Thomas Thomson, John Dalton, and Luke Howard criticized The Royal Society's neglect of meteorological research in general and its sloppiness in handling meteorological registers in particular. In 1823 the *British Register* 55, 70, deplored the non-existence of a meteorological society and dryly announced the lack of "the general principles of the science of meteorology," a claim that the meteorologist Thomas Hopkins entertained as late as 1860 (see his *On Winds and Storms*, London, 1860, 15). In a heyday of his "declinist" phase, David Brewster reprobated "the discontinuance of the hourly observations at Leith Fort." Meteorological observations were made there by non-commissioned officers from 1824 to 1828 when the army ceased to cooperate with the Royal Society of Edinburgh. Brewster to Forbes, July 10, 1830, in Jack Morrell and Arnold Thackray, eds, *Gentlemen of Science. Early Correspondence of the British Association for the Advancement of Science* (London, 1984), 27. See also J. M. C. Burton, "Meteorology and the public health movement in London during the late nineteenth century," *Weather* 45 (1990): 300–7.

86 Quoted in Schneider-Carius, *Weather Science*, 238.

87 Perhaps the most telling criticism of old meteorological practices was voiced by another German, L. F. Kaemtz who observed that naturalists too frequently attach themselves to the phenomena "that present anything marvelous." From the time of Galileo's laws of the falling bodies, explained Kaemtz, this tendency has been gradually challenged [sic!], but "meteorology still feels its baneful influence. Where are the observers who are occupying themselves in the regular succession of the modifications of the atmosphere, and the philosophers who are seeking to recognize their laws? It is a trouble to them to look at their instruments." But when the barometer displays an *extraordinary* rise or fall, meteorologists begin to flatter themselves that they "will draw some beautiful results from it. I have many times had occasion to show that those extraordinary facts *teach us nothing* precisely because they are observed isolatedly, without any trouble being taken as to what precedes them and without examination of what follows." L. F. Kaemtz, *A Complete Course of Meteorology*. Translated by C. V. Walker (London: Hippolyte Baillière Publisher, 1845), 464.

88 Ibid., 9.

89 Read, *A Summary View of the Spontaneous Electricity of the Earth and Atmosphere*, 35.

90 Golinski, *Science as Public Culture*, 285; Theodore M. Porter, *Trust in Numbers*, 85.

Notes to conclusion

1 J. D. Forbes, "The Report on Meteorology," *The Report of the British Association for the Advancement of Science, 1832* (London: John Murray, 1833), 7.

2 James D. Forbes, "Supplementary Report on Meteorology," *The Report of the BAAS, 1840* (London: John Murray, 1841), 38; John Herschel, "Address to the Fourteenth Meeting of the BAAS," *The Report of the BAAS, 1845* (London: John Murray, 1846), 85.

3 Vladimir Jankovic, "Ideological Crests versus Empirical Troughs: John Herschel's and William Radcliffe Birt's Research on Atmospheric Waves," *British Journal for the History of Science* 31 (1998): 21–40.

4 Susan Faye Cannon, *Science in Culture: The Early Victorian Period* (New York: Dawson and Science History Publication, 1978), 105.

5 Martin Rudwick, *The Great Devonian Controversy: The Shaping of Natural Knowledge Among Gentlemanly Specialists* (Chicago: The University of Chicago Press, 1985), 41.

Bibliography

Primary sources

Abstract of the Papers Printed in Philosophical Transactions 1800–1830. London: Richard Taylor, 1851.

An Account of Explosions in the Atmosphere, or Airquakes. Their Distinction from True Earthquakes. London: A. Dodd, 1750.

"An Account of Toaldo's System Respecting the Probability of a Change of Weather at the Different Changes of the Moon," *Philosophical Magazine* 3 (1799): 120–7.

Adams, George. *A Short Dissertation on the Barometer, Thermometer, and other Meteorological Instruments. Together with an Account of Prognostic Signs of Weather.* London: the Author, 1790.

Adams, George. *Lectures on Natural and Experimental Philosophy.* London: J. Dillon and Co., 1798.

Adams, Henry. *A Royal Almanac and Meteorological Diary for the Year of Our Lord 1778.* London: T. Carnan, 1778.

Adams, John. *Extract from a Meteorological Journal Kept at Edmonton, Middlesex, together with a Collection of such Observations as have been Considered the Surest Guides whereby to Foretell the Weather.* London: the Author, 1814.

Alexander of Aphrodisias. *On Aristotle's Meteorology 4.* Translated by Eric Lewis. Ithaca: Cornell University Press, 1996.

Annely, Bernard. *A Theory of the Wind. Shewing by a New Hypothesis the Physical Causes of all Winds in General.* London: Jeremy Batley, 1729.

[Appletree, Thomas]. "The 1703 Weather Diary," MS., National Meteorological Archives, Bracknell, Berkshire.

Arderon, William. "Remains," British Library Manuscript Collection, MS. Addit. 27966.

Arderon, William. "Concerning an Improvement of the Weather Cord," *Phil. Trans.* 44 (1746–47): 169–73.

Arderon, William. "On the Hot Weather in July 1750," *Phil. Trans.* 46 (1749–50): 571–2.

Arderon, William. "Observations on the Late Severe Cold Weather," *Phil. Trans.* 48 (1753–54): 507–8.

Arderon, William. "Of the Rain Fallen in a Foot-Square at Norwich, 1749–1762," *Phil. Trans.* 52 (1761–62): 9.

Aristotle. *Meteorologica.* Translated by H. D. P. Lee. Loeb Classical Library. Cambridge, Mass.: Harvard University Press, 1952.

Atkinson, A. D. "William Derham, FRS (1657–1735)," *Annals of Science* 8 (1952): 368–92.

Atkyns, Robert. *The Ancient and Present State of Glostershire.* London: W. Bowyer, 1712.

Aubert, Alexander. "An Account of Meteors," *Phil. Trans.* 73 (1783): 112.

Aubrey, John. *Natural History and Antiquities of the County of Surrey.* London: E. Curll, 1719.

Aubrey, John. *Aubrey's Natural History of Wiltshire.* Reprint; Ponting, New York: Augustus M. Kelley, 1969.

Baker, Henry. "Abstract of Several Observations of Aurorae Boreales," *Phil. Trans.* 46 (1749–50): 499–502.

Baker, Henry. "An Account of the [Meteor of July 22, 1750]," *Phil. Trans.* 47 (1751–52): 3.

Barham, Henry. "A Letter giving a Relation of a Fiery Meteor Seen by [Barham], in Jamaica, to Strike Into the Earth," *Phil. Trans.* 30 (1717–19): 837–8.

Barker, Thomas. "An Account of an Extraordinary Meteor seen in the County of Rutland, which Resembled a Water-Spout," *Phil. Trans.* 46 (1749–50): 248–50.

Barker, Thomas. "An Account of a Remarkable Halo," *Phil. Trans.* 52 (1761–62): 3–4.

Barker, Thomas. "[C]oncerning Observations of the Quantities of Rain at [Lyndon] for Several Years," *Phil. Trans.* 59 (1769): 221.

Barker, Thomas. *The Duty, Circumstances, and Benefits of Baptism.* London: B. White, 1771.

Barker, Thomas. "Extract of Mr. Barker's Meteorological Register at Lyndon in Rutland [1771]," *Phil. Trans.* 62 (1772): 42.

Barker, Thomas. "Extract of a Register of the Barometer, Thermometer, and Rain, at Lyndon in Rutland, 1772," *Phil. Trans.* 63 (1773): 221.

Barker, Thomas. *On Prophesies Relating to the Messiah.* London: B. White, 1780.

Barker, Thomas. *On the Nature and Circumstances of the Demoniacks in the Gospels.* London: B. White, 1783.

Barlow, Edward. *Meteorological Essays Concerning the Origin of Springs, Generation of Rain, and Production of Wind.* London: J. Hooke, 1715.

Barlow, Peter. *A New Mathematical Dictionary.* London: J. Nichols, 1814.

Bate, John. *The Mysteries of Nature and Art.* London, 1635.

Bayley, Edward. "A Narrative of the Earthquake," *Phil. Trans.* 39 (1735–36): 362–7.

Bédoyère, Guy de la, ed., *The Diary of John Evelyn.* Woodbridge: The Boydell Press, 1995.

Bennet, Abraham. *New Experiments on Electricity.* Derby: John Drewry, 1789.

Bent, William. *A Meteorological Journal of the Year 1793, Kept in London.* London: W. Bent, 1794.

Bentham, Jeremy. *Chrestomathia.* Edited by M. J. Smith and W. H. Burton. Oxford: Clarendon Press, 1983.

Bibliotheca Topographica Britannica. London: J. Nichols, 1780–90.

Biographia Britannica. London: W. Innys, 1747.

Birch, Thomas. *The History of the Royal Society of London.* A Facsimile of the London edition, 1756–57. New York, London: Johnson Reprint Corporation, 1968.

Bisset, Charles. *An Essay on Medical Constitution of Great Britain to which are Added Observations on the Weather.* London: A. Millar, 1762.

Blagden, Charles. "An Account of Some Late Meteors," *Phil. Trans.* 74 (1784): 201–31.

Blome, Richard. *The Gentleman's Recreations.* London: S. Rotcroft, 1686.

Boate, Gerard. *Ireland's Natural History.* London: John Wright, 1652.

Bohun, Ralph. *A Discourse Concerning the Origins and Properties of Wind.* Oxford: W. Hall, 1671.

Bonajutus, Vincentius. "An Account of the Earthquakes in Sicilia," *Phil. Trans.* 17–18 (1693–94): 2–10.

Borlase, William. "An Account of a Storm of Thunder and Lightning, near Ludgvan in Cornwall," *Phil. Trans.* 48 (1753–54): 86–98.

Borlase, William. *Observations on the Antiquities, Historical and Monumental, of the County of Cornwall.* Oxford: W. Jackson, 1754.

Borlase, William. "Of the Earthquake in the West Parts of Cornwall, July 1757," *Phil. Trans.* 50 (1757–58): 499–503.

Borlase, William. *Natural History of Cornwall, the Air, Climate, Waters, Rivers, Lakes, Sea, and Tides etc.* Oxford: W. Jackson, 1758.

Borlase, William. "On the Late Mild Weather in Cornwall," *Phil. Trans.* 53 (1763): 27–9.

Borlase, William. "Of the Quantity of Rain Fallen at Mount Bay in Cornwall and of the Weather in that Place," *Phil. Trans.* 54 (1764): 59–60.

Borlase, William. "Meteorological Observations at Ludgvan in Mount's Bay, Cornwall," *Phil. Trans.* 58 (1768): 89.

Borlase, William. "Meteorological Observations for 1769, made at Bridgewater, Somersetshire; and at Mount's Bay, Cornwall," *Phil. Trans.* 60 (1770): 228.

Borlase, William. "Meteorological Observations for 1770 at Ludgvan in Mount's Bay, Cornwall," *Phil. Trans.* 61 (1771): 195.

Borlase, William. "Meteorological Observations for 1771, at Ludgvan in Mount's Bay," *Phil. Trans.* 62 (1772): 365.

Botanista, Theophilus, M. D. *Rural Beauties, or the Natural History of the Four following Counties: Cornwall, Dorsetshire, Devonshire and Somersetshire.* London: William Fenner, 1757.

Boyle, Robert. "An Experimental History of Cold," *Phil. Trans.* 1–2 (1665–67): 8–14.

Boyle, Robert. "A New Frigorific Experiment," *Phil. Trans.* 1–2 (1665–67): 255–62.

Boyle, Robert. *The Works.* Edited by Thomas Birch (1772). Reprinted by Hildesheim: Georg Olms Verlagsbuchhandlung, 1966.

Bradbury, Thomas. *God's Empire over the Wind, Consider'd in a Sermon on the Fast-Day, January 19, 1703/4.* London: Jonathan Robinson, 1704.

Bradley, R. *The Natural History of Cambridgeshire and Essex.* c. 1735.

Brand, John. *Observations on Popular Antiquities.* London: Vernon et al., 1810.

Brayley, E. W. Jr., "Introductory Sketch of the Objects and Uses of Meteorological Science," *The Magazine of Natural History and Journal of Zoology, Botany, Mineralogy, Geology, and Meteorology.* London: Longman, 1836, 153–4.

Brewster, David, ed. *Letters of Euler on Different Subjects in Natural Philosophy.* New York: J. J. Harper, 1833.

Brinkmair, L. *The Warnings of Germany, by Wonderful Signes.* London: J. Horton, 1638.

Brooke, John Hedley. *Science and Religion.* Cambridge: Cambridge University Press, 1991.

Brydone, Patrick. "Of a Fiery Meteor," *Phil. Trans.* 63 (1773): 163–5.

Budgen, Richard. *The Passage of the Hurricane.* London: the Author, 1733.

Burnet, Gilbert. *History of My Time.* London: J. Hopkins, 1723.

Burrow, Reuben. *The Lady's and Gentleman's Diary.* London: T. Carnan and G. Robinson, 1755–58.

Burton, William. *The Description of Leicester Shire.* London: John White, [1622].

Burton, William. *The Surprizing Miracles of Nature and Art.* London, 1683.

Camden, William. *Britain, Or a Chorographicall Description of England, Scotland and Ireland.* London: F. K. R. G. & I., 1637.

Campbell, Duncan. *Time's Telescope: Universal and Perpetual.* London: J. Wilcox, J. Oswald, 1734.

Capper, James. *Observations on the Winds and Monsoons Illustrated on a Chart and Accompanied with Notes, Geographical and Meteorological.* London: J. Debrett, 1801.

Capper, James. *Meteorological and Miscellaneous Tracts, Applicable to Navigation, Gardening and Farming, with Calendar of Flora for Greece, France, England, and Sweden.* Cardiff: J. D. Bird, 1810.

Carew, Richard. *The Survey of Cornwall.* S[tafford]: J. Jaggard, 1602.

Casaubon, Meric. *A Treatise Concerning Enthusiasm.* London, 1655.

Cavallo, Tiberius. "Description of a Meteor, Aug. 18, 1783," *Phil. Trans.* 73 (1783): 108–15.

Mr. Chalmers. "Of an Extraordinary Fireball Bursting at Sea," *Phil. Trans.* 46 (1749–50): 366.

Chambers, Ephraim. *Cyclopædia: or an Universal Dictionary of Arts and Sciences.* London: James and John Knapton, 1728.

Chambers, E. and Rees, Abraham. *Encyclopaedia, or an Universal Dictionary of Arts and Science.* London: J. F. and C. Rivington, 1784.

Chambers's Encyclopaedia: A Dictionary of Universal Knowledge for the People. Philadelphia: J. B. Lippincott & Co., 1881.

Chapman, Richard. *The Necessity of Repentance Asserted: In Order to Avert Those Judgments which the Present War and Strange Unseasonableness of the Weather At Present seem to Threaten this Nation with in Sermon preached on Wednesday the 26th of May 1703 being the Fast-Day Appointed by Her Majesty's Proclamation.* London: M. Wotton, 1704.

Cheyne, George. *Philosophical Principles of Religion Natural and Revealed.* Fourth edition. London: George Strahan, 1734.

Childrey, Joshua. *Britannia Baconica, or The Natural Rarities of England, Scotland, & Wales.* London: the Author, 1660.

Clap, Thomas, *Conjectures upon the Nature and Motion of Meteors, which are Above the Atmosphere.* Norwich, Connect.: John Trumbull, 1781.

Clare, John. *The Motion of Fluids, Natural and Artificial; in Particular that of the Air and Water.* Third edition. London: A Ward, 1747.

Claridge, John. *The Shepherd, of Banbury's Rules to Judge the Changes of the Weather.* London: T. Waller, 1723.

Clarke, James. "Meteorological Tables for the Year 1812, at Sidmouth," *Annals of Philosophy* 1 (1813): 265.

Clutterbuck, Rev. Robert. *Notes on the Parishes of Fyfield, Kimpton, Penton Mewsey, Weyhill and Wherwell in the County of Hampshire.* Salisbury: Bennet Brothers, 1898.

Cock, William. *Meteorologiae, or the True Way of Foreseeing the Weather.* Edinburgh: the Author, 1671.

Cockin, William. "Account of an Extraordinary Appearance in a Mist," *Phil. Trans.* 70 (1780): 157–62.

Cole, William. "On the Grains Resembling Wheat which Fell Lately in Wiltshire," *Phil. Trans.* 16 (1686–92): 281.

Collinson, Peter. "An Observation of an Uncommon Gleam of Light," *Phil. Trans.* 44 (1746–47): 456.

Comenius, Johannes Amos. *Naturall Philosophie Reformed by Divine Light, or a Synopsis of Physics.* London: Robert and William Leybourn, 1651.

Comenius, Johannes Amos. *Orbis Sensualium Pictus. A World of Things Obvious to the Senses Drawn in Pictures* (1659). Menston: The Scolar Press Limited, 1970.

A Companion to the Weather-Glass: or the Nature, Construction, and Use, of the Barometer, Thermometer, and Hygrometer, with a Short Account of Watery Meteors, the Form of a Register of the Weather, etc. Edinburgh: T. Ross, J. Guthrie, 1796.

Cook, Benjamin. "A Letter . . . Concerning a Ball of Sulphur Supposed to be Generated in the Air," *Phil. Trans.* 40 (1737–38): 427–8.

Cookson, Dr. "A Further Account of the Extraordinary Effects of the Lightning at Wakefield," *Phil. Trans.* 38 (1733–34): 74–5.

Cooper, William. "Observation on a Remarkable Meteor Seen on the 18th of August 1783," *Phil. Trans.* 74 (1783): 116–17.

Copland, Alexander. "Meteorological Observations and Remarks on the Weather at Dumfries," *Manchester Literary and Philosophical Society Memoirs* 4 (1793): 243–51.

"A Copy of a Letter from R. P. Vicar of Kildwick in Yorkshire, to a Friend of his in those parts, wherein he gives an Account of an Extraordinary Eruption of Water, which happened in June, 1698," *Phil. Trans.* 20 (1698): 382–5.

Costard, George. "Concerning a Fiery Meteor Seen in the Air on July 14, 1745," *Phil. Trans.* 43 (1744–45): 522–4.

Cotes, Roger. "A Description of the Great Meteor which was on the 6th of March 1715/6," *Phil. Trans.* 31 (1720): 66–7.

Cotte, Louis. *Traité de météorologie.* Paris, 1772.

Cotte, L. "Meteorological Axioms, reprinted from *Neues Journal der Physik*," *Philosophical Magazine* 6 (1800): 146–7.

[Crocker Mr.], "An Account of a Meteor Seen in the Air in the Day-Time on Dec. 8, 1733," *Phil. Trans.* 41 (1739–41): 346–7.

Crouch, Nathaniel. *Admirable Curiosities, Rarities and Wonders in England, Scotland and Ireland.* London: Thomas Snowden, 1682.

Cruwys, Samuel. "A Relation of the [Aurora Borealis], seen at Cruwys Morchard in Devonshire," *Phil. Trans.* 30 (1717–19): 1101–2.

Cullum, John. "Account of Extraordinary Frost, 23 June 1783," *Phil. Trans.* 74 (1784): 416.

Cunningham, P., ed., *Letters of Horace Walpole.* London: Richard Bentley, 1857.

"Cursory View of Some of the Late Discoveries in Science," *Philosophical Magazine* 1 (1798): 208.

Dalton, John. *Meteorological Observations and Essays.* Manchester: Harrison, 1793.

Daniell, John Frederick. *Meteorological Essays and Observations.* London: T. & G. Underwood, 1823.

Daniell, John Frederick. *Elements of Meteorology. Being the Third edition of Meteorological Essays, Revised and Enlarged.* Edited by W. A. Miller and C. Pemlison. London: John W. Parker, 1845.

Daval, Peter. "Of an Extraordinary Rainbow," *Phil. Trans.* 46 (1749–50): 193.

Davies, Evan. "An Account of what happened from Thunder in Carmarthenshire; . . . communicated to the Royal Society, By John Eames, FRS as he received it in a Letter from Mr Evan Davies," *Phil. Trans.* 36 (1729–30): 444–8.

De Brahm, John Gerard William, *The Levelling Balance and Counter Balance; or, the Method of Observing, by the Weight and Height of Mercury.* London: T. Spilsbury, 1774.

Defoe, Daniel. *The Novels and Miscellaneous Works.* London: George Bell and Son, 1911.

[Defoe, Daniel]. *The Storm: or, a Collection of the Most Remarkable Casualties and Disasters Which Happened in the Late Dreadful Tempest, both by Sea and Land.* London: G. Sawbridge, 1704.

Deluc, Jean Andre. "Barometrical Observations on the Depth of the Mines in the Hartz," [transl. from the French], *Phil. Trans.* 67 (1777): 388–91.

Deluc, Jean Andre. "An Essay on Pyrometry and Aerometry, and on Physical Measures in General," *Phil. Trans.* 67 (1777): 419–30.

Deluc, Jean Andre. "A Second Paper on Hygrometry," *Phil. Trans.* 81 (1791): 1–9.

Deluc, Jean Andre. "On Evaporation," *Phil. Trans.* 82 (1792): 283–87.

Denston, R. L. *A New Theory of the Atmosphere.* Birmingham: Knott and Lloyd, 1807.

Derham, William. "Observations on the Late Storm," *Phil. Trans.* 23 (1702–03): 1530.

Derham, William. "A Letter concerning the late Storm," *Phil. Trans.* 24 (1704–5): 1530–4.

Derham, William. "Theory of Storms," (1705). *Classified Papers of the Royal Society,* IV (1) 53.

Derham, William. "The History of the Great Frost in the last Winter, 1708 and 1708/9," *Phil. Trans.* 26 (1708–09): 477.

Derham, William. *Physico-Theology.* London: W. Innys, 1716.

Derham, William. "Observations on the Lumen Boreale," *Phil. Trans.* 35 (1727–28): 245–6.

Derham, William. "Uncommon Appearances Observed in an Aurora Borealis," *Phil. Trans.* 36 (1729): 137.

Dingley, Robert. *Vox Coeli* (London, 1658).

Dobbs, Richard. "An Account of an Aurora Borealis seen in Ireland in September 1725,"*Phil. Trans.* 34 (1726–27): 366–7.

Dorby, John. "Of a Terrible Whirlwind," *Phil. Trans.* 41 (1739–41): 230–2.

Downie, Murdo. *Observations upon the Nature and Properties of the Atmosphere, to which are Added Observations on the Moon's Influence upon the Atmosphere, and the Rise and Fall in the Barometrical Tube.* Aberdeen: J. Chalmers and Co., 1801.

Drebbel, Cornelius. *A Dialogue Philosophicall. Wherein Natures Secret is Opened and the Cause of All Motion in Nature Shewed out of Matter and Form.* London: C. Knight, 1612.

Dunn, Samuel. "Of a Remarkable Meteor," *Phil. Trans.* 53 (1763): 351.

Dunton, John. *The Life and Errors of John Dunton.* London: J. Nichols, 1818.

Dupleix, M. Scipion. *La Physique* (1603). Edited by Roger Ariew. Paris: Librarie Arthème, Fayard, 1990.

Dyer, James. "An Account of the Effects of a Storm of Thunder and Lightning, in the Parishes of Looe and Lanreath, in the County of Cornwall, on the 27th Day of June, 1756," *Phil. Trans.* 50 (1757–58): 104–7.

Eeles, Henry. "Letter concerning the Cause of the Ascent of Vapour and Exhalation," *Phil. Trans.* 49 (1755–56): 124–9.

Eeles, Henry. "On the Cause of Thunder," *Phil. Trans.* 47 (1751–52): 324–7.

Encyclopedia [Britannica]: or a Dictionary of Arts, Sciences, and Misc. Literature. Philadelphia, 1798.

The English Chapmans and Travellers Almanack for the Year of Christ 1697. London: Thomas James, 1697.

"An Extraordinary Phenomenon," *Gentleman's Magazine* 20 (1750): 136.

Fairfax, Nathaniel. "Hail Stones of Unusual Size," *Phil. Trans.* 1–2 (1665–67): 481.

Farrer, Isaac. *A Sermon or the Occasional Discourse on the Late Great Flood in the North of England, Nov. 16 1771.* Newcastle: T. Slack, 1772.

Folkes, Martin. "An Account of the Aurora Boeralis, Seen at London, the 30th of March Last," *Phil. Trans.* 30 (1717–19): 586–7.

Forbes, J. D. "The Report on Meteorology," *The Report of the British Association for the Advancement of Science, 1832.* London: John Murray, 1833.

Forbes, James D. "Supplementary Report on Meteorology," *The Report of the BAAS, 1840.* London: John Murray, 1841.

Forster, M. "Of a Meteor Seen at Shefford in Berkshire," *Phil. Trans.* 51 (1759–60): 299.

Forster, Thomas Ignatius Maria. *Researches about Atmospheric Phenomena.* London: J. Moyes 1815.

Forster, Thomas Ignatius Maria. *The Pocket Encyclopaedia of Natural Phenomena; being a Compendium of Prognostication of Weather, Signs of the Seasons.* London: John Nichols, 1827.

Forth, Henry. "An Account of the Storm, Jan. 8, 1734–5," *Phil. Trans.* 39 (1735–36): 285–7.

Franklin, Benjamin. "Concerning the Effects of Lightning," *Phil. Trans.* 47 (1751–52): 289–97.

Fromondi, Liberti. *Meteorologicum Libri Sex.* Londini: Typis E. Tyler, 1656.

Frostiface, Icedore, of Freesland, *An Account of all the Principal Frosts for above an Hundred Years Past: with Political Remarks and Poetical Descriptions.* London: C. Corbettt, 1740.

Fulke, William. *A Goodly Gallerye. Book of Meteors* (1563). Philadelphia: The American Philosophical Society, 1979.

Fulke, William. *Meteors: or a Plain Description of all Kinds of Meteors*. London: William Leake, 1654.

Fuller, John. "A Letter Concerning a Strange Effect of the Late Great Storm in that Country," *Phil. Trans.* 24 (1704–4): 1530.

Fuller, Stephen. "A Letter . . . Concerning a Violent Hurricane," *Phil. Trans.* 41 (1739–41): 854–6.

"A Further Relation of the Same Appearance as Seen at Dublin, Communicated to the Publisher by an Unknown Hand," *Phil. Trans.* 30 (1717–19): 1104–5.

Gabry, Peter. "Observation of a Fiery Meteor Made at Hague," *Phil. Trans.* 53 (1763): 300.

Gadbury, John. *De Cometis*. London: L. Chapman, 1665.

Gadbury, John. *Ephemeris: or, a Diary Astronomical, Astrological, Meteorological, for the Year 1706*. London: The Company of Stationers, 1706.

Garden, George. "An Account of the Effects of a Very Extraordinary Thunder," *Phil. Trans.* 19 (1695–97): 311–12.

Garnett, Thomas. "Meteorological Observations, Collected and Arranged by Thomas Garnett," *Memoirs of the Literary and Philosophical Society of Manchester* 4 (1793): 517.

Gassendi, Petrus. *Opera Omnia*. Reprint. Stuttgart-Bad Cannstatt, 1964.

Gibbons, Thomas. *A Sermon [on Habbak. iii, 2] Preached on Occasion of the Earthquake in Lisbon, November 1, 1758*. Second edition; London, 1758.

Gifford, A. *A Sermon in Commemoration of the Great Storm, Commonly called the High Wind, in the year 1703, preached at the CHAPEL in Little Wilde Street, London November 27, 1733, with an Account of the Damage done by it*. London: Aaron Ward, 1733.

Gilbert, C. S. *An Historical Survey of the County of Cornwall*. London: J. Congdon, 1817.

Gilbert, William. *De Mundo Nostro Sublunari Philosophia Nova*. Amstelodami: Ludovicum Elzevirium, 1651.

Godfridus, *The New Book of Knowledge Shewing the Effects of Planetary and other Astronomical Constellations*. London: A. Wilde, 1758.

Gordon, William. "An Account of the Fireball," *Phil. Trans.* 42 (1742–43): 58.

Gostling, William. "Concerning the Meteor Seen in Kent," *Phil. Trans.* 41 (1739–41): 20–1.

Gray, Stephen. "An Unusual Parhelion and Halo," *Phil. Trans.* 22 (1700–01): 535.

Greenwood, Isaac. "A New Method for Composing a Natural History of Meteors Communicated in a Letter to Dr Jurin," *Phil. Trans.* 35 (1727–28): 390–402.

Gregory, G. *The Economy of Nature: Explained and illustrated on the Principles of Modern Philosophy*. London: J. Johnson, 1796.

Grew, Dr. Nehemiah. "Observations on the Nature of Snow," *Phil. Trans.* 7–8 (1672–73): 5193–6.

Hale, Matthew. *Difficiles Nugae. Or Observations Touching the Torricellian Experiment*. London: W. Goodbid, 1674.

Hales, Stephen. *Statical Essays*. London: T. Woodward, 1733.

Hales, Stephen. *Some Considerations on the Causes of Earthquakes*. London: R. Manby, 1750.

Hales, Stephen. *Vegetable Staticks* London: 1727.

Hales, Stephen. "A Collection of Various Papers Presented to the Royal Society Concerning Several Earthquakes," *Phil. Trans.* 46 (1749–50): 602–17.

Hall, Allen. *Observation on the Weather*. London: Drury's Office, 1788.

Hall, A. R. and Hall, M. B., eds, *The Correspondence of Henry Oldenburg*. 6 vols. Madison, 1965–86.

[Haller, Albert von.] "An Historical Account of the Wonderful Discoveries Made in Germany Concerning Electricity," *Gentleman's Magazine* 15 (1745): 193–7.

Hallet, John. "A Letter to Dr Henry Pemberton, FRS. On the Same Subject [Aurora Borealis]," *Phil. Trans.* 34 (1726–27): 143–4.

Halley, E. "An Estimate of the Quantity of Vapour raised out of the Sea by the Warmth of the Sun," *Phil. Trans.* 16 (1686–92): 366.

Halley, Edmund. "On the Circulation of the Watery Vapours at the Sea," *Phil. Trans.* 16 (1686–92): 468–81.

Halley, Edmund. "Account of the Evaporation of Water," *Phil. Trans.* 17–18 (1693–94): 183–90.

Halley, Edmund. "Account of an Extraordinary Iris," *Phil. Trans.* 20 (1698): 193.

Halley, Edmund. "An Account of the Late Surprizing Appearance of the Lights Seen in the Air, on the sixth of March last; with an Attempt to Explain the Principal Phenomena thereof," *Phil. Trans.* 29 (1714–16): 406–28.

Halley, Edmund. "An Account of the Extraordinary Meteor Seen all over England, on the 19th of March 1718–9," *Phil. Trans.* 30 (1717–19): 978–92.

Halley, E. "Observations on Falling Dew, Made at Middleburg in Zealand," *Phil. Trans.* 42 (1742–43): 112.

Hamilton, Hugh. A Dissertation on the Nature of Evaporation and Several Phenomena of Air, Water, and Boiling Vapors," *Phil. Trans.* 55 (1765): 146–52.

Hamilton, Hugh. *Philosophical Essays on the Following Subjects: 1. On the principles of mechanics. 2. On the ascent of vapours, the formation of clouds, rain and dew, and of several other phenomena of air and water, 3. Observations and conjectures on the nature of the aurora borealis, and the tails of comets.* Third edition. London: J. Nourse, 1772.

Hamilton, Sir William. "An Account of the Effects of a Thunder-Storm, on the 15th of March 1773, upon the House of Lord Tylney at Naples," *Phil. Trans.* 63 (1773): 324–32.

Harris, John. *Lexicon Technicum, or An Universal Dictionary of Arts and Sciences* (1704). Reprint; New York and London: Johnson Reprint Corporation, 1966.

Harward, Simon. *A Discourse of the Several Kind and Causes of Lightnings.* London: John Windet, 1607.

Hazlitt, W. Carew. *Faiths and Folklore of the British Isles.* London: Murray, 1846.

Henry, Thomas. "An Account of the Earthquake which was Felt at Manchester and Other Places, on the 14th day of September, 1777," *Phil. Trans.* 67 (1777): 221–5.

Henry, William. "An Account of an Extraordinary Stream of Wind, which Shot thro' Part of the Parishes of Termonomungan and Urney, in the County of Tyrone, on Wednesday October 11, 1752," *Phil. Trans.* 48 (1753–54): 1–4.

Heylen, Peter. *Microcosmus, or a Little Description of the Great World. A Treatise Historicall, Geographicall, etc.* Oxford: J. Lichfield, 1621.

Hill, Sir John. *The Gardener's New Kalendar: Divided according to the Twelve Months of the Year.* Dublin: J. Exsham, 1763.

Hill, Thomas. *Mysteries of Nature.* London, 1646.

Hill, Thomas. *The Contemplation of Mysteries: Contayning the Rare Effects and Significations of Certayine Comets, and a Briefe Rehearsall of Sundrie Hystorical Examples.* London: Henry Denham, 1571.

Hilliard, John. *Fire From Heaven.* London, 1613.

Hinde, M., et al. *New Royal and Universal Dictionary of Arts and Sciences.* London: J. Cooke, 1771.

Hirst, William. "Of a Fireball Seen at Hornsey," *Phil. Trans.* 48 (1753–54): 773.

Hobbes, Thomas. *Leviathan* (1652). Edited by Michael Oakeshott. New York: Collier Books, 1962.

Hone, William. *The Table Book.* London: Hunt and Clarke, 1827, 1512–21.

Hone, William. *The Every-day Book and Table-book, or Everlasting Calendar of Popular Amusements, Sports, Pastimes, Ceremonies, Manners etc., including Accounts of Weather, Rules for Health and Conduct.* 3 vols. London: T. Tegg, 1830–39.

Hopkins, Thomas. *On Winds and Storms.* London, 1860.

Hopkinson, Samuel. *Causes of the Scarcity Investigated: also an Account of the Most Striking Variations in the Weather, from October 1789 to September, 1800.* Stamford: R. Newcomb, 1800.

Hopton, Arthur. *Speculum Topographicum.* Amsterdam: Theatrum Orbis Terrarum, reprinted, 1974.

Horsley, Samuel. "An Abridged State of the Weather at London in the Year 1774," *Phil. Trans.* 65 (1775): 167–8.

Howard, Luke. "Meteorological Journals," *Annals of Philosophy* 1 (1813): 79.

Howard, Luke. *Seven Lectures on Meteorology.* London: Harvey and Darton, 1837.

Howell, James. *Dendrologia. Dodona's Grave, or the Vocall Forest.* London: T. B[adger], 1640.

Howie, John. *An Alarm Unto a Secure Generation.* Kilmarnock: H. & S. Crawford, 1809.

Humble Adorer of God. *Stars and Planets the Best Barometers and Truest Interpreters of all Airy Vicissitudes. With some Brief Rules of Knowledg [sic] of the Weather at all Times.* London: John Nutt, 1701.

Hunter, Rev. Joseph. *The Diary of Ralph Thoresby, FRS.* London: Henry Colburn and Richard Bentley, 1830.

Hussey, Joseph. *A Warning From The Winds. A Sermon preach'd upon Wednesday, January XIX, 1703/4. Being the Day of Public Humiliation for the Late Terrible and Awakening Storm of Wind, Sent in Great Rebuke upon this Kingdom.* London: William and Joseph Marshal, 1704.

Hutchins, John. *The History and Antiquities of the County of Dorset with some Remarkable Particulars of Natural History.* 2 vols. London: W. Bowyer and J. Nichols, 1774.

Hutchinson, Benjamin. *A Kalendar of the Weather, for the Year 1781. With an Introductory Discourse on the Moon's Influence at Common Lunations in general, and on the Winds at the Eclipses in particular, founded on Observations at Kimbolton.* London: for J. Fielding, 1782.

Huxham, John. "[*Aurora Borealis*] Described in a Letter to the Publisher," *Phil. Trans.* 34 (1726–27): 137–9.

Jenyns, Leonard. *Observations in Natural History: with an Introduction on Habits of Observing, as Connected with the Study of That Science.* London: J. Murray, 1846.

Johnson, Samuel. *The History of Rasselas, Prince of Abissinia.* London: Oxford University Press, 1971 (1759).

Jones, William. *Physiological Disquisitions; or Discourses on the Natural Philosophy of the Elements.* London: J. Rivington and Sons, 1781.

Jurin, James. "Invitatio ad Observatione Meteorologicas Communi Consilio Instituendas," *Phil. Trans.* 31–2 (1720–23): 422–7.

Justice, James. *The British Gardener's Calendar, chiefly Adapted to the Climate of North-Britain etc.* Edinburgh: R. Fleming, 1759.

J. W. *O-Yes from the Court of Heaven to the Northern Nations, by the Streaming Lights that have Appeared of the Late Years in the Air; or Mathematical Reasons Shewing, that the said Lights &c, are no less than Supernatural. Presented to the Royal Society and Others.* London: the Author, 1741.

Kaemtz, L. F. *A Complete Course of Meteorology.* Translated by C. V. Walker. London: Hippolyte Baillière Publisher, 1845.

Kirby, Thomas. *An Analysis of the Electrical Fire.* Chatham: the Author, 1777.

Kirwan, Richard. *An Estimate of the Temperature of Different Latitudes.* London: J. Davis, 1787.

Kirwan, Richard. *A Comparative View of Meteorological Observations Made in Ireland Since the Year 1789, with some Hints towards Forming Prognostics of the Weather.* Dublin: George Bonham, 1794.

Laertius, Diogenes. *Lives of Eminent Philosophers.* Translated by R. D. Hicks. Loeb Classical Library; Cambridge, Mass.: Harvard University Press, 1979.

Lambarde, William. *A Perambulation of Kent, Containing the Description, Hystories, and Customes of that Shyre.* London: Ralphe Newberrie, 1576.

Langwith, Benjamin. "An Account of the Aurora Borealis that appeared October 8, 1726," *Phil. Trans.* 34 (1726–27): 132.

Laurence, John. *Gardening Improved: The Clergyman's Recreation Shewing the Pleasure and Profit of Gardening.* London, 1718.

Leigh, Charles. *The Natural History of Lancashire, Cheshire, and the Peak in Derbyshire, with an Account of the British, Phoenician, Armenian, Greek and Roman Antiquities in those Parts.* Oxford: the Author, 1700.

"Letter to Dr. Sloane, Concerning the Effects of a Great Storm of Thunder and Lightning at Everdon in Northamptonshire (wherein divers Persons were killed)," *Phil. Trans.* 20 (1698): 5–7.

van Leuwenhoek, Anthony. "Observations on the Late Storm," *Phil. Trans.* 24 (1704–05): 1535–37.

[Lhwyd, Edward], "Extracts of Several Letters from Mr. Edward Lhwyd Containing Observations Made on His Travels through Wales and Scotland," *Phil. Trans.* 28 (1713): 99–105.

Lister, Martin. "On the Nature of Earthquakes," *Phil. Trans.* 13 (1683): 512–19.

Lister, Martin. "Some Experiments about Freezing," *Phil. Trans.* 74 (1784): 836–9.

Lord Petre, James. 'A Letter Concerning Some Extraordinary Effects of Lightning," *Phil. Trans.* 42 (1742–43): 136–40.

Lovell, Thomas Lord. "Of a Meteor Seen Near Holkam in Norfolk," *Phil. Trans.* 42 (1742–43): 184–5.

Lover of the Sciences. *A Succint Treatise of Popular Astronomy to which are Subjoined Prognostics of the Weather.* Edinburgh, *c.* 1780.

Lucretius, *De Natura Rerum*. Translated by W. H. D. Rouse. Loeb Classical Library. Cambridge, Mass.: Harvard University Press, 1982.

Lux, William. *Poems on Several Occasions: viz. The garden. The phaenomenon, a Poem on the Late Surprising Meteor, Seen in the Sky March 19th, 1719*. Oxford: Leon Lichfiled, 1719.

Lyon, John. *Account of Several New and Interesting Phenomena, Discovered in [bodies] Killed by Lightning. With Remarks on the Insufficiency of Electricity to Explain Them*. London: James Phillips, 1796.

Marshall, William. *Experiments and Observations Concerning Agriculture and Weather*. London: for J. Dodsley, 1779.

Marshall, William. *Minutes of Agriculture; with Experiments and Observations Concerning Agriculture and Weather*, etc. London: J. Dodsley, 1783.

Marsham, Robert. "Naturalist Journal, 1736–1788," *Phil. Trans.* 79 (1789): 154–6.

Martin, Benjamin. *The Philosophical Grammar; Being a View of the Present State of Experimented Physiology or Natural Philosophy*. London: J. Noon, 1762.

Maunder, William. "A Relation of [the Luminous Appearance], seen at Cruwys Morchard in Devonshire," *Phil. Trans.* 30 (1717–19): 1101–4.

Mawgridge, Robert. "A True and Exact Relation of the Dismal and Surprizing Effects of a Terrible and Unusual Clap of Thunder with Lightning," *Phil. Trans.* 19 (1695–97): 782–3.

Mayne, Zachary. "Concerning a Spout of Water that Happened at Topsham," *Phil. Trans.* 17–18 (1693–94): 28–31.

Mayow, John. *Medico-Physical Works. Being a Translation of Tractatus Quinque Medico-Physici*. Reprint, Edinburgh: Alembic Club, 1907.

The Meteorologist's Assistant in Keeping Diary of the Weather; or the Atmospherical Register of the State of the Barometer etc. London: R. Baldwin, J. Owen, 1793.

"Meteorology from the Intelligence of the Learned Societies," *Philosophical Magazine* 1 (1798): 428.

Miles, Henry. "A Representation of the Parhelia seen in Kent, December 19, 1741," *Phil. Trans.* 42 (1742–43): 16–17.

Miles, Henry. "Of the Storm of Thunder which happened June 12, 1748," *Phil. Trans.* 45 (1748): 383–5.

Miles, Henry. "On Thermometers and the Weather," *Phil. Trans.* 46 (1749–50): 1–6.

Miles, Henry. "Concerning Aurora Borealis, Seen February 16, 1749," *Phil. Trans.* 46 (1749–50): 319.

Miles, Henry. "On the Heat of the Weather at Tooting in July and September Last," *Phil. Trans.* 46 (1749–50): 571–2.

Milles, Jeremiah. "Meteorological Observations for 1768, Made at Bridgewater, Somerset, and Ludgvan, Cornwall," *Phil. Trans.* 59 (1769): 234.

Mills, John, Esq. *An Essay on the Weather; with Remarks on the Shepherd of Banbury Rules for Judging of it's Changes: Intended Chiefly for the Use of Husbandmen*. London: S. Hooper, 1770.

Mitchell, Sir Andrew. "Of an Extraordinary Shower of Black Dust," *Phil. Trans.* 50 (1757–58): 297.

Molyneux, Thomas. "A Relation of the Strange Effects of Thunder and Lightning," *Phil. Trans.* 26 (1708–9): 36.

Moray, Robert. "Extraordinary Tides in the West Isles of Scotland," *Phil. Trans.* 1–2 (1665–67): 53.

More, Henry. *Enthusiasmus Triumphatus*. Reprint; Los Angeles: William Andrews Clark Memorial Library, 1966.

Mortimer, John. *The Whole Art of Husbandry. The Way of Managing and Improving the Land*. London: R. Robinson, 1721.

Morton, Charles. *Compendium Physicae* (1687). *Publications of the Colonial Society of Massachusetts*, Boston. 33 (1940): 3–237.

Morton, John. *Natural History of Northamptonshire with Some Account of the Antiquities, to Which is Annexed a Transcript of Doomsday-Book so Far as It Relates to That County*. London: R. Knaplock, 1712.

Murphy, Patrick. *Meteorology Considered in its Connection with Astronomy, Climate, and the Geographical Distribution of Animals and Plants, equally with the Seasons and Changes of Weather*, London: J. B. Baillière, 1836.

"A Narrative of Several Odd Effects of a Dreadful Thunderclap," *Phil. Trans.* 5–6 (1670–71): 2084.

Neale, Thomas. "A Sad Effect of Thunder and Lightning," *Phil. Trans.* 1–2 (1665–67): 247.

Neve, Richard. *Baroscopologia, or, a Discourse of the Baroscope, or Quicksilver Weather-Glass.* London: W. Keble, 1708.

Neve, Timothy. "[Concerning] the Aurora Australis of March 18, 1738," *Phil. Trans.* 41 (1739–41): 843.

The Newest, Best, and Very Much Esteemed Book of Knowledge. Together with the Husbandman's practice, Also a Brief Discourse of the Natural Causes of Meteors, and Observations on the Weather. London: A. Wilde, 1764.

Newton, Isaac. *Opticks.* Reprint of Third Edition [1723]; New York: Dover, 1979.

Nicholson, J. "An Account of a Storm of Lightning Observed March 1, 1776," *Phil. Trans.* 64 (1774): 350–1.

Nicholson, William. *An Introduction to Natural Philosophy.* 2 vols. London: J. Johnson, 1796.

Nicholson, William. *British Encyclopedia or the Dictionary of Arts and Sciences.* London: C. Whittingham, 1809.

Norden, John the Elder. *Speculum Britanniae. An Historical Description of Middlesex.* [London], 1593.

North Country Diaries. Durham: Surtees Society, 1915.

"Of the Causes of Earthquakes," *Gentleman's Magazine* 20 (1750): 89.

"Of Four Suns and Two Uncommon Rainbows observed in France," *Phil. Trans.* 1–2 (1665–67): 219.

"On a Whirlwind," *Phil. Trans.* 17–18 (1693–94): 192.

Palmer, Samuel, V. D. M., *A Sermon Preach'd on the XIXth day of January. Being a Day of Solemn Fasting and Humiliation, appointed by Her Majesty, on Account of Nation's Sins, the Calamities of the War, and the Late Amazing Storm.* London: John Lawrence, 1704.

"Part of a Letter from Sir John Clark to Roger Gale, Esq.," *Phil. Trans.* 41 (1739–41): 235.

Paul, James B., ed., *The Diary of George Ridpath 1755–1761.* London, 1922.

Pedini, Pasqual R. "An Account of the Earthquakes Felt in Leghorn," *Phil. Trans.* 43 (1742–43): 90–2.

Pennant, Thomas. *The Literary Life of the Late Thomas Pennant, Esq, by Himself.* London: Benjamin and John White, 1793.

The Penny Encyclopedia of the Society for the Diffusion of Useful Knowledge. London: Charles Knight, 1839.

Percival, Philip. "An Account of a Luminous Appearance in the Air at Dublin," *Phil. Trans.* 31–2 (1720–23): 21–2.

Percival, Thomas. "On the Different Quantities of Rain which Fall, at Different Heights, over the Same Spot of Ground; with a Letter from Benjamin Franklin," *Memoirs of the Literary and Philosophical Society of Manchester* 2 (1787): 106–13.

Perry, Mr. "A Letter of Mr. Perry on the Earthquake in Sumatra," *Phil. Trans.* 50 (1757–58): 491–3.

Philotheus, *A True and Particular History of Earthquakes containing A Relation of that Dreadful Earthquake which Happened at Lima and Callao in Peru, October 28, 1746.* London: the Author, 1748.

Pliny the Elder. *Natural History.* Translated by H. Rackham. Loeb Classical Library. Cambridge, Mass.: Harvard University Press, 1986.

Plot, Robert. *The Natural History of Oxfordshire, Being Essay Toward the Natural History of England.* Oxford: Theatre, 1677.

Plot, Robert. *Natural History of Staffordshire.* Oxford: Theatre, 1686.

Pointer, John. *A Rational Account of the Weather. Shewing the Signs of its Several Changes and Alterations, together with the Philosophical Reasons of them.* London: Aaron Ward, 1723.

Posidonius. Edited by L. Edelstein and I. G. Kidd. Cambridge: Cambridge University Press, 1972.

Pringle, John. "Of an Unusual Agitation of the Sea," *Phil. Trans.* 49 (1755–56): 642–4.

Pringle, John. "Several Accounts of the Fiery Meteor, which Appeared on Sunday Nov. 26, 1758 between 8 and 9 at Night," *Phil. Trans.* 51 (1759–60): 218–59.

Pringle, John. "Some Remarks upon the Several Accounts of the Fiery Meteor," *Phil. Trans.* 51 (1759–60): 265–81.

Pryme, Abraham De la. "Concerning a Waterspout Observed by Him in Yorkshire," *Phil. Trans.* 23 (1702–03): 1248–51.

Pryme, Abraham de la. *The Diary*. Durham: Surtees Society, 1869.

Read, John. *A Summary View of the Spontaneous Electricity of the Earth and Atmosphere*. London: the Author, 1795.

Rees, Abraham. *The Cyclopaedia; or Universal Dictionary of Arts, Sciences, and Literature*. London: Longman, 1819.

Renwick, James. *First Principles of Natural Philosophy*. New York, 1856.

Reverend Divine, *A Practical Discourse on the Late Earthquakes with An Historical Account of Prodigies and their Various Effects*. London: J. Dunton, 1692.

Richardson, Richard. "An Account of a Wonderful Fall of Water from a Spout in Lancashire," *Phil. Trans.* 30 (1717–19): 1097–101.

Ringsted, Josiah. *A Diary for Gentlemen, Sportsmen, Gardeners, Grasiers, Game-keepers, Cow-keepers, Horse dealers, Carriers, etc., etc.. with a Column for Observations on the Weather*. London: J. Dixwell, *c.* 1780.

Robertson, Henry. *A General View of the Natural History of the Atmosphere and of its Connection with the Sciences of Medicine and Agriculture*. Edinburgh, 1808.

Robinson, Thomas. *New Observations on the Natural History of the World of Matter and this World of Life. Being a Philosophical Discourse, Grounded Upon Mosaick System of Creation, and the Flood. To which is Added a Treatise on Meteorology*. London: John Newton, 1696.

Rooke, Hayman. *A Meteorological Register, kept at Mansfield Woodhouse. With Probable Indications of Weather*. Nottingham: S. Tupman, 1795.

Rooke, Hayman. *A Continuation of Meteorological Register (1794–1805)*. Nottingham: S. Tupman, 1796.

[Rooke, Hayman.], *A Continuation of the Register of Weather, 1801–2*. Nottingham: S. Tupman, 1802.

Rowning, John. *A Compendious System of Natural Philosophy*. London: John Rivington, 1779.

Ruskin, John. "Remarks on the Present State of Meteorological Science," *Transactions of the Meteorological Society* i (1839): 56–7.

Rutherforth, T. *A System of Natural Philosophy*. Cambridge: J. Bentham, 1748.

Rutty, John. *A Chronological History of the Weather and Seasons and of the Prevailing Diseases in Dublin*. London: Robinson and Roberts, 1770.

Rutty, John. *An Essay Toward a Natural History of the County of Dublin, Accomodated to the Noble Designs of the Dublin Society*. Dublin: W. Sleater for the Author, 1772.

Sarotti, Signore. "Of Red Snow seen in Genoa," *Phil. Trans.* 11–12 (1676–78): 863.

Saul, Edward. *An Historical and Philosophical Account of the Barometer, or Weather-Glass*. London: A. Bettesworth and C. Hitch, 1730.

The Scots Gardiner [sic] for the Climate of Scotland; Together with Gardiner's Calendar [and] Observations on the Weather. Edinburgh: James Reid, 1755.

Seneca. *Naturales Quaestiones*. Translated by Thomas H. Corcoran. Loeb Classical Library. Cambridge, Mass.: Harvard University Press, 1971.

Sheldrake, T[imothy]. *The Cause of Heat and Cold in the Several Climates and Situations of this Globe*. London: the Author, 1756.

Sheldrake, Timothy, *The Gardener's Best Companion in a Greenhouse*. London, 1756.

A Short Essay upon the Cause and Usefulness of the W. and S. W. Wind's Frequent Blowing in England, Liverpool: the Author, 1742.

Short, Thomas. "An Account of Several Meteors, Communicated in a Letter from Thomas Short," *Phil. Trans.* 41 (1739–41): 625–30.

Short, Thomas. *A General Chronological History of the Air, Weather, Seasons, Meteors etc in Sundry Places and Different Times; with Some of their Most Remarkable Effects on Animal (especially Human) Bodies, and Vegetables* 2 vols. London: T. Longman and A. Millar, 1749.

Short, Thomas. *New Observations, Natural, Moral, Civil, Political and Medical on City, Town, and Country Bills of Mortality*. London: T. Longman, 1750.

Short, Thomas. "A Meteorological Discourse," in idem., *A Comparative History of the Increase and Decrease of Mankind in England and Several Countries Abroad*. London: W. Nicoll, 1767.

Short, Thomas. *A Comparative History of the Increase and Decrease of Mankind in England and Several Countries Abroad*. London: W. Nicoll, 1767.

Sinclair, George. *Natural Philosophy; Improven by New Experiments. Touching the Mercurial Glass, the Hygroscope, Eclipsis, Conjuction of Saturn and Jupiter.* Edinburgh: the Author, 1683.

Smellie, John. *The Philosophy of Natural History.* Edinburgh: the Heirs of C. Elliot, 1790–99.

Smith, John. *A Compleat Discourse of the Nature, Use, and Right Managing of that Wonderful Instrument, the Baroscope, or Quick-silver Weather-Glass.* No publisher cited, *c.* 1700.

Spedding James, et. al., eds, *The Works of Francis Bacon.* 8 vols. London: Longmans and Co., 1870.

Spinoza, Baruch. *Tractatus Theologico-Politicus.* Leiden: E. J. Brill, 1989.

Sprat, Thomas. *The History of Royal Society* (1667). Reprint. St Louis: Washington University Press, 1966.

St Clair, Robert. "On an Erruption of Fire," *Phil. Trans.* 21 (1699): 378.

Stanley, Thomas. *The History of Philosophy.* 3 vols. London: Humphrey Moseley, 1660.

Stars and Planets the best Barometers and Truest Interpreters of all Airy Vicissitudes. With some Brief Rules of Knowledg [sic] of the Weather at all Times. By an humble Adorer of God in his Word and Works. London: John Nutt, 1701.

Steinmetz, Andrew. *Sunshine and Showers.* London: Reeve and Co, 1867.

Stennet, Samuel DD. *A Sermon in Commemoration of the Great Storm of Wind on November 27, 1703 and of the More Dreadful Storm Which Threatened the Destruction of British Freedom at the Eve of the Revolution, Preached in Little-Wilde Street, Nov. 27, 1788.* London: J. Buckland, 1788.

Stephens, John. "Of an Uncommon Phenomenon on Dorsetshire," *Phil. Trans.* 52 (1761–62): 108.

Stillingfleet, Benjamin. *Literary Life and Selected Works of Benjamin Stillingfleet,* 2 vols. London: J. Nichols, 1811.

Stillingfleet, Benjamin. *Miscellaneous Tracts Relating to Natural History, Husbandry, and Physick.* London: R. and J. Dodsley, 1762.

Stirling, James. "Of a Remarkable Darkness at Detroit, in America," *Phil. Trans.* 53 (1763): 63.

"Strange Frost about Bristol," *Phil. Trans.* 7–8 (1672–73): 5138.

Stukeley, William. "A Collection of Various Papers Concerning Several Earthquakes," *Phil. Trans.* 46 (1749–50): 602.

Swift, Jonathan. 'A Description of a City Shower,' in Robert A. Greenberg and William Bowman Piper, eds, *The Writings of Jonathan Swift.* New York, London: W. W. Norton, 1973, 518–19.

Swinton, John. "An Account of a Remarkable Meteor seen at Oxford," *Phil. Trans.* 52 (1761–62): 99–101.

Swinton, John. "An Account of an Anthelion Observed Near Oxford," *Phil. Trans.* 52 (1761–62): 94.

Swinton, John. "Of a Remarkable Meteor ["a resplendant whiteness of the heaven"] seen at Oxford, March 5, 1764," *Phil. Trans.* 54 (1764): 326–9.

Swinton, John. "An Account of Two Aurorae Boreales Observed in Oxford," *Phil. Trans.* 59 (1769): 367–70.

Taylor, Joseph. *The Complete Weather Guide: A Collection of Practical Observations for Prognostication the Weather, Drawn from Plants, Animals, Inanimate Bodies, and also by Means of Philosophical Instruments; Including Shepherd and Banbury Rules.* London: John Harding, 1812.

Taylor, Robert. "Account of a Great Hail-Storm in Herefordshire," *Phil. Trans.* 19 (1695–97): 577–9.

Temple, Henry. "Concerning an Earthquake at Naples," *Phil. Trans.* 41 (1739–41): 340–3.

Thomson, David Purdie. *Introduction to Meteorology.* Edinburgh and London: W. Blackwood, 1849.

Thomson, Thomas. *History of the Royal Society.* London: Murray, 1812.

Thoresby, Ralph. "An Account of a Young Man Slain with Thunder and Lightning," *Phil. Trans.* 21 (1699): 51–4.

Thoresby, Ralph. "Of an Accident by Thunder and Lightning at Leeds," *Phil. Trans.* 22 (1700–01): 577–9.

Thoresby, Ralph. "A Letter Giving an Account of a Lunar Rainbow Seen in Darbishire, and of a Storm of Thunder and Lightning which Happened near Leeds in Yorkshire," *Phil. Trans.* 27 (1710–12): 320–1.

Thoresby, Ralph. "An Account of the Damage done by a Storm of Hail, which Happened near Rotherham in Yorkshire on June 7, 1711," *Phil. Trans.* 27 (1710–12): 514.

Thoresby, Ralph. *Ducatus Leodiensis: Or the Topography of the Ancient and Populous Town and Parish of Leeds, and Parts Adjacent in the West-Riding of the County of York.* London: Maurice Atkins, 1715.

Thoresby, Ralph. "The Effects of a Violent Shower of Rain in Yorkshire," *Phil. Trans.* 31–2 (1720–23): 101.

Time's Telescope Universal and Perpetual [and] a Brief Discourse of all Kinds of Meteors, or Appearances in Heavens. Natural Prognosticks of Weather. London: J. Wilcox, J. Oswald, 1734.

Toaldo, M. "On the Signs Exhibited by Animals which Indicate Changes of the Weather, with Remarks on other Prognostications," *Philosophical Magazine* 4 (1799): 367–75.

Todd, Thomas Philomath. *Perpetuum Kalendarum Astronomicum: or, a Perpetual Astronomical Kalendar.* Edinburgh: Thomas Lumisden and John Robertson, 1738.

Tonge, Ezekiel. "Some Observations Concerning the Variety of the Running of Sap in Trees," *Phil. Trans.* 5–6 (1670–71): 2070–1.

Townley, Richard. *A Journal Kept in the Isle of Man, Giving an Account of the Wind, Weather, and Daily Occurences, for Upwards of* 11 *months.* 2 vols. Oxford: J. and J. Merrill, 1791.

Virgil. *The Works of Virgil.* Philadelphia: Claxton, 1870.

Virgil. Translated by H. Rushton Fairclough. Cambridge, Mass.: Harvard University Press, 1940.

Walker, Adam. *A System of Familiar Philosophy.* London: the Author, 1802.

Walker, John. *Lectures on Geology, Including Hydrography, Mineralogy, and Meteorology, with an Introduction to Biology. Edited with Notes and Introduction by Harold W. Scott.* Chicago: The University of Chicago Press, 1966.

Wallis, John. "Relation of an Accident by Thunder and Lightning at Oxford," *Phil. Trans.* 1–2 (1665–67): 222.

Wallis, John. "Of an Unusual Meteor," *Phil. Trans.* 11–12 (1676–78): 863.

Wallis, John. "The Effects of a Great Storm of Thunder and Lightning at Everdon in Northamptonshire," *Phil. Trans.* 20 (1698): 5.

Wallis, John. "On the Production and Effects of Hail, Thunder, and Lightning," *Phil. Trans.* 19 (1695–97): 653–65.

Ward, Edward. *British Wonders: or, A Poetical Description of the Several Prodigies and Most Remarkable Accidents that have Happened in Britain since the Death of Queen Anne.* London: John Morphew, 1717.

Wasse, Joseph. "Two Letters on the Effects of Lightning," *Phil. Trans.* 33 (1724–25): 366–70.

Whiston, William, M. A. *An Account of a Surprizing Meteor, Seen in the Air, March 1715/6 at Night.* London: J. Senex, 1716.

Whiston, William. *An Account of a Surprizing Meteor, Seen in the Air March 19, 1718/1719, at Night.* London: W. Taylor, 1719.

Whiston, William. "An Account of Two Mock-Suns, and an Arc of a Rainbow Inverted, with a Halo," *Phil. Trans.* 31 (1720–23): 212.

White, Gilbert. *Natural History of Selborne* (1790). Harmondsworth, Middlesex: Penguin, 1977.

Widdowes, Daniel. *Naturall Philosophy, or A Description of the World.* London: J. Dowson, 1621.

Williams, Anthony. "Of a Remarkable Thunder-Storm," *Phil. Trans.* 61 (1771): 71–2.

Williams, John. *The Climate of Great Britain; or Remarks on the Change it has Undergone, Particularly within the Last Fifty Years.* London: C. and R. Baldwin, 1806.

Willsford, Thomas. *Natures Secrets, or, the Admirable and Wonderfull History of the Generation of Meteors.* London: Brooke, 1658.

Wilson, Alexander. "Of a Remarkable Cold Observed at Glasgow," *Phil. Trans.* 61 (1771): 326.

Winthorp, John. "Of a Meteor Seen in New England," *Phil. Trans.* 52 (1761–62): 6–8.

Winthorp, John. "Of Several Fiery Meteors," *Phil. Trans.* 54 (1764): 185–6.

W. K., *News from Hereford; or, a Wonderful and Terrible Earthquake.* London, 1661.

Woodward, John. *An Essay Toward a Natural History of the Earth and Terrestrial Bodies.* London: Richard Wilkin, 1695.

Woodward, John. *Brief Instructions for Making Observations.* London: Richard Wilkin, 1696.

Woodward, John. *The Natural History of the Earth.* London: Thomas Edlin, 1726.

Yorkshire Diaries and Autobiographies in the Seventeenth and Eighteenth Centuries. Durham: the Surtees Society, 1877.

Young, Arthur. *Arthur Young's Tour in Ireland 1776–7.* London and New York: G. Bell and Sons, 1892.

Zuccolo, Vitale. *Dialogo delle Cose Meteorologiche.* Venetia: Paolo Megietti, 1590.

Secondary sources

Allen, Don Cameron. *The Star-Crossed Renaissance*. Durham, N. C.: Duke University Press, 1941.

Allen, D. E. "Natural History in Britain in the Eighteenth Century," *Archives of Natural History* 20 (1993): 333–47.

Allen, David E. *The Naturalist in Britain: A Social History*. Princeton N. J.: Princeton University Press, 1994.

Allen, Phyllis. "Scientific Studies in the English Universities of the Seventeenth Century," *Journal of the History of Ideas* 10 (1949): 219–53.

Anderson, J. L. "Combined Operations and the Protestant Wind: Some Maritime Aspects of the Glorious Revolution of 1688," *The Great Circle* 5 (1983): 13–23.

Armitage, Angus. *Edmund Halley*. London: Nelson, 1966.

Arnold, Ralph. *Northern Lights: the Story of Lord Derwentwater*. London: Constable, 1959.

Ashworth, William B. Jr. "Natural History and the Emblematic World View," in David C. Lindberg and Robert S. Westman, eds, *Reappraisals of the Scientific Revolution*. Cambridge: Cambridge University Press, 1990, 303–33.

Asmis, Elizabeth. *Epicurus' Scientific Method*. Ithaca and London: Cornell University Press, 1984.

Atkinson, A. D. "William Derham, FRS (1657–1735)," *Annals of Science* 8 (1952): 368–92.

Atkinson, D. H. *Ralph Thoresby the Topographer; his Town and Times*. Leeds: Walker and Laycock, 1885.

Averley, Gwendoline. *English Scientific Societies of the Eighteeenth and Early Nineteeenth Centuries*. Ph.D. thesis. Teesside Polytechnic, 1989.

Baker, J. N. L. *The History of Geography*. New York: Barnes and Noble, Inc., 1963.

Barrell, John. *The Idea of Landscape and the Sense of Place 1730–1840*. Cambridge: Cambridge University Press, 1972.

Baskerville, Stephen. *Not Peace but a Sword: The Political Theology of the English Revolution*. London: Routledge, 1993.

Baudrillard, Jean. *In the Shadow of the Silent Majorities, or the End of the Social*. New York: Semiotext(e), 1983.

Beier, Lucinda McCray. "Experience and Experiment: Robert Hooke, Illness and Medicine," in Michael Hunter and Simon Schaffer, eds, *Robert Hooke. New Studies*. Woodbridge: The Boydell Press, 1989.

Benedict, Barbara. "The 'Curious Attitude' in Eighteenth-Century Britain: Observing and Owning," *Eighteenth-Century Life* 14 (1990): 59–98.

Bennett, G. V. *The Tory Crisis in Church and State*. Oxford: Oxford University Press, 1975.

Beresford, John, ed., *The Diary of a Country Parson, The Rev. James Woodforde, vol. 2, 1782–1787*. London: Humphrey Milford, 1926.

Birket, Kirsten and Oldroyd, David. "Robert Hooke, Physico-Mythology, Knowledge of the World of the Ancients and Knowledge of the Ancient World," in Stephen Gaukgroger, ed., *The Uses of Antiquity*. Kluwer Academic Press, 1991, 145–70.

Black, Jeremy. *Natural and Necessary Enemies: Anglo-French Relations in the Eighteenth Century*. Athens: University of Georgia Press, 1986.

Blair, Ann. "Humanist Methods in Natural Philosophy: The Commonplace Book," *Journal of the History of Ideas* 53 (1992): 541–51.

Borsay, Peter. *The English Urban Renaissance: Culture and Society in the Provincial Town 1660–1770*. Oxford: Clarendon Press, 1989.

Bourde, Andre. *The Influence of England on the French Agronomes, 1750–1789*. Cambridge: Cambridge University Press, 1953.

Boyer, Carl B. *The Rainbow: From Myth to Mathematics*. London: Thomas Yoseloff, 1959.

Bowen, Margarita. *Empiricism and Geographical Thought from Francis Bacon to Alexander von Humboldt*. Cambridge: Cambridge University Press, 1981.

Brekke, Asgeir and Egeland, Alv. *The Northern Light: From Mythology to Space Research*. Berlin: Springer Verlag, 1983.

Briggs, J. Morton, Jr. "Aurora and Enlightenment: Eighteenth-Century Explanations of Aurora Borealis," *Isis* 58 (1967): 491–503.

Brownell, Morris R. *Alexander Pope and the Arts of Georgian England.* Oxford: Clarendon Press, 1978.

Buell, Llewellyn M. "Elizabethan Portents: Superstition or Doctrine?" in idem., *Essays Critical and Historical.* Berkeley and Los Angeles: University of California Press, 1950, 27–41.

Burke, John G. *Cosmic Debris. Meteorites in History.* Berkeley: University of California Press, 1986.

Burnham, Frederick B. "The More-Vaughan Controversy: The Revolt Against Philosophical Enthusiasm," *Journal of the History of Ideas* 35 (1974): 33–49.

Burton, J. M. C. "Meteorology and the Public Health Movement in London During the Late Nineteenth Century," *Weather* 45 (1990): 300–7.

Bushaway, Bob. *By Rite: Custom, Ceremony, and Community in England, 1700–1800.* London: Junction Books, 1982.

Camden, Carroll Jr. "Elizabethan Almanacs and Prognostications," *The Library* 12 (1931), 100–8.

Campbell, Colin. "Understanding Traditional and Modern Patterns of Consumption in Eighteenth-century England: A Character-Action Approach," in John Brewer and Roy Porter, eds, *Consumption and the World of Goods,* London: Routledge, 1994, 40–58.

Cannon, Susan Faye. *Science in Culture: The Early Victorian Period.* New York: Dawson and Science History Publication, 1978.

Capp, Bernard. *English Almanacs, 1500–1800: Astrology and the Popular Press.* Ithaca, NY: Cornell University Press, 1979.

Carter, Harold B. *Sir Joseph Banks, 1743–1820.* London: British Museum, 1988.

Clark, Jerome. *The UFO Encyclopedia. The Emergence of a Phenomenon: UFOs from the Beginning through 1959.* Detroit: Omnigraphics Inc, 1993.

Clark, J. C. D. *English Society, 1688–1832: Ideology, Social Structure and Political Practice during the Ancien Régime.* Cambridge: Cambridge University Press, 1985.

Clark, J. C. D. *Revolution and Rebellion. State and Society in England in the Seventeenth and Eighteenth Centuries.* Cambridge University Press, 1986.

Clark, J. C. D. *Samuel Johnson: Literature, Religion, and English Cultural Politics from the Restoration to Romanticism.* Cambridge: Cambridge University Press, 1994.

Clark, Peter. *Small Towns in Early Modern Europe.* Cambridge: Cambridge University Press, 1995.

Clark, Sandra. *The Elizabethan Pamphleteers: Popular Moralistic Pamphlets: 1580–1640.* Rutherford, N. J.: Humanities Press, 1983.

Clark, Stuart. "The Rational Witchfinder: Conscience, Demonological Naturalism and Popular Superstitions," in Stephen Pumfrey, et al., eds, *Science, Culture and Popular Belief in Renaissance Europe.* Manchester: Manchester University Press, 1991, 222–49.

Clark, Stuart. *Thinking with Demons: The Idea of Witchcraft in Early Modern Europe.* Oxford: Clarendon Press, 1997.

Cooter, Roger and Pumfrey, Stephen. "Separate Spheres and Public Places: Reflections on the History of Scientific Popularizations and Science in Popular Culture," *History of Science* 32 (1994): 237–67.

Corbin, Alain. *The Lure of the Sea. The Discovery of the Seaside in the Western World 1750–1840.* London: Polity Press, 1994.

Costello, William T. *The Scholastic Curriculum at Early Seventeenth-Century Cambridge,* Cambridge, Mass.: Harvard University Press, 1958.

Cressy, David *Literacy and the Social Order: Reading and Writing in Tudor and Stuart England.* Cambridge: Cambridge University Press, 1980.

Crosland, Maurice P. "The Development of Chemistry in the Eighteenth Century," *Studies on Voltaire and the Eighteenth Century* 24 (1963): 369–441.

Cruickshanks, Eveline, ed., *Ideology and Conspiracy: Aspects of Jacobitism, 1689–1759.* Edinburgh: John Donald Publishers, 1982.

Curry, Patrick. *Prophecy and Power. Astrology in Early Modern England.* Cambridge: Polity Press, 1989.

Curry, Patrick. "Astrology in Early Modern England: The Making of a Vulgar Knowledge," in Stephen Pumfrey et al., eds, *Science, Culture and Popular Belief in Renaissance Europe.* Manchester: Manchester University Press, 1991, 274–9.

Dahm, John J. "Science and Apologetics in the Early Boyle Lectures," *Church History* 39 (June 1970): 176.

Daniels, Stephen. *Fields of Vision: Landscape Imagery and National Identity in England and the United States*. Cambridge: Cambridge University Press, 1993.

Daston, Lorraine. "Historical Epistemology," in James Chandler, Arnold I. Davidson, and Harry Harootunian, eds, *Questions of Evidence. Proof, Practice, and Persuasion across the Disciplines*. Chicago: The University Chicago Press, 1994, 282–9.

Daston, Lorraine. "Marvelous Facts and Miraculous Evidence in Early Modern Europe," *Critical Inquiry* 18 (1991): 93–124.

Davies, Gordon L. *The Earth in Decay. A History of British Geomorphology 1578–1878*. New York: American Elsevier Publishing Company, Inc, 1969.

Davies, Horton. *Like Angels from a Cloud: The English Metaphysical Preachers, 1588–1645*. San Marino: Huntington Library, 1986.

Dawson, Warren R., ed., *The Banks Letters*. London: British Museum, 1958.

Dear, Peter. "*Totius in Verba*. Rhetoric and Authority in the Early Royal Society," *Isis* 76 (1985): 145–61.

Dear, Peter. "Miracles, Experiments, and the Ordinary Course of Nature," *Isis* 81 (1990): 671.

Dear, Peter. *Discipline and Experience: The Mathematical Way in the Scientific Revolution*. Chicago: The University of Chicago Press, 1995.

De Bruyn, Frans. "From Virgilian Georgic to Agricultural Science: An Instance in the Transvaluation of Literature in Eighteenth-Century Britain," in Albert J. Rivero, ed., *Augustan Subjects. Essays in Honor of Martin C. Battestin*, Delaware, 1997.

Debus, Allen G. "The Paracelsian Aerial Niter," *Isis* 55 (1964): 43–61.

Debus, Allen G. *Science and Education in the Seventeenth Century*. London: Macdonald, 1970.

Desaive, J. P., et al., *Médecins, climat and épidémies à la fin du XVIIIe siècle*. Paris: CNRS, 1972.

Dewhurst, Kenneth. *Dr. Thomas Sydenham (1624–1689)*. Berkeley and Los Angeles: University of California Press, 1966.

Dictionary of National Biography, eds, Lord Blake and C. S. Nichols. Oxford: Oxford University Press, 1986.

Dictionary of Scientific Biography. New York: Scribner, 1970.

Dod, Bernard G. "Aristoteles Latinus," in the *Cambridge Companion of Later Medieval Philosophy*. Cambridge: Cambridge University Press, 1982, 45–80.

Donaldson, John. *Agricultural Biography*. London: the Author, 1854.

Donovan, Arthur L. *Philosophical Chemistry in the Scottish Enlightenment: The Doctrines and Discoveries of William Cullen and Joseph Black*. Edinburgh: Edinburgh University Press, 1975.

Douglas, David C. *English Scholars*. London: Jonathan Cape, 1939.

Dundes, Alan. *The Study of Folklore*. New Jersey, 1965.

Duthie, R. E. "English Florists. Societies and Feasts in the Seventeenth and First Half of the Nineteenth Centuries," *Garden History* 10 (1982): 17–35.

Eamon, William. *Science and the Secrets of Nature: Books of Secrets in Medieval and Early Modern Culture*. Princeton: Princeton University Press, 1994.

Eisenstein, Elizabeth. *The Printed Press as an Agent of Change: Communications and Cultural Transformations in Early Modern Europe*. Cambridge: Cambridge University Press, 1979.

Elliade, Mircea, ed., *Encyclopedia of Religion*. New York: Macmillan, 1982.

Encyclopedia Britannica. 15th edn. Chicago: William Benton Publisher, 1979.

Estabrook, Carl. *Urbane and Rustic England*. Manchester: Manchester University Press, 1998.

Everett, Nigel. *The Tory View of Landscape*. New Haven and London: Yale University Press, 1994.

Everitt, Alan. *Landscape and Community in England*. London and Ronceverte: The Hambledon Press, 1985.

Everitt, Alan. "Country, County and Town: Patterns of Regional Evolution in England," in Peter Borsay ed., *The Eighteenth-Century Town*. London: Longman, 1990, 83–116.

Fara, Patricia. "'A Treasure of Hidden Vertues': the Attraction of Magnetic Marketing," *British Journal for the History of Science* 28 (1995): 5–35.

Fara, Patricia. "Lord Derwentwater's Light: Prediction and the Aurora Polaris," *Journal for the History of Astronomy* 27 (1996): 239–58.

Fara, Patricia. *Sympathetic Attractions*. Princeton, N. J.: Princeton University Press, 1996.

Fassig, Oliver L., ed., *Bibliography of Meteorology*. Washington D. C.: US Signal Corps, 1889–91.

Fawcet, T. "Measuring the Provincial Enlightenment: The Case of Norwich," *Eighteenth-Century Life*, New Series 18 (1982): 15–27.

Feingold, Richard. *Nature and Society: Later Eighteenth-Century Uses of the Pastoral and Georgic*. New Brunswick, N. J.: Rutgers University Press, 1978.

Feldman, Theodore S. "The History of Meteorology, 1750–1800: A Study in the Quantification of Experimental Physics." *Diss. Abstr. Int.*, 45 (1984): 922-A.

Feldman, Theodore. "Applied Mathematics and the Quantification of Experimental Physics: The Example of Barometric Hypsometry," *Historical Studies in the Physical Sciences* 15 (1985): 127–97.

Feldman, Theodore S. "Late Enlightenment Meteorology," in Frängsmyr, Tore, Heilbron, J. L. and Rider, Robin E., eds, *The Quantifying Spirit in the 18th Century*. Berkeley/Los Angeles/Oxford: University of California Press, 1990, 143–97.

Ferguson, Arthur B. *Clio Unbound: Perception of the Social and Cultural Past in Renaissance England*. Durham, N. C.: Duke University Press, 1979.

Fleming, James R. *Meteorology in America, 1800–1870*. Baltimore and London: The Johns Hopkins University Press, 1990.

Fletcher, Anthony and Stevenson, John, eds, *Order and Disorder in Early Modern Enlgand*. Cambridge: Cambridge University Press, 1985.

Force, James. *William Whiston. An Honest Newtonian*. Cambridge: Cambridge University Press, 1985.

Force, James E. "Hume and the Relation of Science to Religion among Certain Members of the Royal Society," in John W. Yolton, ed., *Philosophy, Religion and Science in the Seventeenth and Eighteenth Centuries*. Rochester: Rochester University Press, 1990.

Foster, G. M. *Gilbert White and His Records*. London: Christopher Helm, c. 1988.

Fothergill, Robert A. *Private Chronicles. A Study of the English Diaries*. London: Oxford University Press, 1974.

Frank, Robert G. Jr. "Science, Medicine and the Universities of Early Modern England: Background and Sources, Part I," *History of Science* 11 (1973): 194–216.

Frank, R. G. *Harvey and the Oxford Physiologists: A Study of Scientific Ideas and Social Interaction*. Berkeley and Los Angeles: University of California Press, 1980.

Frei, Hans. *Eclipse of Biblical Narrative: A Study in Eighteenth and Nineteenth Century Hermeneutics*. New Haven and London: Yale University Press, 1978.

French, Roger. *Ancient Natural History: Histories of Nature*. London: Routledge, 1994.

Frisinger, H. Howard. "Aristotle's Legacy in Meteorology," *Bulletin of the American Meteorological Society* 54 (1973): 198–204.

Frisinger, H. Howard. *The History of Meteorology: To 1800*. New York: Science History Publications, 1977.

Fritz, Paul and Williams, David, eds, *The Triumph of Culture: 18th Century Perspective*. Toronto: A. M. Hakkert Ltd., 1972.

Froom, Le Roy Edwin. *Prophetic Faith of Our Fathers: The Historical Development of Prophetic Interpretation*. Washington: Review and Herald, 1946–54.

Fulford, Tim. *Landscape, Liberty and Authority*. Cambridge: Cambridge University Press, 1996.

Fussell, G. E. *The Old English Farming Books: From Fitzherbert to Tull*. London: Crosby Lockwood, 1947.

Fussell, G. E. *The Classical Tradition in West European Farming*. Rutherford: Fairleigh Dickinson University Press, 1972.

Gascoigne, John. "A Reappraisal of the Role of the Universities in the Scientific Revolution," in David C. Lindberg and Robert Westman, eds, *Reappraisals of the Scientific Revolution*. Cambridge: Cambridge University Press, 1990, 207–61.

Geneva, Ann. *Astrology and the Seventeenth-Century Mind: William Lilly and the Language of Stars*. Manchester: Manchester University Press, 1995.

Genuth, Sara Schechner. *Comets, Popular Culture and the Birth of Modern Cosmology*. Princeton, N. J.: Princeton University Press, 1997.

Gilbert, E. W. "The Idea of the Region," *Geography* 45 (1960): 157–75.

Gilbert, Otto. *Die Meteorologischen Theorien des Griechischen Altertums*. Leipzig: Teubner, 1907.

Gillispie, Charles Coulston. *Genesis and Geology*. Cambridge, Mass.: Harvard University Press, 1951.

Goldgar, Anne. *Impolite Learning: Conduct and Community in the Republic of Letters 1680–1750*. New Haven: Yale University Press, 1995.

Golinski, Jan. "Utility and Audience in Eighteenth-century Chemistry: Case Studies of William Cullen and Joseph Priestley," *British Journal for the History of Science* 68 (1988): 1–32.

Golinski, Jan. *Science as Public Culture*. Cambridge: Cambridge University Press, 1992.

Golinski, Jan. "Barometers of Change: Meteorological Instruments as Machines of Enlightenment," in William Clark, Jan Golinski, and Simon Schaffer, eds, *The Sciences in Enlightened Europe*. Chicago: The University of Chicago Press, 1999, 69–93.

Goodison, Nicholas. *English Barometers 1680–1860. A History of Domestic Barometers and their Makers*. New York: Clarkson N. Potter, Inc, 1968.

Guerlac, Henry. "The Poet's Nitre," *Isis* 45 (1954): 243–55.

Gunther, R. T. *Early Science in Oxford*. 4 vols. London: Dawson of Pall Mall, 1925.

Hale, J. R. "Gunpowder and the Renaissance: An Essay in the History of Ideas," in idem., *Renaissance War Studies*. London: The Hambledon Press, 1983.

Hall, Marie Boas. "Henry Miles, FRS (1698–1763) and Thomas Birch, FRS (1705–1766)," *Notes and Records of the Royal Society* 18 (1963): 39–44.

Hamlin, Christopher. "Chemistry, Medicine, and the Legitimization of English Spas, 1740–1840," *Medical History*, Suppl. No. 10 (1990): 67–81.

Hammond, Lansing V. "Gilbert White, Poetizer of the Commonplace," in Frederick W. Hilles, ed., *The Age of Johnson. Essays Presented to Chauncey Brewster Tinker*. New Haven: Yale University Press, 1949, 377–83.

Hankins, Thomas L. *Science and the Enlightenment*. Cambridge: Cambridge University Press, 1985.

Harris, Tim. *London Crowds in the Reign of Charles II: Propaganda and Politics from the Restoration until the Exclusion Crisis*. Cambridge: Cambridge University Press, 1987.

Harrison, J. F. C. *The Second Coming. Popular Millenarianism, 1750–1850*. New Brunswick, N. J.: Rutgers University Press, 1979.

Harrison, J. F. C. *The Common People*. London: Croom Helm, 1984.

Hayward, Geoffrey D. "Home as an Environmental and Psychological Concept," *Landscape* 20: 1 (1975).

Heilbron, J. L. "Franklin, Haller and Franklinist History," *Isis* 68 (1977): 539–49.

Helgerson, Richard. *Forms of Nationhood. The Elizabethan Writing of England*. Chicago and London: The University of Chicago Press, 1992.

Hellmann, Gustav. *Repertorium der Deutschen Meteorologie*. Leipzig, 1883.

Hellmann, Gustav. *Beiträge zur Geschichte der Meteorologie*. Berlin: Behrend & Co, 1917.

Hellman, Gustav. *Meteorologische Beobachtungen vom XIV bis XVII Jahrhundert*. Berlin, 1901, Nendeln, Lichenstein: Kraus Reprint, 1969.

Heninger, S. K. *A Handbook of Renaissance Meteorology*. Durham, N. C.: Duke University Press, 1978.

Heyd, Michael. "The New Experimental Philosophy: A Manifestation of 'Enthusiasm' or an Antidote to it?" *Minerva* 25 (1987): 423–40.

Heyd, Michael. *'Be Sober and Reasonable': The Critique of Enthusiasm in the Seventeenth and Early Eighteenth Centuries*. Leiden: E. J. Brill, 1995.

Hill, Christopher. *God's Englishman*. New York: Penguin Books, 1970.

Hill, Christopher. "'Till the Conversion of the Jews'," in Richard Popkin, ed., *Millenarianism and Messianism in English Literature and Thought, 1650–1800*. Leiden: E. L. Brill, 1988, 12–37.

Hindle, Steve. "Custom, Festival and Protest in Early Modern England: The Little Budworth Wakes, St Peter's Day, 1596," *Rural History* 6 (1995): 155–78.

Holmes, Geoffrey. *British Politics in the Age of Anne*. New York: Macmillan, 1967.

Holmes, Geoffrey. "The Sacheverell Riots: The Crowd and the Church in Early Eighteenth Century London," *Past and Present* 72 (1976): 42.

Holmes, Geoffrey. "Science and the English Revolution in the Age of Newton," *The British Journal for the History of Science* 38 (1978): 164–71.

Holmes, Geoffrey. *Augustan England: Professions, State and Society*. London: George Allen and Unwin, 1982.

Holmes, Geoffrey. *Politics, Religion and Society in England 1679–1742*. London and Ronceverte: The Humbledon Press, 1986.

Holmes, Geoffrey and Szechi, Daniel. *The Age of Oligarchy*. London: Longman, 1993.

Holt-White, Rashleigh. *The Life and Letters of Gilbert White of Selborne*. London: John Murray, 1901.

Hoppen, K. Theodore. *The Common Scientist in the Seventeenth Century. A Study of the Dublin Philosophical Society 1683–1708*. Charlottesville: The University Press of Virginia, 1970.

Hoppen, K. Theodore. "The Nature of the Early Royal Society," *British Journal for the History of Science* 9 (1976): 6.

Hoskins, W. G. *Local History in England*. London: Longman, 1959.

Hughes, Arthur. "Science in British Encyclopaedias, 1704–1875. I," *Annals of Science* 7 (1951): 340–70.

Hughes, Arthur. "Science in English Encyclopaedias, 1704–1875. III. Meteorology," *Annals of Science* 9 (1953): 233–64.

Hunter, J. Paul. *Before Novels. The Cultural Contexts of Eighteenth-Century English Fiction*. New York, London: W. W. Norton & Company, 1992.

Hunter, Michael. *John Aubrey and The Realm of Learning*. London: Duckworth, 1975.

Hunter, Michael. *Science and Society in Restoration England*. Cambridge: Cambridge University Press, 1981.

Hunter, Michael and Gregory, Annabel, eds, *An Astrological Diary of the Seventeenth Century. Samuel Jeake of Rye 1652–1699*. Oxford: Clarendon Press, 1988.

Hunter, Michael. *Establishing New Science. The Experience of the Early Royal Society*. Woodbridge: The Boydell Press, 1989.

Hunter, Michael. *Science and the Shape of Orthodoxy. Intellectual Change in Late Seventeenth-Century Britain*. Woodbridge: The Boydel Press, 1995.

Illife, Robert. "'The Puzzleing Problem': Isaac Newton and the Political Physiology of Self," *Medical History* 39 (1995): 433–58.

Israel, Jonathan I. and Parker, Geoffrey. "Of Providence and Protestant Winds: The Spanish Armada of 1688," in Jonathan I. Israel, ed., *The Anglo-Dutch Moment*. Cambridge: Cambridge University Press, 1991, 335–63.

Jacob, Margaret. *The Newtonians and the English Revolution 1689–1720*. Ithaca: Cornell University Press, 1976.

Jacob, Margaret C. "Scientific Culture in the Early English Enlightenment: Mechanisms, Industry, and Gentlemanly Facts," in Alan Charles Kors and Paul J. Korshin, eds, *Anticipations of the Enlightenment in England, France, and Germany*. Philadelphia: University of Pennsylvania Press, 1987.

Jacob, Margaret. *Scientific Culture and the Making of the Industrial West*. Oxford: Oxford University Press, 1997.

Jacob, W. M. *Lay People and Religion in the Early Eighteenth Century*. Cambridge: Cambridge University Press, 1996.

James, Mervyn. *English Politics and the Concept of Honour, 1482–1647*. Oxford: Oxford University Press, 1978.

Jankovic, Vladimir. "Ideological Crests versus Empirical Troughs: John Herschel's and William Radcliffe Birt's Research on Atmospheric Waves," *British Journal for the History of Science* 31 (1998): 21–40.

Jankovic, Vladimir, "The Place of Nature and the Nature of Place: The Chorographic Challenge to the History of British Provincial Science," *History of Science* 38 (2000): 79–113.

Jaritz, Gerhardt and Winiwarter, Verena. "On the Perception of Nature in a Renaissance Society," in Mikulas Teich, Roy Porter and Bo Gustafson, eds, *Nature and Society in Historical Context*. Cambridge: Cambridge University Press, 1997, 91–112.

Jermyn, L. A. S. "Virgil's Agricultural Lore," *Greece and Rome* 53 (1949): 49–69.

Jermyn, L. A. S. "Weather-Signs in Virgil," *Greece and Rome* 58 (1951): 28–59.

Jones, Richard Foster. *Ancients and Moderns*. St Louis: Washington University Press, 1961.

Jones, William Powell. *Thomas Gray, Scholar*. Cambridge, Mass.: Harvard University Press, 1937.

Jordanova, L. J. "Earth Sciences and Environmental Medicine," in Ludmilla Jordanova and Roy Porter, eds, *Images of the Earth: Essays in the History of the Environmental Sciences*. Chalfont St Giles: British Society for the History of Science, 1979, 119–46.

Jordanova, L. J. *Languages of Nature: Critical Essays on Science and Literature*. New Brunswick, NJ: Rutgers University Press, 1986.

Kearney, Hugh. *Scholars and Gentlemen. Universities and Society in Pre-Industrial Britain, 1500–1700*. London: Faber and Faber, 1970.

Keith, W. J. *The Rural Tradition. A Study of the Non-Fiction Prose Writers of the English Countryside*. Toronto and Buffalo: University of Toronto Press, 1974.

Keith, W. J. *Regions of the Imagination: The Development of British Rural Fiction*. Toronto and Buffalo: University of Toronto Press, 1988.

Kelly, Suzanne, O. S. B., *The De Mundo of William Gilbert*. Amsterdam: Menno Hertzberger & Co., 1965.

Kendrick, T. D. *The Lisbon Earthquake*. New York, Philadelphia: J. B. Lippincott Company, 1955.

Kent, Nathaniel. *Hints to Gentlemen of Landed Property*. London: J. Dodsley, 1773, 1793.

Ketton-Cremer, R. W. *Thomas Gray, A Biography*. Cambridge: Cambridge University Press, 1955.

Key, Newton E. "The Localism of the County Feast in Late Stuart Political Culture," *Huntington Library Quarterly* 58 (1995): 211–37.

Khrgian, A. Kh. *Meteorology: A Historical Survey*. The Israel Program for Scientific Translations, Jerusalem, 1970.

Kidd, I. G. "Theophrastus' *Meteorology*, Aristotle and Posidonius," in William Fortenbaugh and Dimitri Gutas, eds, *Theophrastus: His Psychological, Doxographical and Scientific Writings*. New Brunswick and London: Transactions Publishers, 1992, 294–300.

King-Hele, Desmond. "Erasmus Darwin, Grandfather of Meteorology," *Weather* 28 (1973): 240–50.

Kington, J. A. "The Societas Meteorologica Palatina: An Eighteenth-Century Meteorological Society," *Weather* 29 (1974): 416–26.

Kington, J. *The Weather of the 1780s over Europe*. Cambridge: Cambridge University Press, 1978.

Kington, John. *The Weather Journals of a Rutland Squire: Thomas Barker of Lyndon Hall*. Rutland Record Society: Oakham, 1988.

Kuklick, Henrietta and Kohler, Robert E., eds, *Science in the Field*. Ithaca, N. Y.: Cornell University, 1996.

Lake, Peter. "Popular Form, Puritan Content? Two Puritan Appropriations of the Murder Pamphlet from Mid-Seventeenth-century London," in Anthony Fletcher, and Peter Roberts, eds, *Religion, Culture, and Society in Early Modern England*. Cambridge: Cambridge University Press, 1994.

Lamb, Hubert and Frydenhal, Knud. *Historic Storms of the North Sea, British Isles and Northwest Europe*. Cambridge: Cambridge University Press, 1992.

Landau, Norma. *The Justices of Peace, 1679–1760*. Berkeley: University of California Press, 1984.

Landsberg, H. E. "Roots of Modern Climatology," *Journal of the Washington Academy of Sciences* 54 (1964): 130–41.

Langford, Paul. *A Polite and Commerical People. England 1727–1783*. Oxford: Clarendon Press, 1989.

Langford, Paul. *Public Life and the Propertied Englishman 1689–1789*. Oxford: Clarendon Press, 1991.

Laslet, Peter. "The Gentry of Kent in 1640," *The Cambridge Historical Journal* 9 (1947): 148–64.

Latham, Robert and Matthews, William, eds, *The Diary of Samuel Pepys*. 2 vols. Berkeley: University of California Press, 1970.

Latour, Bruno. *We Have Never Been Modern*. Cambridge, Mass.: Harvard University Press, 1993.

Leighly, John. "An Early Drawing and Description of a Tornado," *Isis* 65 (1974): 474–86.

Lepenies, Wolf. *Melancholy and Society*. Cambridge, Mass.: Harvard University Press, 1992.

Levine, Joseph M. *Dr. Woodward's Shield: History, Science, and Satire in Augustan England*. Ithaca and London: Cornell University Press, 1977.

Levine, Joseph M. "Natural History and the History of Scientific Revolution," *Clio* 13 (1983): 57–69.

Levine, Joseph M. *Humanism and History: Origins of Modern English Historiography*. Ithaca and London: Cornell University Press, 1987.

Liddell, Henry George and Scott, Robert. *A Greek-English Lexicon*, Oxford: Clarendon Press, 1968.

Lloyd, G. E. R. *Aristotelian Explorations*. Cambridge: Cambridge University Press, 1996.

Locher, A. "The Structure of Pliny the Elder's Natural History," in Roger French and Frank Greenaway, eds, *Science in the Early Roman Empire: Pliny the Elder, His Sources and Influence*. Totowa, N. J.: Barnes and Noble Books, 1986.

Lord, Carnes. "On the Early History of the Aristotelian Corpus," *American Journal of Philology* 107, No. 2 (1986): 137–61.

Low, Anthony. *The Georgic Revolution*. Princeton, N. J.: Princeton University Press, 1985.

Lowenthal David, and Prince, Hugh C. "English Landscape Tastes," in Paul Ward English and Robert C. Mayfield, eds, *Man, Space, and Environment*. New York: Oxford University Press, 1972, 81–114.

Lowood, Henry. *Patriotism, Profit, and the Promotion of Science in the German Enlightenment. The Economic and Scientific Societies, 1760–1815*. New York: Garland Publishing, Inc., 1991.

Lund, Roger D. "'More Strange than True': Sir Hans Sloane, King's 'Transactioneer', and the Deformation of English Prose," in O. M. Brack Jr., ed., *Studies in Eighteenth Century Culture*, vol. 14. Madison: The University of Wisconsin Press, 1985.

Lux, David S. and Cook, Harold J. "Closed Circles or Open Networks?: Communicating at a Distance During the Scientific Revolution," *History of Science* 26 (1998), 179–211.

Lyons, Sir Henry, FRS. *The Royal Society 1660–1940. A History of its Administration under its Charters*. New York: Greenwood Press, 1968.

Mabey, Richard. *Gilbert White: A Biography of the Author of The Natural History of Selborne*. Century: Vermont, 1986.

Maddox, Lucy B. "Gilbert White and the Politics of Natural History," *Eighteenth Century Life* 10 (1986): 45–57.

Manley, Gordon. "The Weather and Diseases: Some Eighteenth-Century Contributions to Observational Meteorology," *Notes and Records of the Royal Society of London* 9 (1952): 300–7.

Margolin, J. C. "Cardan, interprète dè Aristote," in *Platon et Aristote à la Renaissance*. Paris: Vrin, 1976, 307–34.

Mayer, Robert. *History and the Early English Novel: Matters of Fact from Bacon to Defoe*. Cambridge: Cambridge University Press, 1997.

McClellan III, James E. *Colonialism and Science: Saint Domingue in the Old Regime*. Baltimore and London: The Johns Hopkins University Press, 1992.

McCormack, Lesley. "'Good Fences Make Good Neighbors': Geography as Self-Definition in Early Modern England," *Isis* 82 (1991): 639–61.

Macdonald, Michael. *Mystical Bedlam*. Cambridge: Cambridge University Press, 1981.

MacGregor, Arthur, ed., *Sir Hans Sloane: Collector, Scientist, Antiquary, Founding Father of the British Museum*. London: British Museum Press, 1994.

McKeon, Michael. *The Origins of the English Novel 1600–1740*. Baltimore: The Johns Hopkins University Press, 1987.

Medick, Hans. "Plebeian Culture in the Transition to Capitalism," in Raphael Samuel and Gareth Stedman Jones, eds, *Culture, Ideology and Politics*. London: Routledge, 1981.

Mendyk, A. E. Stan. *"Speculum Britanniae." Regional Study, Antiquarianism, and Science in Britain to 1700*. Toronto: University of Toronto Press, 1989.

Merret, Robert James. "Natural History and the Eighteenth-Century English Novel," *Eighteenth Century Studies* 25 (1991): 145–70.

Middleton, W. E. Knowles. *The History of the Barometer*. Baltimore: The Johns Hopkins Press, 1964.

Middleton, W. E. Knowles. "Chemistry and Meteorology, 1700–1825." *Annals of Science* 20 (1965): 125–41.

Middleton, W. E. Knowles. *A History of the Theories of Rain and Other Forms of Precipitation*. London: Oldbourne Book Co. Ltd., 1965.

Middleton, W. E. Knowles. *A History of the Thermometer and Its Use in Meteorology*. Baltimore: Johns Hopkins Press, 1966.

Miller, David P. "'Into the Valley of Darkness': Reflections on the Royal Society in the Eighteenth Century," *History of Science* 27 (1989): 155–66.

Mitchell, John Cairns. *Chester: Its Situation and Climate with a Record of Remarkable Weather in 1893*. Chester: Courant Press, 1894.

Monk, Samuel Halt. *The Sublime: A Study of Critical Theories in XVIII Century England*. New York: Modern Language Association, 1935.

Monod, Paul Kleber. *Jacobitism and the English People 1688–1788*. Cambridge: Cambridge University Press, 1989.

Mooney, Michael. *Vico in the Tradition of Rhetoric*. Princeton: Princeton University Press, 1985.

Morgan, F. W. "Three Aspects of Regional Consciousness," *The Sociological Review* 31 (1939): 68–88.

Morrell, Jack and Thackray, Arnold. *Gentlemen of Science: Early Years of the BAAS.* Oxford: Clarendon Press, 1981.

Morrell, Jack and Thackray, Arnold, eds, *Gentlemen of Science. Early Correspondence of the British Association for the Advancement of Science.* London, 1984.

Newman, Gerald. *The Rise of English Nationalism 1740–1830: A Cultural History.* London: Weidenfield and Nicolson, 1987.

Nicholson, Marjorie and Rousseau, G. S. *"This Long Disease, My Life:" Alexander Pope and the Sciences.* Princeton: Princeton University Press, 1968.

Norton, C. E. *Thomas Gray as a Naturalist* (Boston, 1903).

Oliver, J. "William Borlase's Contribution to Eighteenth-Century Meteorology and Climatology," *Annals of Science* 25 (1969): 275–317.

Olson, Richard. "Tory-High Church Opposition to Science and Scientism in the Eighteenth Century: The Works of John Arbuthnot, Jonathan Swift, and Samuel Johnson," John G. Burke, ed., *The Uses of Science in the Age of Newton.* Berkeley: University of California Press, 1983.

Olwig, Kenneth Robert. "Sexual Cosmology: Nation and Landscape at the Conceptual Interstices of Nature and Culture; or What does Landscape really Mean?," in Barbara Bender, ed., *Landscape. Politics and Perspectives.* Providence and Oxford: Berg, 1993.

Ophir, Adi and Shapin, Steven, "The Place of Knowledge: A Methodological Survey," *Science in Context,* 4: 1 (1991): 3–21.

Otis, Brooks. *Virgil: A Study in Civilized Poetry.* Oxford: Clarendon Press, 1963.

Packer, J. W. *The Transformation of Anglicanism.* Manchester: Manchester University Press, 1969.

Park, Katharine and Daston, Lorraine. "Unnatural Conceptions: The Study of Monsters in Sixteenth- and Seventeenth-Century France and England," *Past and Present* 92 (1981): 20–54.

Parks, George B. "Travel as Education," in Richard Foster Jones, ed., *The Seventeenth Century: Studies in the History of English Thought and Literature from Bacon to Pope.* Stanford: Stanford University Press, 1951, 264–91.

Patterson Annabel. *Pastoral and Ideology: Virgil to Valéry.* Berkeley: University of California Press, 1987.

Paul, James B., ed., *The Diary of George Ridpath 1755–1761.* London, 1922.

Perkins, Maureen. *Visions of the Future: Almanacs, Time, and Cultural Change, 1775–1870.* Oxford: Oxford University Press, 1996.

Phillipson, N. T. "Culture and Society in the 18th Century Province: The Case of Edinburgh and the Scottish Enlightenment," in Lawrence Stone, *The University in Society.* Princeton: Princeton University Press, 1974, 407–48.

Pickering, Michael. "The Four Angels of the Earth: Popular Cosmology in a Victorian Village," *Southern Folklore Quarterly* 45 (1981): 1–18.

Pickstone, John V. "Ways of Knowing: Towards a Historical Sociology of Science, Technology and Medicine," *British Journal for the History of Science* 26 (1993): 433–58.

Pickstone, John V. *Ways of Knowing.* Manchester: Manchester University Press, 2000.

Piggott, Stuart. *Ancient Britons and the Antiquarian Imagination. Ideas from the Renaissance to the Regency.* New York: Thames and Hudson, 1989.

Piggott, Stuart. *Ruins in the Landscape. Essay in Antiquarianism.* Edinburgh: Edinburgh University Press, 1976.

Piggott, Stuart. *William Stukeley.* Oxford: Clarendon Press, 1950.

Pittock, Murray G. H. *Poetry and Jacobite Politics in Eighteenth-Century Britain and Ireland.* Cambridge: Cambridge University Press, 1994.

Pomian, Krzystof. *Collectors and Curiosities: Paris and Venice, 1500–1800.* Cambridge: Cambridge University Press, 1990.

Pool, P. A. S. *William Borlase.* Truro: The Royal Institution of Cornwall, 1987.

Porteous, Douglas J. "Home: The Territorial Core," *Geographical Review* 66: 4 (1976): 383–90.

Porter, Roy. "Gentlemen and Geology: The Emergence of a Scientific Career, 1660–1920," *The Historical Journal* 21 (1978): 809–36.

Porter, Roy. "Science, Provincial Culture and Public Opinion in Enlightenment England," *The British Journal for Eighteenth-century Studies* 3 (1980): 20–46.

Porter, Roy. "The Terraquaeous Globe," in G. S. Rousseau and Roy Porter, eds, *The Ferment of Knowledge*. Cambridge: Cambridge University Press, 1980.

Porter, Roy. "The People's Health in Georgian England," in Tim Harris, ed., *Popular Culture in England, c. 1500–1850*. New York: St Martin's Press, 1995, 124–42.

Porter, Roy. "The Urban and the Rustic in Enlightenment London," in Mikulas Teich, Roy Porter and Bo Gustafson, eds, *Nature and Society in Historical Context*. Cambridge: Cambridge University Press, 1997, 176–94.

Porter, Theodore M. *Trust in Numbers: The Pursuit of Objectivity in Science and Public Life*. Princeton, N. J.: Princeton University Press, 1995.

Potter, George Reuben. "The Significance to the History of English Natural Science of John Hill's Review of the Works of the Royal Society," *University of California Publications in English* 14 (1943): 157–80.

Prince, Hugh C. "English Landscape Tastes," in Paul Ward English, and Robert C. Mayfield, eds, *Man, Space, and Environment*. New York: Oxford University Press, 1972, 81–114.

The Publications of the Surtees Society

Redwood, John. *Reason, Ridicule and Religion*. London: Thames and Hudson, 1976.

Reed, Arden. *Romantic Weather: The Climates of Coleridge and Baudelaire*. Hanover and London: University Press of New England, 1983.

Reed, Michael. "The Cultural Role of Small Towns in England 1600–1800," in Peter Clark, ed., *Small Towns in Early Modern Europe*. Cambridge: Cambridge University Press, 1995, 121–47.

Reif, Patricia. "The Textbook Tradition in Natural Philosophy, 1600–1650," *Journal of the History of Ideas* 30 (1969): 17–32.

Relph, E. *Place and Placelessness*. London: Pion, 1976.

Riley, James. *The Eighteenth-Century Campaign to Avoid Disease*. New York: St. Martin's Press, 1986.

Rivers, Isabel, ed., *Books and their Readers in Eighteenth-Century Britain*. New York: Leicester University Press, 1982.

Rivero, Albert, ed., *Augustan Subjects: Essays in Honor of Martin C. Battestin*. Newark, Del.: University of Delaware Press, 1997.

Rogers, Nicholas. "Popular Protest in Early Hanoverian London," *Past and Present* 79 (1978): 70–100.

Rollins, Hyder Edward, ed. *The Pack of Autolycus*. Cambridge: Cambridge University Press, 1927.

Rollinson, D. "Property, Ideology and Popular Culture in a Gloucestershire Village 1660–1740," *Past and Present* 93 (1981): 70–97.

Ross, David O. Jr., *Virgil's Elements: Physics and Poetry in the Georgics*. Princeton, N. J.: Princeton University Press, 1987.

Ross, W. D. *Aristotle*. London: Methuen, 1945.

Rossi, Paolo. *Dark Abyss of Time: The History of the Earth and the History of Nations from Hooke to Vico*. Chicago: The University of Chicago Press, 1984.

Rudwick, Martin. *The Great Devonian Controversy: The Shaping of Natural Knowledge Among Gentlemanly Specialists*. Chicago: University of Chicago Press, 1985.

Rusnock, Andrea. ed., *The Correspondence of James Jurin (1684–1750)*. Amsterdam: Rodopi, 1996.

Rusnock, Andrea. "Correspondence Networks and the Royal Society, 1700–1750" *British Journal for the History of Science* 32 (1999): 155–69.

Sandbach, F. H. *Aristotle and the Stoics*. Cambridge: Cambridge Philosophical Society, 1985.

Sargent, Frederick. *Hippocratic Heritage: A History of Ideas about Weather and Human Health*. New York, Oxford: Pergamon Press, 1982.

Schmitt, Carles B. "Towards a Reassessment of Renaissance Aristotelianism," *History of Science* 11 (1973): 159–93.

Schove, D. J. and Reynolds, David. "Weather in Scotland, 1659–1660: The Diary of Andre Hay," *Annals of Science* 30 (1973): 165–77.

Serres, Michel. *Naissance de la physique dans le texte de Lucrèce: Fleuves et Turbulences*. Paris: Editions de Minuit, 1975.

Schaffer, Simon. "Natural Philosophy and Public Spectacle in the Eighteenth Century," *History of Science* 21 (1983): 1–43.

Schaffer, Simon. "Making Certain," *Social Studies of Science* 14 (1984): 137–52.

Schaffer, Simon, "Scientific Discoveries and the End of Natural Philosophy," *Social Studies of Science* 16: 3 (1986): 387–421.

Schaffer, Simon. "Newton's Comets and the Transformation of Astrology," in Patrick Curry, ed., *Astrology, Science, and Society: Historical Essays*. Woodbridge: The Boydell Press, 1987, 219–45.

Schaffer, Simon. "A Social History of Plausibility: Country, City and Calculation in Augustan Britain," in Adrian Wilson, ed., *Rethinking Social History: English Society 1750–1920 and its Interpretation*. Manchester: Manchester University Press, 1993, 128–57.

Schaffer, Simon. "The Consuming Flame: Electrical Showmen and Tory Mystics in the World of Goods," in John Brewer and Roy Porter, eds, *Consumption and the World of Goods*. London: Routledge, 1994, 489–526.

Schneider-Carius, Karl. *Weather Science, Weather Research*. The Indian National Scientific Documentation Centre, New Delhi, 1975.

Secord, James A. "Natural History in Depth," *Social Studies of Science* 15 (1985): 181–200.

Secord, James A. *Controversy in Victorian Geology: The Cambrian-Silurian Dispute*. Princeton, N. J: Princeton University Press, 1986.

Shaaber, Matthias A. *Some Forerunners of the Newspaper in England 1476–1622*. Philadelphia: University of Pennsylvania Press, 1929.

Shapin, Steven. "Social Uses of Science," in George Rousseau and Roy Porter, eds, *The Ferment of Knowledge: Studies in the Historiography of Eighteenth Century Science*. Cambridge: Cambridge University Press, 1980, 93–139.

Shapin, Steven. "Of Gods and Kings: Natural Philosophy and Politics in the Leibniz-Clarke Dispute," *Isis* 72 (1981): 187–215.

Shapin, Steven. *A Social History of Truth. Civility and Science in Seventeenth-Century England*. Chicago and London: The University of Chicago Press, 1994.

Shapin, Steven. *The Scientific Revolution*. Chicago: The University of Chicago Press, 1996.

Shapin, Steven. "Rarely Pure and Never Simple: Talking About Truth," *Configurations* 7 (Winter 1999), 1–14.

Shapin, Steven and Schaffer, Simon. *Leviathan and the Air Pump: Hobbes, Boyle and the Experimental Life*. Princeton: Princeton University Press, 1985.

Shapin, Steven and Thackray, Arnold. "Prosopography as a Research Tool in History of Science: The British Scientific Community 1700–1900," *History of Science* 12 (1974): 1–28.

Shapiro, Barbara. "History and Natural History in 16th and 17th-century England: An Essay on the Relationship between Humanities and Science," in idem., ed., *English Scientific Virtuosi in the 16th and 17th Centuries*. Los Angeles: William Andrews Clarke Memorial Library, 1979, 3–55.

Shapiro, Barbara. *Probability and Certainty in Seventeenth-Century England*. Princeton: Princeton University Press, 1983.

Shapiro, Barbara. "The Concept 'Fact': Legal Origins and Cultural Diffusion," *Albion* 26 (1994): 1–26.

Sharpe, J. A. " 'Last Dying Speeches': Religion, Ideology, and Public Execution in Seventeenth-Century England," *Past and Present* 107 (1985): 144–67.

Shaw, William Napier, and Austin, Elaine. *Manual of Meteorology: Volume I: Meteorology in History*. Cambridge: Cambridge University Press, 1932.

Shiach, Morag. *Discourse on Popular Culture. Class, Gender and History in Cultural Analysis, 1730 to the Present*. Stanford: Stanford University Press, 1989.

Shields, Lisa. "The Beginnings of Scientific Weather Observation in Ireland (1684–1708)," *Weather* 40 (1984): 304–11.

Simmons, Jack, ed., *English County Historians*. East Ardsley: E. P. Publishing, 1978.

Simms, J. G. *William Molyneux of Dublin, 1656–1695*. Blackrock: Irish Academy Press, 1982.

Sopher, David E. "The Landscape of Home. Myth, Experience, Social Meaning," in D. W. Meinig, ed., *The Interpretation of Ordinary Landscapes*. Oxford: Oxford University Press, 1979.

Sorensen, Conner W. *Brethren on the Net: American Entomology 1840–1880*. Tuscaloosa and London: The University of Alabama Press, 1995.

Soupel, Serge. "Science and Medicine and the Mid-Eighteenth-Century Novel: Literature and the Language of Science," in idem., *Literature and Science and Medicine*. Los Angeles: William Andrews Clark Memorial Library, 1982.

Speck, William. "Whigs and Tories Dim Their Glories: English Political Parties under the First Two Georges," in John Cannon, ed., *The Whig Ascendancy: Colloquies on Hanoverian England*. New York: St. Martin Press, 1981, 51–71.

Spufford, Margaret. *Small Books and Pleasant Histories: Popular Fiction and its Readership in Seventeenth-Century England*. Oxford: Oxford University Press, 1981.

Spurr, John. "'Latitudinarianism' and the Restoration Church," *The Historical Journal* 31 (1988): 61–82.

Spurr, John. "'Virtue, Religion and Government': The Anglican Uses of Providence," in Tim Harris, Paul Seaward and Mark Goldie, eds, *The Politics of Religion in Restoration England*. London: Basil Blackwell, 1990.

Stephen, Leslie. *History of English Thought in the Eighteenth Century*. New York: Putnam's Sons, 1927.

Stevenson, John. *Popular Disturbances in England, 1700–1832*. New York and London: Longman, 1992.

Stewart, Larry. "Samuel Clarke, Newtonianism, and the Factions of Post-Revolutionary England," *Journal of the History of Ideas* 42 (1981): 53–72.

Stewart, Larry. "Public Lectures and Private Patronage in Newtonian England," *Isis* 77 (1986): 47–58.

Stewart, Larry. *The Rise of Public Science*. Cambridge: Cambridge University Press, 1992.

Stilgoe, John R. *Common Landscape of America, 1580–1845*. New Haven: Yale University Press, 1982.

Stokes, Francis Griffin. *The Bletchley Diary of the Rev. William Cole, 1765–67*. London: Constable, 1931.

Stone, L. and Stone, J. C. F. *An Open Elite? England 1540–1880*. Oxford: Oxford University Press, 1984.

Stubbs, Mayling. "John Beale, Philosophical Gardener of Herefordshire. Part Two: The Improvement of Agriculture and Trade in the Royal Society," *Annals of Science* 46 (1989): 323–63.

Symons, G. J. "The History of English Meteorological Societies, 1823 to 1880," *Quarterly Journal of the Meteorological Society* 7 (1881): 65–68.

Szechi, Daniel. *The Jacobites. Britain and Europe 1688–1788*. Manchester: Manchester University Press, 1994.

Taylor, E. G. R. *Late Tudor and Early Stuart Geography 1583–1650*. New York: Octagon Books, 1968.

Thackray, Arnold. "The Business of Experimental Philosophy: The Early Newtonian Group at the Royal Society," *Actes du XIIᵉ Congrès International d'Histoire des Sciences*, 1970–1, 3B: 155–9.

Thackray, Arnold. "John Frederick Daniell," *Dictionary of Scientific Biography*, vol. 3, 556–8.

Thomas, Keith. *Religion and the Decline of Magic*. New York: Charles Scribner's Sons, 1971.

Thomas, Keith. *Man and the Natural World*. Harmondsworth: Penguin Books, 1984.

Thompson, E. P. "Anthropology and the Discipline of Historical Context," *Midland History* 1 (1972): 52.

Thompson, E. P. "Patrician Society, Plebeian Culture," *Journal of Social History* 7 (1973): 382–405.

Thompson, E. P. "Peculiarities of the English," in idem., ed. *The Poverty of Theory and Other Essays*. 1978, 35–91.

Thompson, E. P. "Eighteenth-Century English Society: Class Struggle without Class?" *Social History* 3 (1978): 157.

Thompson, E. P. *Customs in Common*. New York: The New Press, 1991.

Thomson, Derick. *Edward Lhuyd in the Scottish Highlands 1699–1700*. Oxford: Oxford University Press, 1963.

Thorndike, Lynn. *History of Magic and Experimental Science*, 8 vols. London, New York: Macmillan, 1923–58.

Tiller, Kate. *English Local History, An Introduction*. Wolfeboro Falls, N. H.: Alan Sutton, 1992.

Tindal Hart, A. *The Eighteenth Century Country Parson*. Shrewsbury: Wilding and Son, Ltd, 1955.

Todd, Dennis. *Imagining Monsters: Miscreation of the Self in Eighteenth-Century England*. Chicago: The University of Chicago Press, 1995.

Trevelyan, George Macaulay. *Blenheim. England under Queen Anne*. London: Longman, Green and Company, 1959.

Tuan, Yi-Fu. *The Hydrologic Cycle and the Wisdom of God*. Toronto: University of Toronto Press, 1968.

Tuan, Yi-Fu. *Topophilia. A Study of Environmental Perception, Attitudes, and Values.* Englewood Cliffs, N. J.: Prentice Hall, 1974.

Tuan, Yi-Fu. "Place: An Experiential Perspective," *Geographical Review* 65 (1975): 151–6.

Tuan, Yi-Fu. "Geopiety A Theme in Man's Attachment to Nature and to Place," in F. Lowenthal and Martyn Bowden, eds, *Geographies of the Mind.* New York: Oxford University Press, 1976.

Tunbridge, Paul A. "Jean Andre DeLuc, FRS (1727–1817)," *Notes and Records of the Royal Society of London* 26 (1971): 15–23.

Turner, Frank M. *Contesting Cultural Authority: Essays in Victorian Intellectual Life.* Cambridge: Cambridge University Press, 1993.

Turner, James. *The Politics of Landscape: Rural Scenery and Society in English Poetry 1630–1660.* Cambridge, Mass.: Harvard University Press, 1979.

Ultee, Maarten. "Sir Hans Sloane, Scientist," *The British Journal.* London: British Library, 1988, 1–21.

Vickers, Ilse. *Defoe and the New Sciences.* Cambridge, Mass.: Cambridge University Press, 1996.

Viner, Jacob. *The Role of Providence in the Social Order.* Philadelphia: American Philosophical Society, 1972.

Wahrman, Dror. "National Society, Communal Culture: an Argument About the Recent Historiography of Eighteenth-Century Britain," *Social History* 17 (1992): 43–72.

Walker, J. M. "A Meteorological Curio," *Weather* 25 (1970): 30–2.

Walsh, John, Haydon, Colin and Taylor, Stephen, eds, *The Church of England c.1689–c.1833. From Toleration to Tractarianism.* Cambridge: Cambridge University Press, 1993.

Walsham, Alex. "Aspects of Providentialism in Early Modern England." Ph.D., Cambridge University, 1995.

Walters, Alice. "Tools of Enlightenment: The Material Culture of Science in Eighteenth-Century England." Ph.D. Dissertation, University of California at Berkeley, 1992.

Warnecke, Sara. "A Taste for Newfangledness: The Destructive Potential of Novelty in Early Modern England," *Sixteenth-Century Journal* 26 (1995): 881–96.

Watt, Tessa. *Cheap Print and Popular Piety.* Cambridge: Cambridge University Press, 1989.

Wear, Andrew. "William Harvey and the 'Way of Anatomists'," *History of Science* 21 (1983): 220–49.

Webster, Charles. "Richard Towneley (1629–1707), the Towneley Group and Seventeenth-Century Science," *Transactions of the Historical Society of Lancashire and Cheshire* 118 (1967): 51–76.

Webster, Charles. *The Great Instauration. Science, Medicine, and Reform 1626–1660.* New York: Holmes & Meier Publishers, 1975.

Westfall, Richard S. *Science and Religion in Seventeenth-century England.* New Haven: Yale University Press, 1958.

Westrum, Ron. "Science and Social Intelligence about Anomalies: The Case of Meteorites," *Social Studies of Science* 8 (1978): 461–93.

White, Andrew Dickson. *A History of the Warfare of Science with Theology in Christendom* (1896). Gloucester, Mass.: Peter Smith, 1978.

Wilkinson, L. P. "The Intention of Virgil's Georgics," *Greece and Rome* 55 (1950): 19–28.

Wilkinson, L. P. *The Georgics of Virgil.* Cambridge: Cambridge University Press, 1969.

Willmoth, Frances. "John Flamsteed's Letter concerning the Natural causes of Earthquakes," *Annals of Science* 44 (1987): 23–70.

Wilson, Kathleen. *The Sense of the People. Politics, Culture and Imperialism in England, 1715–1785.* Cambridge: Cambridge University Press, 1995.

Withers, Charles. "Geography, Natural History, and the Eighteenth-Century Enlightenment: Putting the World in Place," *History Workshop Journal* 39 (1995): 137–64.

Withers, Charles. "Toward a History of Geography in the Public Sphere," *History of Science* 36 (1998): 45–78.

Wood, P. B. "Methodology and Apologetics: Thomas Sprat's History of the Royal Society," *The British Journal for the History of Science* 13 (1980): 1–26.

Woolf, D. R. "The 'Common Voice': History, Folklore, and Oral Tradition in Early Modern England", *Past and Present* 120 (1988): 26–52.

Woolf, D. R. *The Idea of History in Early Stuart England. Erudition, Ideology, and 'The Light of Truth' from the Accesion of James I to the Civil War.* Toronto: University of Toronto Press, 1990.

Worden, Blair. "Providence and Politics in Cromwellian England," *Past and Present* 109 (1985): 55.

Worster, Donald. *Nature's Economy: A History of Ecological Ideas*. Cambridge: Cambridge University Press, 1977.

Wrightson, K. *English Society 1580–1680*. New Brunswick, N. J.: Rutgers University Press, 1982.

Broadsheets, pamphlets, and sermons dealing with extreme weather phenomena, 1600–1800

A Wonderful and Strange News which Happened in the County of Suffolk and Essex the first of February, where it rained wheat the space of six or seven miles compass. London, 1583.

Strange Fearful & True News, which Happened at Carlstadt, in the Kingdome of Croatia. George Vincent and William Blackwall, 1606.

A Wonderfull and most Lamentable Declaration of the Great Hurt done, and Mighty Losse sustained by Fire, Winde, Thunder, Lightning, Haile, and Raine. London, 1613.

Looke Up and see Wonders. A miraculous Apparition in the Ayre, lately seen in Barke-shire at Bawlkin Greene neere Hatford, April 9, 1628. London: Roger Michell, 1628.

A Second and Most Exact Relation of those Sad and Lamentable Accidents, which happened in and about the Parish Church of Wydecombe neere Dartmoores in Devonshire on Sunday the 21. of October last, 1638. London: R. Harford, 1638.

A Report from Abbington towne in Berkshire, being a relation of what harme Thunder and Lightning did on Thursday last upon the body of Humphrey Richardson, a rich miserable farmer. London: William Bowde, 1641.

A Strange Wonder or The Cities Amazement. Being a Relation occasioned by a Wonderful and unusual Accident that happened in the River of Thames, Friday Feb. 4, 1641. London: John Thomas, 1641.

A Strange and Wonderful Relation of the Miraculous Judgment of God in the late Thunder and Lightning on sat. 23 of this august being the next day after Mr, Love and Mr. Gibbons were beheaded. With certificate and the names of divers persons of quality. London: Bernard Alsop, n.d., c. 1620–50.

Charles Hammond, *A Warning-peece [sic] for England. By that sad and fearfull Example that happened to Men, Women and Children, all sorts of cattle and fowles; by Stormes, Tempests, Hail-Stones, Lightning and Thunder, June 25 1652*. London, 1652.

More Warning Yet. Being a True Relation of A Strange and most Dreadful Apparition Which was seen in the Air By several persons at Hull, the third day of this present Sept, 1654. London: J. Cottrel, 1654.

Mirabilis Annus (Eniagtos Terastios) or the year of Prodigies and Wonders. London, 1661.

A True and Perfect Relation, of A Strange and Wonderful Apparition in the Air, 1664. London: Thomas Leach, 1664.

A True Relation of what happened at Bedford, On Monday last, Aug. 19 instant while Thundering, Lightning, and Tempestuous Winds tore up the Trees by the Roots. London: Fra. Smith, 1672.

This Winter Wonders: or A True Relation of a Calamitous Accident at Bennenden in the County of Kent. London, 1673.

A True and Perfect Narrative of the Great and Dreadful Damages Sustained in Several Parts of England by the Late Extraordinary Snows, London, 1674.

A True and Perfect Relation of the Great Harm done by a Dreadful Storm of Wind, Thunder and Lightning, mixed with Rain and Hail. London: A. Maxwell, 1674.

News from the Sea: Or a true Relation brought from Dover of a Terrible Tempest of Thunder and Lightning, which on the 26th of July last kill'd outright one William Eaton of Dover, on board a Ship called the Henry and Mary etc. The Truth whereof is well known to most Inhabitants of Dover, ther being many Eye Witnesses etc. London: for D. M. 1676.

A True Relation of the Terrible Earthquake at West-Brummidge in Staffordshire And the places adjacent on Tuesday the 4th of this instant January 1675/6. London: D. M. 1676.

A Strange and Wonderful Relation of a Clap of Thunder which lately set Fire to the Dwelling-House of one Widdow Rosingrean living in the town of Ewloe, in the parish of Howerden in the County of Flint. London: W. Harris, 1677.

Strange and Terrible News from Sea. Or, A True Relation of a Most Wonderful Violent Tempest of Lightning and Thunder on Friday, the 18th of this Instant Jan. 1678. Westsmithfield: A. P. and T. H. for John Clarke, 1678.

A Brief Relation of a Wonderful Account of a Wonderful Accident a Dissolution of the Earth in the Forest of Charnwood. London: two Lovers of Art, I. C. and I. W., 1679.

Strange News from Berkshire, of an Apparition of Several Ships and Men in the Air . . . near Abbington, on Tuesday 26th August 1679. London, 1679.

An Account of a Strange and Prodigious Storm of Thunder, Lightning, hail, which happened in and about London, etc. London: N. I. 1680.

A Full and True Relation of the Death and Slaughter of a Man and his Son at Plough together with Four Horsese, in the Parish of Cookham in the County of Berks, Sept. 1680. Slain by the Thunder and Lightning. London: John Harding, 1680.

Memento to the World; Or An Historical Collection of Divers Wonderful Comets and Prodigious Signs in Heaven. London: T. Haly, 1680.

A True Relation of the Extraordinary Thunder & Lightning which lately happened in the North of Ireland as it was sent to Dublin in several Letters to Persons of Quality. Dublin, 1680.

Wonderful Signs of Strange Times, 1680.

A Strange and True Account from Shrewsbury of a Dreadful Storm, which happened on the 4th of May last. London, 1681.

Strange News from Oxfordshire: Being a true and faithful Account of a Wonderful and Dreadful Earthquake that happened in those Parts on Monday the 17th of this present September 1683. Oxford, 1683.

A Strange and Wonderful Account of the Great Mischiefs Sustained by the late Dreadful Thunder, Lightening and Terrible Land-Floods Caused by the Immoderate Rain in England, Scotland & Holland. London: J. Wright, 1683.

A Strange and Wonderful Account of the late Dreadful Thunder, Lightening, and Terrible Land-Floods. Giving an Exact Relation of the Men, Cattle, Houses, etc that have been Thunderstruck. London, 1683.

Modest Observations on the Present Extraordinary Frost: Containing a Brief Description thereof, and its Natural Celestial Causes inquired into. London: George Larkin, 1684.

Strange and Terrible News from Alton in Hampshire: Being a full Account of a Dreadful Tempest which happened there by Thunder and Lightning, December 19th, 1686. London: for S. M., 1687.

A True Relation of the Great Thunder, Lightning, Rain, Great Wind, and Prodigious Hail, that Happened at Alvanley in the Parish of Frodsham in Cheshire, on Sunday the 19th of June. London: D. Mallet for G. P., 1687.

A True Relation of the Wonderful Apparition of a Cross in the Moon, visible at Wexford and other places, kingdom of Ireland upon April the 4th, 1688. Dublin, 1688.

An Account of the wicked Design of Poysoning the Prince of Orange before he came out of Holland. Also a relation of a strange Meteor, representing a Crown of Light that was there seen in the air, May 6th 1688. London, 1689.

A True Account of the damages done by the Late Storm which happened between the Hours of 12 and 4 of the Clock on Sunday Morning, January 12, 1689. London, 1689.

A True and Impartial Account of the Strange Earthquake which happened in most parts of the City of London, on this September 8, 1692. London: J. Gerrard, 1692.

God's Marvellous Wonders in England in 1694. London: P. Brocksby, 1694.

Strange and Wonderful News from Ireland, Giving a Dreadful Relation of a Prodigious Motion of the Earth near Charleville, Limerick, June 7, 1697. London: J. Wilkins, 1697.

Strange and Wonderful News from Chippingnorton in the County of Oxon. Of certain dreadful Apparition, which were seen in the Air, on 28 July. n.d., c. 1700.

An Account of the Sad and Dreadful Accidents that was Done about the Cities of London and Westminster, and the Liberties thereof, by the Prodigious Hurricane of Wind, on 27 November, 1703. etc. Edinburgh: heirs of Andrew Anderson, 1703.

A Pindarick Poem upon the Hurricane in November 1703, referring to this text in Psalm 148 v8 Winds and Storms fulfilling his word, With a Hymn, compos'd of the 148 Psalm paraphrased, by the author of the Poem upon the Spleen. MS, National Meteorological Archive, Bracknell, Berkshire.

An Exact Relation Of the Late Dreadful Tempest. Faithfully collected by an Ingenious Hand, to preserve the Memory of so Terrible a Judgment, 1704. London, 1704.

The Terrible Stormy Wind and Tempest November 27th 1703 Consider'd, Improved, and Collected to be had in Everlasting Rememberance. London: W. Freeman, 1705.

To begin Harvest Three Days too soon rather than Two Days too late, or Sentences of the Dissenters containing, Relations of Pretended Judgments, Prodigies and Apparitions in [sic] Behalf of the Nonconformists, in Opposition to the Established Church. With Timely Remarks. London: B. Bragg, 1708.

The Age of Wonders: or, a Particular Description of the Remarkable, and Fiery Apparition that was Seen in the Air, on Thursday, May the 11th 1710. London: J. Read, 1710.

A True and Particular Account of a Storm of Thunder and Lightning which Fell at Richmond, Surrey, on White Sunday last in the Afternoon, being May 20th, 1711 etc. London: John Morphew, 1711.

God's Judgment's and Mercies shown to Mankind. Being a Particular Account of a Great and Dismal Rain that fell in Scotland, on the 24th of September 1712, which has laid under Water above eight hundred Houses and twelve hundred Families. London: Edward Midwinter, 1712.

S., J. *The Shepherd's Kalender: or the citizen's and country man's daily companion.* Sixth edition, London: L. Hawes, Clark and Colins, and S. Crowder, 1715.

A Full and True Relation of the Strange and Wonderful Apparitions which were Seen in the Clouds upon Tuesday Evening at Seven of the Clock . . . the 6th and 7th of March 1715/6, n.d.

Their Vision at Deal; being an Account of most Wonderful Apparitions of Ships, Castles and Other Figures in the Air, to which is added an Account of the late strange Appearance in the Sky over City of London. London: S. Popping, 1716.

A Strange and Wonderful Account of the Appearance of a Fiery Meteor in the Air: which was seen by many Hundreds of Spectators, at the Town of Boston in Lincolnshire, etc. Stamford, Lincolnshire: Henry Wilson, 1719.

The Uxbridge Wonder, Being a true and particular Account of a sad and dreadful Storm of Hail, Thunder and Lightning. London, 1738.

O-Yes from the Court of Heaven to the Northern Nations, by the Streaming Lights that have appeared of the Late Years in the Air. London: the Author, 1741.

A True and Particular Relation of the Dreadful Earthquake, with an Enquiry into Causes of Earthquakes. London: T. Osborne, 1748.

A True and Wonderful Narrative of Two Particular Phenomena, which were seen in the sky in Germany. Philadelphia: Anthony Armbruster, 1764.

Narrative of the Great Flood in the Rivers Tyne, Tease, Wear, etc on the 16th and 17th of Nov. 1771. Newcastle, T. Slack 1772.

God's Awful Warnings to a Giddy, Careless, Sinful World. London: G. Riebau, 1795.

The Thunder Bolt: or a Merciful Preservation from the Awful Effects of Lightning. London: James Nisbeth, 1832.

Select bibliography of newspapers, magazines and periodicals

Annals of Philosophy; or Magazine of Chemistry, Mineralogy, Mechanics, Natural History, Agriculture and Arts.

Annals of Science, or Magazine of Chemistry, Mineralogy, Mechanics, Natural History, Agriculture and Arts

The Anti-Jacobin Review and Magazine

The Athenian Oracle

The Magazine of Natural History and Journal of Zoology, Botany, Mineralogy, Geology, and Meteorology.

Daily Courant

Flying Post

Freethinker

Gentleman's Magazine

Historical Register

The London Gazette

Memoirs of the Literary and Philosophical Society of Manchester

Philosophical Magazine

Philosophical Transactions of the Royal Society of London

Index

Figures and boxes are indicated by page numbers in *italics*; notes by the annotation "n".